NORTH-HOLLAND MATHEMATICS STUDIES 132
Annals of Discrete Mathematics (31)

General Editor: Peter L. HAMMER
Rutgers University, New Brunswick, NJ, U.S.A.

Advisory Editors
C. BERGE, Université de Paris, France
M. A. HARRISON, University of California, Berkeley, CA, U.S.A.
V. KLEE, University of Washington, Seattle, WA, U.S.A.
J.-H. VAN LINT, California Institute of Technology, Pasadena, CA, U.S.A.
G.-C. ROTA, Massachusetts Institute of Technology, Cambridge, MA, U.S.A.

NORTH-HOLLAND – AMSTERDAM · NEW YORK · OXFORD · TOKYO

SURVEYS IN COMBINATORIAL OPTIMIZATION

Edited by

Silvano MARTELLO
DEIS - University of Bologna
Italy

Gilbert LAPORTE
École des H.É.C.
Montréal, Québec, Canada

Michel MINOUX
CNET
Paris, France

Celso RIBEIRO
Catholic University of Rio de Janeiro
Brazil

1987

NORTH-HOLLAND – AMSTERDAM · NEW YORK · OXFORD · TOKYO

© *Elsevier Science Publishers B.V., 1987*

All rights reserved. No part of this publication may be reproduced, stored in a retrieval system, or transmitted, in any form or by any means, electronic, mechanical, photocopying, recording or otherwise, without the prior permission of the copyright owner.

ISBN: 0 444 70136 2

Publishers:
ELSEVIER SCIENCE PUBLISHERS B.V.
P.O. Box 1991
1000 BZ Amsterdam
The Netherlands

Sole distributors for the U.S.A. and Canada:
ELSEVIER SCIENCE PUBLISHING COMPANY, INC.
52 Vanderbilt Avenue
New York, N.Y. 10017
U.S.A.

Library of Congress Cataloging-in-Publication Data

Surveys in combinatorial optimization.

 (North-Holland mathematics studies ; 132) (Annals
of discrete mathematics ; 31)
 "Based on a series of tutorial lectures given at the
School on Combinatorial Optimization, held at the Federal
University of Rio de Janeiro, Brazil, July 8-19, 1985"--
CIP fwd.
 1. Combinatorial optimization. I. Martello, Silvano.
II. Series. III. Series: Annals of discrete
mathematics ; 31.
QA402.5.S85 1987   519   86-24128
ISBN 0-444-70136-2

PRINTED IN THE NETHERLANDS

FOREWORD

This book is based on a series of tutorial lectures given at the School on Combinatorial Optimization, held at the Federal University of Rio de Janeiro, Brazil, July 8 - 19, 1985.

Over 100 participants benefitted from the high quality of the tutorial lectures and technical sessions. This event was the first of its kind held in Latin America and undoubtedly contributed to the diffusion of knowledge in the field of Combinatorial Optimization.

I would like to take this opportunity to acknowledge with pleasure the efforts and assistance provided by Professors Ruy Eduardo Campello, Gerd Finke, Peter Hammer, Gilbert Laporte, Silvano Martello, Michel Minoux and Celso Ribeiro in the organization of the School. I am also very grateful to CNPq – Conselho Nacional de Desenvolvimento Científico e Tecnológico and to FINEP – Financiadora de Estudos e Projetos of the Brazilian Ministry of Science and Technology, who provided financial support for this stimulating meeting.

Nelson Maculan, chairman
Rio de Janeiro, 1985

PREFACE

Ever since the field of Mathematical Programming was born with the discovery of the Simplex method by G.B. Dantzig in the late 1940s, researchers have devoted considerable attention to optimization problems in which the variables are constrained to take only integer values, or values from a finite or discrete set of numbers.

After the first significant contributions of R.E. Gomory to the field of Integer Programming, discrete optimization problems were to attract growing interest, partly because of the fact that a large number of practical applications of Operations Research techniques involved integrality constraints, and partly because of the theoretical challenge.

During the same period, Graph Theory had been developing, providing a natural and effective conceptual framework for expressing and sometimes solving combinatorial problems commonly encountered in applications, thus giving rise to network flow theory, matroid theory and complexity theory.

What is nowadays referred to as *Combinatorial Optimization* therefore derives from the combination and cross-fertilization of various research streams, all developed somewhat independently over the past two or three decades (such as Graph Theory, Integer Programming, Combinatorics and Discrete Mathematics).

The present volume is an attempt to provide a synthetic overview of a number of important directions in which Combinatorial Optimization is currently developing, in the form of a collection of survey papers providing detailed accounts of recent progress over the past few years.

A number of these papers focus more specifically on theoretical aspects and fundamental tools of Combinatorial Optimization: «Boolean Programming» and «Probabilistic Analysis of Algorithms». From the computational point of view, a good account of recent work on the use of vector processing and parallel computers for implementing algorithms will be found in the paper on «Parallel Computer Models and Combinatorial Algorithms».

In addition, substantial space has been allocated to a number of well-known problems, which have some relation with applications, but which are addressed here as prototypes of hard-to-solve combinatorial problems, and which, as such, have traditionally acted as stimulants for research in the field: we refer, in particular, to the papers on «The Linear Assignment Problem», «The Quadratic Assignment Problem», «The Knapsack Problem» and «The Steiner Problem in Graphs».

Besides these, it also seemed important that the wide applicability of the techniques and algorithmic tools developed in the field of Combinatorial Optimization be illustrated with a number of more complicated models chosen for their technical, industrial or economic importance. Accordingly, the reader will find a number of applications oriented papers devoted to combinatorial problems arising in such practical contexts as: communication networks («Network Synthesis and Dynamic Network Optimization»), location and routing («Single Facility Location Problems on Networks» and «The Vehicle Routing Problem»), manufacturing and planning in production systems («Scheduling Problems»).

All the survey papers included herein have been written by well-known specialists in the field, with particular emphasis on pedagogical quality (in this respect, special mention should be given to the bibliography which contains almost 1000 titles) and, as far as possible, completeness.

The Editors wish to thank the participants of the School on Combinatorial Optimization for their valuable comments and criticisms on preliminary versions of the papers. The financial support of the Federal University of Rio de Janeiro, the National Research Council of Italy and the École des Hautes Études Commerciales de Montréal is gratefully acknowledged.

We hope that this volume will be considered as a milestone in the rich and fast evolving field of Combinatorial Optimization.

Silvano Martello
Gilbert Laporte
Michel Minoux
Celso Ribeiro

CONTENTS

Preface
1. J. BŁAŻEWICZ, Selected topics in scheduling theory 1
2. G. FINKE, R.E. BURKARD, F. RENDL, Quadratic assignment problems 61
3. P.L. HAMMER, B. SIMEONE, Order relations of variables in 0-1 programming 83
4. P. HANSEN, M. LABBÉ, D. PEETERS, J.-F. THISSE, Single facility location on networks 113
5. G. LAPORTE, Y. NOBERT, Exact algorithms for the vehicle routing problem 147
6. N. MACULAN, The Steiner problem in graphs 185
7. S. MARTELLO, P. TOTH, Algorithms for knapsack problems 213
8. S. MARTELLO, P. TOTH, Linear assignment problems 259
9. M. MINOUX, Network synthesis and dynamic network optimization 283
10. C.C. RIBEIRO, Parallel computer models and combinatorial algorithms 325
11. A.H.G. RINNOOY KAN, Probabilistic analysis of algorithms 365

SELECTED TOPICS IN SCHEDULING THEORY

Jacek BŁAŻEWICZ

1. Introduction

This study is concerned with deterministic problems of scheduling tasks on machines (processors), which is one of the most rapidly expanding areas of combinatorial optimization. In general terms, these problems may be stated as follows. A given set of tasks is to be processed on a set of available processors, so that all processing conditions are satisfied and a certain objective function is minimized (or maximized). It is assumed, in contrast to stochastic scheduling problems, that all task parameters are known *a priori* in a deterministic way. This assumption, as will be pointed out later, is well justified in many practical situations. On the other hand, it permits the solving of scheduling problems having a different practical interpretation from that of the stochastic approach. (For a recent survey of stochastic scheduling problems see [117, 118]). This interpretation is at least a valuable complement to the stochastic analysis and is often imposed by certain applications as, for example, in computer control systems working in a hard-real-time environment and in many industrial applications. In the following we will emphasize the applications of this model in computer systems (scheduling on parallel processors, task preemptions). However, we will also point out some other interpretations, since tasks and machines may also represent ships and dockyards, classes and teachers, patients and hospital equipment, or dinners and cooks. To illustrate some notions let us consider the following example.

Let us assume that our goal is to perform five tasks on three parallel identical processors in the shortest possible time. Let the processing times of the tasks be 3, 5, 2, 1, 2, respectively, and suppose that the tasks are to be processed without preemptions, i.e. each task once started must be processed until completion without any break on the same processor. A solution (an optimal schedule) is given in Fig. 1.1, where a Gantt chart is used to present the assignment of the processore to the tasks in time. We see that the minimum processing time for these five tasks is 5 units and the presented schedule is not the only possible solution. In this case, the solution has been obtained easily. However, this is not generally the case and, to construct an optimal schedule, one has to use special algorithms depending on the problem in question.

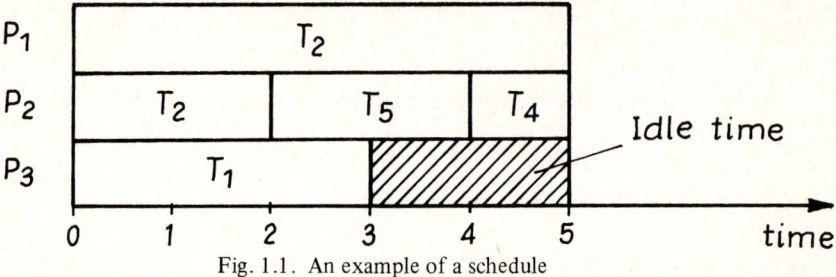
Fig. 1.1. An example of a schedule

In this paper we would like to present some important topics in scheduling theory. First, and interpretation of its assumptions and results, especially in computer systems, is described and some other applications are given. Then a general approach to the analysis of scheduling problems is presented in detail and illustrated with an example. In the next two sections, the most important results, in our opinion, concerning the classical scheduling problems with parallel machines and in job systems, are summarized. These include NP-hardness results and optimization as well as approximation algorithms with their accuracy evaluation (mean or in the worst case). Because of the limited space we have not been able to describe all the existing results. Our choice has been motivated first by the importance of a particular result, e.g. in the case of algorithms, their power to solve more than one particular scheduling problem. This is, for example, the case with the level algorithm, network flow or linear programming approaches. Secondly, our choice has been influenced by the relevance of the results presented to a computer scheduling problem. Hence, not much attention has been paid to enumerative methods. However, in all the considered situations we refer to the existing surveys which cover the above issues in a very detailed way. New directions in scheduling theory are then presented, among which special attention is paid to scheduling under resource constraints and scheduling in microprocessor systems. An Appendix for the notations of scheduling problems can be found at the end of this study.

2. Basic Notions

2.1. Problem formulation

In general, we will be concerned with two sets: a set of n tasks $\mathcal{T} = \{T_1, T_2, \ldots, T_n\}$ and a set of m processors $\mathcal{P} = \{P_1, P_2, \ldots, P_m\}$. There are two general rules to be followed in classical scheduling theory. Each task is to be processed by at most one processor at a time and each processor is capable of processing at most one task at a time. In Section 6 we will show some new applications in which the first constraint will be removed.

We will now characterize the processors. They may be either *parallel*, performing the same functions, or *dedicated*, i.e. specialized for the execution of certain tasks. Three types of parallel processors are distinguished depending on their speeds. If all processors from set $\mathscr{P}$ have equal task-processing speeds, then we call them *identical* processors. If the processors differ in their speeds, but the speed of each processor is constant and does not depend on the tasks of $\mathscr{T}$, then they are called *uniform*. Finally, if the speeds of the processors depend on the particular task which is processed, then they are called *unrelated*.

In the case of dedicated processors, there are three modes of processing: flow shop, open shop and job shop systems. In a *flow shop* system each task from set $\mathscr{T}$ must be processed by all the processors in the same order. In an *open shop* system each task must also be processed by all processors, but the order of processing is not given. In a *job shop* system the subset of processors which are to process a task and the order of the processing are arbitrary, but must be specified *a priori*.

We will now describe the task set in detail. In the case of parallel processors each task can be processed by any given processor. When tasks are to be processed on dedicated processors, then each $T_j \in \mathscr{T}$ is divided into operations $0_{j1}, 0_{j2}, \ldots, 0_{jk_j}$, each of which may require a different processor. Moreover, in a flow shop system, the number of operations per task is equal to m and their order of processing is such that 0_{j1} is processed on processor P_1, 0_{j2} on P_2, and so on. Moreover, the processing of $0_{j,i-1}$ must always precede the processing of 0_{ji}, $j = 1, 2, \ldots, n$, $i = 2, \ldots, m$. In the case of an open shop system, each task is also divided into m operations with the same allocation of the operations to processors, but the order of processing operations is not specified. In a job shop system, the number of operations per task, their assignment to processors and the order of their processing, are arbitrary but must be given in advance.

Task $T_j \in \mathscr{T}$ is characterized by the following data.

1. *A vector of processing times* $- \bar{p}_j = [p_{j1}, p_{j2}, \ldots, p_{jm}]$, where p_{ji} is the time needed by processor P_i to complete T_j. In the case of identical processors we have $p_{ji} = p_j$, $i = 1, 2, \ldots, m$. If the processors in $\mathscr{P}$ are uniform then $p_{ji} = = p_j/b_i$, $i = 1, 2, \ldots, m$, where p_j is a *standard processing time* (usually on the slowest processor) and b_i is a processing speed factor of processor P_i.

2. *An arrival time* (a ready time) $- r_j$, which is the time at which T_j is ready for processing. If for all tasks from $\mathscr{T}$ the arrival times are equal, then it is assumed that $r_j = 0, j = 1, 2, \ldots, n$.

3. *A due-date* $- d_j$. If the processing of T_j must be completed at time d_j, then d_j is called the deadline.

4. *A weight* (priority) $- w_j$, which expresses the relative urgency of T_j.

Below, some definitions concerning task preemptions and precedence constraints among tasks are given.

The mode of processing is called *preemptive* if each task (each operation in the case of dedicated processors) may be preempted at any time and restarted later at no cost, perhaps on another processor. If the preemption of any task is not allowed we will call the scheduling *nonpreemptive*.

In set $\mathcal{T}$ *precedence constraints* among tasks may be defined. $T_i < T_j$ means that the processing of T_i must be completed before T_j can be started. In other words, set $\mathcal{T}$ is ordered by a binary relations $<$. The tasks in set $\mathcal{T}$ are called *dependent* if at least two tasks in $\mathcal{T}$ are ordered by this relation. Otherwise, the tasks are called *independent*. A task set ordered by the precedence relation is usually represented as a directed graph (a digraph) in which nodes correspond to tasks and arcs to precedence constraints. An example of a dependent task set is shown in Fig. 2.1 (nodes are denoted by T_j/p_j). Let us notice that in the case of dedicated processors (except in the open shop system) operations that constitute a task are always dependent, but tasks can be either independent or dependent. It is also possible to represent precedence constraints as a task-on-arc graph, which is also called an activity network representation. Usually, the first approach is used, but sometimes the second presentation may be useful and we will mention these cases in what follows. Task T_j will be called *available* at moment t if $r_j \leq t$ and all its predecessors (with respect to the precednece contraints) have been completed at time t.

Now we will give the definitions concerning schedules and optimality criteria.

A *schedule* is an assignment of processors from set $\mathcal{P}$ to tasks from set $\mathcal{T}$ in time such that the following conditions are satisfied:

— at every moment each processor is assigned to at most one task and each task is processed by at most one processor;

— task T_j is processed in time interval $[r_j, \infty)$;

— all tasks are completed;

— for each pair T_i, T_j, such that $T_i < T_j$, the processing of T_j is started after the completion of T_i;

— in the case of nonpreemptive scheduling no task is preempted (the schedule is called *nonpreemptive*), otherwise the number of preemptions of each task is

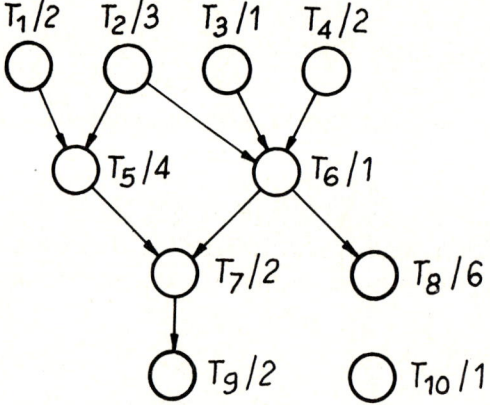

Fig. 2.1. An example of a task set.

finite (the schedule is called *preemptive*) ([1]). An example schedule for the task set of Fig. 2.1, is given in Fig. 2.2.

The following parameters can be calculated for each task $T_j, j = 1, 2, \ldots, n$, processed in a given schedule:
- a completion time — c_j:
- a flow time — f_j, being the sum of waiting and processing times

$$f_j = c_j - r_j:$$

- a lateness — l_j

$$l_j = c_j - d_j:$$

- a tardiness — t_j

$$t_j = \max \{c_j - d_j, 0\}.$$

A schedule for which the value of a particular performance measure is at its minimum will be called *optimal*. To evaluate schedules we will use three main criteria.

Schedule length (makespan)

$$C_{\max} = \max \{c_j\}.$$

Mean flow time

$$\bar{F} = \frac{1}{n} \sum_{j=1}^{n} f_j$$

or *mean weighted flow time*

$$\bar{F}_w = \sum_{j=1}^{n} w_j f_j / \sum_{j=1}^{n} w_j.$$

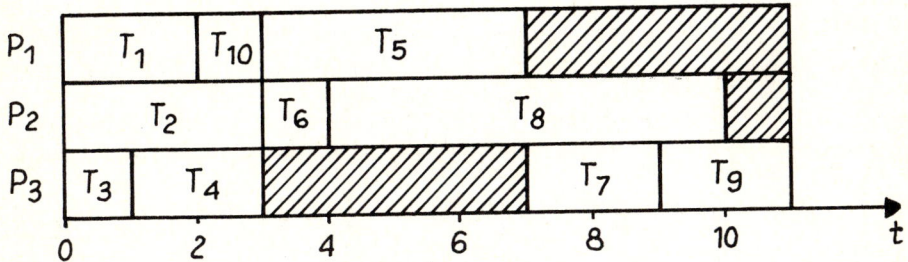

Fig. 2.2. A schedule for the task set given in Fig. 2.1.

([1]) The last condition is imposed by practical considerations only.

Maximum lateness

$$L_{max} = \max_j \{l_j\}.$$

In some applications, other related criteria are also used, as for example: mean tardinees $\bar{T} = 1/n \; \Sigma_{j=1}^{n} \, t_j$, mean weighted tardiness $\bar{T}_w = \Sigma_{j=1}^{n} w_j \, t_j / \Sigma_{j=1}^{n} w_j$, or number of tardy jobs $\bar{U} = \Sigma_{j=1}^{n} u_j$, where $u_j = 1$, if $c_j > d_j$, and 0 otherwise.

We may now define the *scheduling problem* Π as a set of parameters described in this subsection not all of which have numerical values, together with a criterion. An *instance* I of problem Π is obtained by specifying particular values for all the problem parameters. As we see, there is a great variety of scheduling problems and to describe a particular one we need several sentences. Thus, having a short notation of problem types would greatly facilitate the presentation and discussion of scheduling problems. Such a notation has been given in [70, 102] and we present it in the Appendix and use it throughout the paper.

2.2. Interpretation of assumptions and results

In this subsection, an analysis of the assumptions and results in deterministic scheduling theory is presented taking especially computer applications into account. However, also other practical applications, as mentioned in the Introdiction, should not be forgotten.

Let us begin with an analysis of processors. Parallel processors may be interpreted as central processors which are able to process every task (every program). Uniform processors differ from each other by their speeds, but they do not prefer any type of task. Unrelated processors, on the contrary, are specialized in the sense that they prefer certain types of tasks for example numerical computations, logical programs or simulation procedures, since the processors have different instruction lists. Of course, they can process tasks of each type, but at the expense of longer processing time.

A different type are the dedicated processors which may process only certain types of tasks. As an example let us consider a computer system consisting of an input processor, a central processor and an output processor. It is not difficult to see that the above system corresponds to a flow shop with $m = 3$. On the other hand, a situation in which each task is to be processed by an input/output processor, then by a central one and at the end again by the input/output processor, can be easily modelled by a job shop system with $m = 2$. As far as an open shop is concerned, there is no obvious computer interpretation. But this case, like the other shop scheduling problems, has great significance in other applications.

Let us now consider the assumptions associated with the task set. As mentioned in Subsection 2.1. in deterministic scheduling theory *a priori* knowledge of ready times and processing times of tasks, is assumed. Ready times are obviously known in computer systems working in an off-line mode. They are also often known in computer control systems in which measurement samples are taken from sensing devices at fixed time moments.

As far as processing times are concerned, they are usually not known *a priori* in computer systems, unlike in many other applications. Despite this fact the solution of a deterministic scheduling problem is also important in these systems. Firstly, when scheduling tasks to meet deadlines, the only approach (when the task processing times are not known) is to solve the problem with assumed upper bounds on the processing times. (Such a bound for a given task may be implied by the worst case complexity function of an algorithm connected with that task). Then, if all deadlines are met with respect to the upper bounds, no deadline will be exceeded for the real task processing times [2]. This approach is often used in a broad class of computer control systems working in a hard--real-time environment, where a certain set of control programs must be processed before taking the next sample from the same sensing device.

Secondly, instead of exact values of processing times one can take their mean values and, using the procedure described in [37], calculate an optimistic estimate of the mean value of schedule length.

Thirdly, one can measure the processing times of tasks after processing a task set scheduled according to a certain algorithm A. Taking these values as an input in the deterministic scheduling problem, one may construct an optimal schedule and compare it with the one produced by algorithm A, thus evaluating the latter.

Apart from the above, optimization algorithms for deterministic scheduling problems give some indications for the construction of heuristics for weaker assumptions than those made in stochastic scheduling problems (c.f. [31, 32]).

The existence of precedence constraints also requires an explanation. In the simplest case the results of certain programs may be the input data for others. Moreover, precedence constraints may also concern parts of the same program. A conventional serially written program may be analyzed by a special procedure looking for parallel parts in it (see for example [122, 130, 146]). These parts may also be defined by the programmer who can use special programming languages (see [66]). Apart from this, a solution of certain reliability problems in operating systems, for example the *determinacy problem* [2, 6, 10], requires an introduction of additional precedence constraints. In other applications, the existence of precedence constraints is more obvious and follows, for example, from technological constraints.

We will not analyze the importance of particular criteria for scheduling problems. Minimizing schedule length is important from the viewpoint of the owner of a set of processors (machines), since it leads to a maximization of the processor utilization factor. This criterion may also be of importance in a computer control system in which a task set arrives periodically and is to be

---

[2] However, one has to take into account list scheduling anomalies which will be explained in Section 3.

processed in the shortest time.

The mean flow time criterion is important from the user's viewpoint since its minimization yields a minimization of the mean response time.

The maximum lateness criterion is of great significance in computer control systems working in the hard-real-time environment since its minimization leads to the construction of a schedule with no task late whenever such a schedule exists (i.e. when $L^*_{max} \leqslant 0$ for an optimal schedule). Other criteria involving deadlines are of importance in some economic applications.

The three criteria mentioned above are basic in the sense that they concern three main applications and require three different approaches to the designing of optimization algorithms.

2.3. *An approach to the analysis of scheduling problems*

Deterministic scheduling problems are a part of a much broader class of combinatorial optimization problems. Thus, the general approach to the analysis of these problems can follow similar lines, but one should take into account their peculiarities, which are evident in computer applications. It is rather obvious that in these applications the time we can devote to solving scheduling problems is seriously limited so that only low order polynomial-time algorithms may be used. Thus, the examination of the complexity of these problems should be the basis of any further analysis.

It has been known for some time [45, 85] that there exists a large class of combinatorial optimization problems for which most probably there are no *efficient optimization* algorithms ([3]). These are the problems whose decision counterparts (i.e. problems formulated as questions with «yes» or «no» answers) are *NP-complete*. The optimization problems are called NP-hard in this case. We refer the reader to [58] for a comprehensive treatment of the NP-completeness theory, and in the following we assume knowledge of its basic concepts like NP-completeness, NP-hardness, polynomial-time transformation, etc. It follows that the complexity analysis answers the question whether or not an analyzed scheduling problem may be solved (i.e. an optimal schedule found) in time bounded from above by a polynomial in the input length of the problem (i.e. in polynomial time). If the answer is positive, then an optimization, polynomial--time algorithm must have been found. Its usefulness depends on the order of its worst-case complexity function and on a particular application. Sometimes, when the worst-case complexity function is not low enough, although polynomial, a mean complexity function of the algorithm may be sufficient. This issue is discussed in detail in [4]. On the other hand, if the answer is negative, i.e.

([3]) By an efficient (polynomial-time) algorithm we mean one whose worst-case complexity function can be bounded from above by a polynomial in the problem input length, and by an optimization algorithm one which finds an optimal schedule for all instances of a problem.

when the decision version of the analyzed problem is NP-complete, then there are several ways of further analysis.

Firstly, one may try to relax some constraints imposed on the original problem and then solve the relaxed one. The solution of the latter may be a good approximation to the solution of the original one. In the case of scheduling problems such a relaxation may consist of the following:
— allowing preemptions, even if the original problem dealt with non-preemptive schedules,
— assuming unit-length tasks, when arbitrary-length tasks were considered in the original problem,
— assuming certain types of precedence graphs, e.g. trees or chains, when arbitrary graphs were considered in the original problem, etc.

In computer applications, especially the first relaxation can be justified in the case when parallel processors share a common primary memory. Moreover, such a relaxation is also advantageous from the viewpoint of certain criteria.

Secondly, when trying to solve hard scheduling problems one often uses approximation algorithms which try to find an optimal schedule but do not always succeed. Of course, the necessary condition for these algorithms to be applicable is practice is that their worst-case complexity function is bounded from above by a low-order polynomial in the input length. Their sufficiency follows from an evaluation of the distance between the solution value they produce and the value of an optimal solution. This evaluation may concern the worst case or a mean behavior. To be more precise, we give here some definitions, starting with the worst case analysis [58].

If Π is a minimization (maximization) problem, and I is any instance of it, we may define the ratio $R_A(I)$ for an approximation algorithm A as

$$R_A(I) = \frac{A(I)}{\text{OPT}(I)} \qquad \left(R_A(I) = \frac{\text{OPT}(I)}{A(I)}\right),$$

where $A(I)$ is the value of the solution constructed by algorithm A for instance I, and OPT(I) is the value of an optimal solution for I. The *absolute performance ratio* R_A for an approximation algorithm A for Π is given as

$$R_A = \inf\{r \geqslant 1 : R_A(I) \leqslant r \text{ for all instances of } \Pi\}.$$

The *asymptotic performance ratio* R_A^∞ for A is given as

$$R_A^\infty = \inf\left\{r \geqslant 1 : \begin{array}{l}\text{for some positive integer } K,\ R_A(I) \leqslant r \text{ for all}\\ \text{instances of } \Pi \text{ satisfying OPT}(I) \geqslant K\end{array}\right\}.$$

The above formulae define a measure of the «goodness» of approximation algorithms. The closer R_A^∞ is to 1, the better algorithm A performs (4). However,

---

(4) One can also consider several possibilities of the worst case behavior of an approximation algorithm for which $R_A^\infty = 1$, and we refer the reader to [58] for detailed treatment of the subject.

for some combinatorial problems it can be proved that there is no hope of finding an approximation algorithm of a certain accuracy (i.e. this question is as hard as finding a polynomial-time algorithm for any NP-complete problem).

Analysis of the worst-case behavior of an approximation algorithm may be complemented by an analysis of its mean behavior. This can be done in two ways. The first consists in assuming that the parameters of instances of the considered problem Π are drawn from a certain distribution D and then one analyzes the *mean performance* of algorithm A. One may distinguish between the *absolute error* of an approximation algorithm, which is the difference between the approximate and optimal solution values and the *relative error*, which is the ratio of the two. Asymptotic optimality results in the stronger (absolute) sense are quite rare. On the other hand, asymptotic optimality in the relative sense is often easier to establish [86, 124, 135].

It is rather obvious that the mean performance can be much better than the worst case behavior, thus justifying the use of a given approximation algorithm. A main obstacle is the difficulty of proofs of the mean performance for realistic distribution functions. Thus, the second way of evaluating the mean behavior of approximation algorithms, consisting of simulation studies, is still used very often. In the latter approach one compares solutions, in the sense of the values of a criterion, constructed by a given approximation algorithm and by an optimization algorithm. This comparison should be made for a large representative sample of instances. There are some practical problems which follow from the above statement and they are discussed in [134].

The third and last way of dealing with hard scheduling problems is to use exact enumerative algorithms whose worst-case complexity function is exponential in the input length. However, sometimes, when the analyzed problem is not NP-hard in the strong sense, it is possible to solve it by a pseudopolynomial optimization algorithm whose worst-case complexity function is bounded from above by a polynomial in the input length and in the maximum number appearing in the instance of the problem. For reasonably small numbers such an algorithm may behave quite well in practice and it can be used even in computer applications. On the other hand, «pure» exponential algorithms have probably to be excluded from this application, but they may be used sometimes for other scheduling problems which may be solved by off-line algorithms.

The above discussion is summarized in a schematic way in Fig. 2.3. To illustrate the above considerations, we will analyze an example of a scheduling problem in the next section. (In the following we will use a term complexity function instead of worst-case complexity function).

3. An example of a scheduling problem analysis

In this section an examplary scheduling problem will be analyzed along the

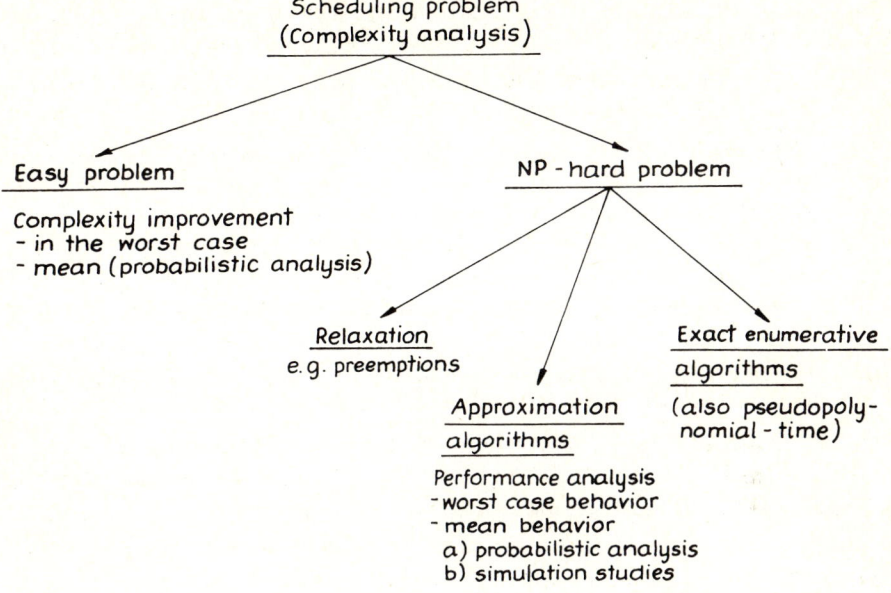

Fig. 2.3. An analysis of a scheduling problem - schematic view.

lines described in Section 2. We have chosen problem $P \| C_{max}$, i.e. the problem of nonpreemptive scheduling of independent tasks on identical processors with the objective of minimizing schedule length. A reason for this choice was the fact that this is probably one of the most thoroughly analyzed problems. We will start with the complexity analysis and the considered problem appears to be not an easy one, since a problem with two processors is already NP-hard.

Theorem 3.1. *Problem $P2 \| C_{max}$ is NP-hard.*

Proof. The proof is very easy. As a known NP-complete problem we take PARTITION [85] which is formulated as follows.

Instance: Finite set A and a size $s(a_i) \in N$ (5) for each $a_i \in A$.

Question: Is there a subset $A' \subseteq A$ such that $\Sigma_{a_i \in A'} s(a_i) = \Sigma_{a_i \in A - A'} s(a_i)$?

Given any instance of PARTITION defined by the set of positive integers $\{s(a_i) : a_i \in A\}$, we define a corresponding instance of the decision counterpart of $P2 \| C_{max}$ by assuming $n = |A|$, $p_j = s(a_j)$, $j = 1, 2, \ldots, n$ and the threshold value of schedule length $y = (1/2) \Sigma_{a_i \in A} s(a_i)$. It is obvious that there exists a subset A' with the desired property for the instance of PARTITION if, for the corresponding instance of $P2 \| C_{max}$, there exists a schedule with $C_{max} \leq y$ (cf. Fig. 3.1), and the theorem follows. □

---

(5) N denotes the set of positive integers.

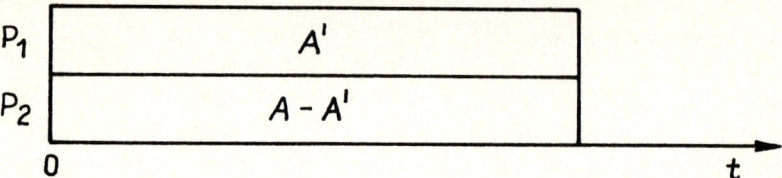

Fig. 3.1. A schedule for Theorem 3.1.

Since there is no hope of finding an optimization polynomial-time algorithm for $P\|C_{max}$, one may try to solve the problem along the lines presented in Subsection 2.3. Firstly, we may relax some constraints imposed on problem $P\|C_{max}$ and allow preemptions of tasks. It appears that problem $P|\text{pmtn}|C_{max}$ may be solved very efficiently. It is easy to see that the length of a preemptive schedule cannot be smaller than the maximum of two values: the maximum processing time of a task and the mean processing requirement on a processor [116], i.e.:

$$C^*_{max} = \max\left\{\max_j\{p_j\},\ \frac{1}{m}\sum_{j=1}^n p_j\right\}. \tag{3.1}$$

The following algorithm given by Mc Naughton [116] constructs a schedule whose length is equal to C^*_{max}.

Algorithm 3.1. (Problem $P|\text{pmtn}|C_{max}$)
1. Assign any task to the first processor at time $t = 0$.
2. Schedule any nonassigned task on the same processor on which the last assigned task has been completed, at the moment of completion. Repeat this step until either all tasks are scheduled or $t = C^*_{max}$. In the latter case go to step 3.
3. Schedule the remaining part of the task that exceeds C^*_{max} on the next processor at time $t = 0$, and go to step 2.

Note that the above algorithm is an optimization procedure, since it always find a schedule whose length is equal to C^*_{max}. It is complexity is $O(n)$ (⁶).

We see that by allowing preemptions we now have an easy problem. However, there still remains the question of the practical applicability of the solution obtained in this way. It appears that in multiprocessor systems with a common primary memory, the assumptions of task preemptions can be justified and preemptive schedules can be used in practice. If this is not the case, one may try to find an approximation algorithm for the original problem and evaluate its worst case as well as its mean behavior. We will present such an analysis below.

(⁶) The notation «$q(n) = O(p(n))$» means that there exists a constant $c \geq 0$ such that $|q(n)| \leq cp(n)$ for all $n > 0$.

One of the most often-used general approximation strategies for solving scheduling problems is *list scheduling*, whereby a priority list of the tasks is given and at each step the first available processor is selected to process the first available task on the list [67]. The accuracy of a given list scheduling algorithm depends on the order in which tasks appear on the list. On the other hand, this strategy may result in the unexpected behavior of constructed schedules, since the schedule length for problem $P \mid \text{prec} \mid C_{\max}$ may increase if:
- the number of processors increases,
- task processing times decrease,
- precedence constraints are weakened, or
- the priority list changes.

Figures 3.2 through 3.6 indicate the effects of changes of the above-mentioned parameters [69]. These list scheduling anomalies have been discovered by Graham [67], who has also avaluated the maximum change in schedule length that may be induced by varying one or more problem parameter. We will quote this theorem since its proof is one of the shortest in that area and illustrates well the technique used in other proofs of that type. Let there be defined a task set $\mathcal{T}$ together with precedence constraints $<$. Let the processing times of the tasks be given as vector $\bar{p}$, and let $\mathcal{T}$ be scheduled on m processors using list L, and the obtained value of schedule length be equal to $C_{\max}$. On the other hand, let the above parameters be changed: a vector of processing times $\bar{p}' \leq \bar{p}$ (for all the components), precedence constraints $<' \subseteq <$, priority list L' and the number of processors m'. Let the new values of schedule length be $C'_{\max}$. Then the following theorem is valid.

Theorem 3.2. [67]. *On the above assumptions*

$$\frac{C'_{\max}}{C_{\max}} \leq 1 + \frac{m-1}{m'}. \tag{3.2}$$

Proof. Let us consider the schedule S' obtained by processing the task set $\mathcal{T}$ with primed parameters. Let an interval $[0, C'_{\max})$ be divided into two subsets, A and B, defined in the following way:
$A = \{t \in [0, C'_{\max}) : \text{all processors are busy at time } t\}$,
$B = [0, C'_{\max}) - A$.

Notice that both A and B are unions of disjoint half-open intervals. Let T_{j_1} denote a task completed in S' at time $C'_{\max}$, i.e. $c_{j_1} = C'_{\max}$. Two cases may occur.

1. The starting time of T_{j_1}, s_{j_1}, is an interior point of B. Then by the definition of B there is some processor P_i which for some $\epsilon > 0$ is idle during interval $[s_{j_1} - \epsilon, s_{j_1})$. Such a situation may only occur if for some T_{j_2} we have $T_{j_2} <' T_{j_1}$ and $c_{j_2} = s_{j_1}$.

2. The starting time of T_{j_1} is not an interior point of B. Let us also suppose that $s_{j_1} \neq 0$. Let $x_1 = \sup\{x : x < s_{j_1} \text{ and } x \in B\}$ or $x_1 = 0$ if this set is empty. By the construction of A and B, we see that $x_1 \in A$ and for some $\epsilon > 0$ processor

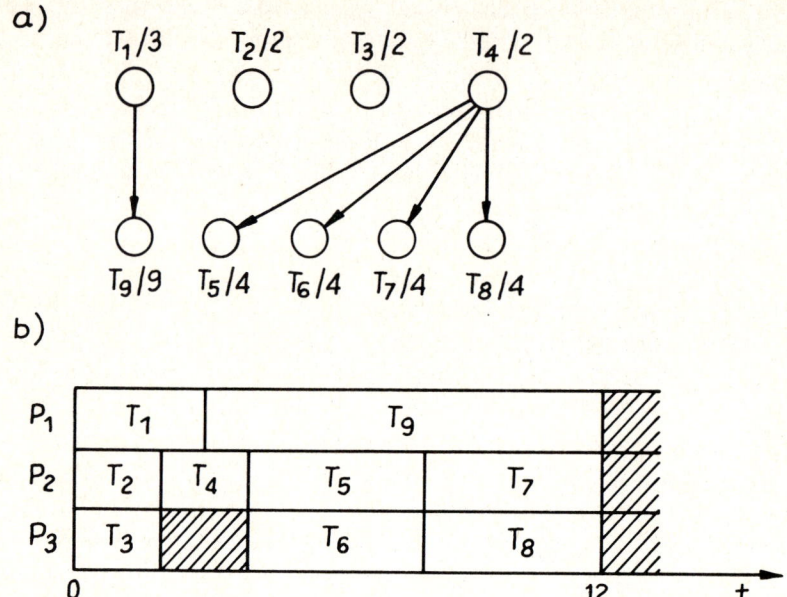

Fig. 3.2. A task set (a) and an optimal schedule (b); $m = 3$, $L = (T_1, T_2, T_3, T_4, T_5, T_6, T_7, T_8, T_9)$.

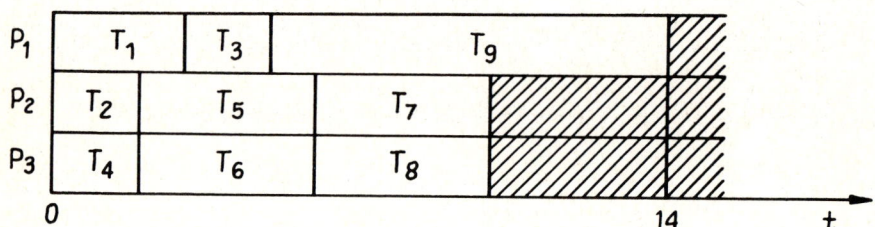

Fig. 3.3. Priority list changed: a new list $L' = (T_1, T_2, T_4, T_5, T_6, T_3, T_9, T_7, T_8)$.

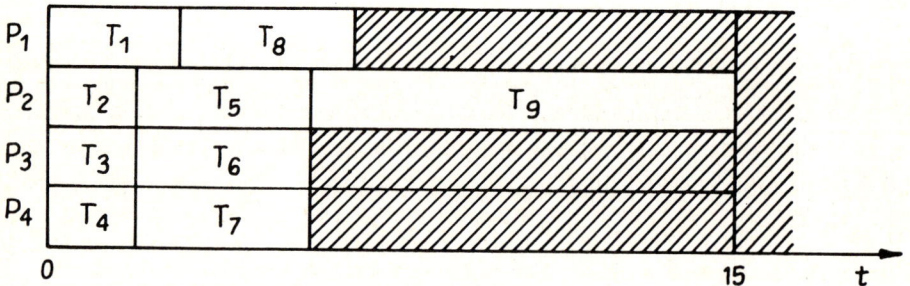

Fig. 3.4. A number of processors increased: $m' = 4$.

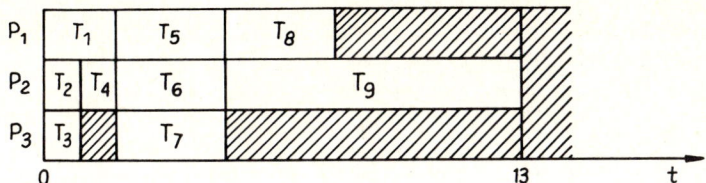

Fig. 3.5. Processing times decreased: $p'_j = p_j - 1, j = 1, 2, \ldots, n$.

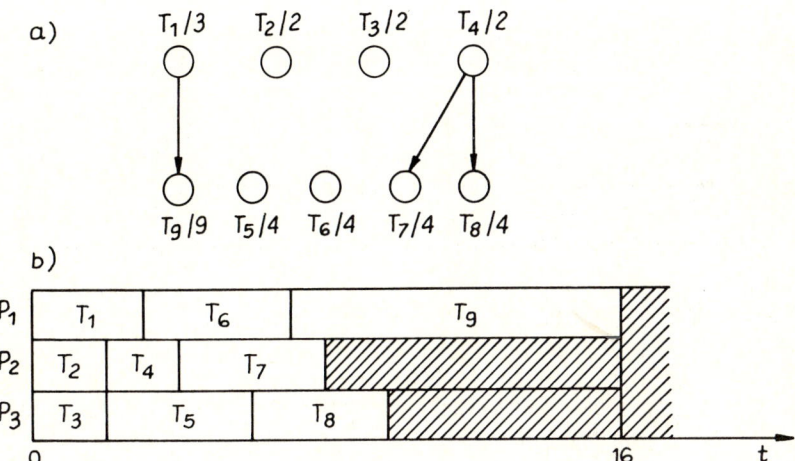

Fig. 3.6. Precedence constraints weakened (a), a resulting list schedule (b).

P_i is idle in time interval $[x_1 - \epsilon, x_1)$. But again, such a situation may only occur if some task $T_{j_2} <' T_{j_1}$ is processed during this time interval.

It follows that either there exists a task $T_{j_2} <' T_{j_1}$ such that $y \in [c_{j_2}, s_{j_1})$ implies $y \in A$ or we have: $x < s_{j_1}$ implies either $x \in A$ or $x < 0$.

The above procedure can be inductively repeated, forming a chain $T_{j_3}, T_{j_4}, \ldots$, until we reach task T_{j_r} for which $x < s_{j_r}$ implies either $x \in A$ or $x < 0$. Hence there must exist a chain of tasks

$$T_{j_r} <' T_{j_{r-1}} <' \ldots <' T_{j_2} <' T_{j_1} \tag{3.3}$$

such that in S' at each moment $t \in B$, some task T_{j_k} is being processed. This implies that

$$\sum_{\phi' \in S'} p'_{\phi'} \leq (m' - 1) \sum_{k=1}^{r} p'_{j_k} \tag{3.4}$$

where the sum of the left-hand side is made over all empty ϕ' tasks in S'. But by

(3.3) and the hypothesis $<' \subseteq <$ we have

$$T_{j_r} < T_{j_{r-1}} < \ldots < T_{j_2} < T_{j_1}. \tag{3.5}$$

Hence,

$$C_{\max} \geq \sum_{k=1}^{r} p_{j_k} \geq \sum_{k=1}^{r} p'_{j_k}. \tag{3.6}$$

Furthermore, by (3.4) and (3.6) we have

$$\begin{aligned} C'_{\max} &= \frac{1}{m'} \left(\sum_{k=1}^{n} p'_k + \sum_{\phi' \in S'} p'_{\phi'} \right) \\ &\leq \frac{1}{m'} (m \, C_{\max} + (m'-1) \, C_{\max}). \end{aligned} \tag{3.7}$$

It follows that

$$\frac{C'_{\max}}{C_{\max}} \leq 1 + \frac{m-1}{m'} \tag{3.8}$$

and the theorem is proved. □

Using the above theorem, one can prove the absolute performance ratio for an arbitrary list scheduling algorithm solving problem $P \| C_{\max}$.

Corollary 3.3. [67]. *For an arbitrary list scheduling algorithm LS for $P \| C_{\max}$ we have*

$$R_{LS} = 2 - \frac{1}{m}. \tag{3.9}$$

Proof. The upper bound of (3.9) follows immediately from (3.2) by taking $m' = m$ and by considering the list leading to an optimal schedule. To show that this bound is achievable let us consider the following example: $n = (m-1)m+1$, $\overline{p} = [1, 1, \ldots, 1, 1, m]$, $<$ is empty, $L = (T_n, T_1, T_2, \ldots, T_{n-1})$ and $L' = (T_1, T_2, \ldots, T_n)$. The corresponding schedules for $m = 4$ are shown in Fig. 3.7. □

It follows from the above considerations that an arbitary list scheduling algorithm can produce quite bad schedules, twice as long as an optimal one. An improvement can be obtained if we order the task list properly. The simplest algorithm orders tasks on the list in order of nonincreasing p_j (the so-called *longest processing time* (LPT) rule). The absolute performance ratio for the LPT algorithm is given in Theorem 3.4.

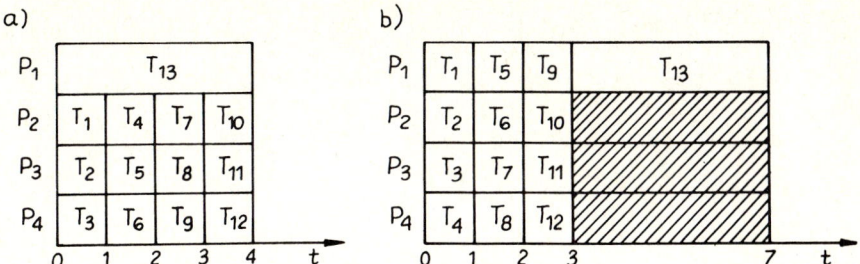

Fig. 3.7. Schedules for Corollary 3.3.: an optimal one (a), an approximate one (b).

Theorem 3.4. [68]. *If the LPT scheduling algorithm is used to solve problem $P \| C_{max}$, then*

$$R_{LPT} = \frac{4}{3} - \frac{1}{3m}.$$ (3.10)

Space limitations prevent us from including here the proof of the upper bound in the above theorem. However, we will give an example showing that this bound can be achieved. Let $n = 2m + 1$, $p = [2m - 1, 2m - 1, 2m - 2, 2m - 2, \ldots, m + 1, m + 1, m, m, m]$. Fig. 3.8 shows an optimal and an LPT schedule for $m = 4$.

We see that an LPT schedule can be longer in the worst case by 33% than an optimal one. However, one is led to expect better performance from the LPT algorithm than is indicated by (3.10), especially as the number of tasks becomes large. In [43] another absolute performance ratio for the LPT rule was proved, taking into account the least number k of tasks on any processor.

Theorem 3.5. [43]. *For the assumptions stated above, we have*

$$R_{LPT}(k) \leq 1 + \frac{1}{k} - \frac{1}{mk}.$$ (3.11)

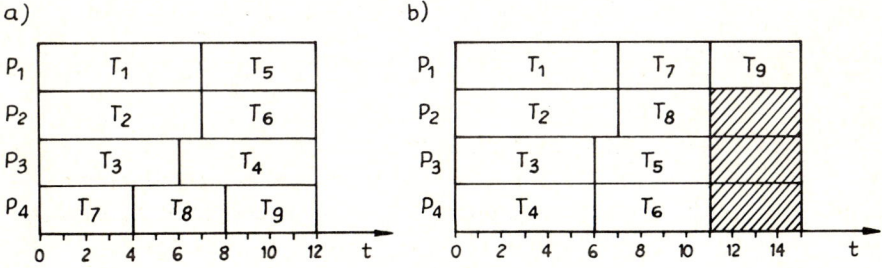

Fig. 3.8. Schedules for Theorem 3.4: on optimal one (a), LPT one (b).

This result shows that the worst-case performance bound for the LPT algorithm approaches unity approximately as $1 + 1/k$.

On the other hand, one can be interested in how good the LPT algorithm is on the average. Recently such a result was obtained [38] where the relative error was found for two processors on the assumption that task processing times are independent samples from the uniform distribution on [0, 1].

Theorem 3.6. [38]. *For the assumptions already stated, we have*

$$\frac{n}{4} + \frac{1}{4(n+1)} \leqslant E(C_{\max}^{\mathrm{LPT}}) \leqslant \frac{n}{4} + \frac{e}{2(n+1)}. \tag{3.12}$$

Taking into account that $n/4$ is a lower bound on $E(C_{\max}^*)$ we get $(E(C_{\max}^{\mathrm{LPT}})/E(C_{\max}^*)) < 1 + 0(1/n^2)$. Therefore, as n increases, $E(C_{\max}^{\mathrm{LPT}})$ approaches the optimum no more slowly than $1 + 0(1/n^2)$ approaches 1. The above bound can be generalized to cover also the case of m processors for which we have [39]:

$$E(C_{\max}^{\mathrm{LPT}}) \leqslant \frac{n}{2m} + 0\left(\frac{m}{n}\right).$$

Moreover, it is also possible to prove [49, 50] that $C_{\max}^{\mathrm{LPT}} - C_{\max}^*$ almost surely converges to 0 as $n \to \infty$ if the task processing time distribution has a finite mean and a density f satisfying $f(0) > 0$. It is also shown that if the distribution is uniform or exponential, the rate of convergence is $O(\log(\log n)/n)$. This result, obtained by a complicated analysis, can also be guessed by simulation studies. Such an experiment was reported in [88] and we present the summary of the results in Table 3.1. The last column presents the ratio of schedule lengths obtained by the LPT algorithm and the optimal preemptive one. Task processing times were drawn from the uniform distribution of the given parameters.

To conclude the above analysis we may say that the LPT algorithm behaves quite well and may be used in practice. However, if one wants to have better performance guarantees, other approximation algorithms should be used, as for example MULTIFIT [41] or the algorithms proposed in [72] or in [84]. A comprehensive treatment of approximation algorithms for this and related problems is given in [42].

We may now pass to the third way of analyzing our chosen problem $P \| C_{\max}$. Theorem 3.1. gave a negative answer to the question about the existence of an optimization polynomial-time algorithm for solving $P2 \| C_{\max}$. However, we have not proved that our problem is NP-hard in the strong sense and we may try to find a pseudopolynomial optimization algorithm. It appears that such an algorithm may be constructed using ideas presented in [129, 104, 109]. It is based on a dynamic programming approach and its formulation for $P \| C_{\max}$ is follows. Let

Table 3.1. Mean performance of the LPT algorithm.

n, m	Parameters of task processing time distribution	C^*_{max}	C^{LPT}_{max}/C^*_{max}
6,3	1,20	20	1.00
9,3	1,20	32	1.00
15,3	1,20	65	1.00
6,3	20,50	59	1.05
9,3	20,50	101	1.03
15,3	20,50	166	1.00
8,4	1,20	23	1.09
12,4	1,20	30	1.00
20,4	1,20	60	1.00
8,4	20,50	74	1.04
12,4	20,50	108	1.02
20,4	20,50	185	1.01
10,5	1,20	25	1.04
15,5	1,20	38	1.03
20,5	1,20	49	1.00
10,5	20,50	65	1.06
15,5	20,50	117	1.03
25,5	20,50	198	1.01

$$x_j(t_1, t_2, \ldots, t_m) = \begin{cases} true \text{ if tasks } T_1, T_2, \ldots, T_j \text{ can be scheduled on processors} \\ P_1, P_2, \ldots, P_m, \text{ in such a way that } P_i \text{ is busy in time} \\ \text{interval } [0, t_i], i = 1, 2, \ldots, m \\ false \text{ otherwise} \end{cases}$$

with

$$x_0(t_1, t_2, \ldots, t_m) = \begin{cases} true \text{ if } t_i = 0, i = 1, 2, \ldots, m, \\ false \text{ otherwise.} \end{cases}$$

Then one defines the recursive equation in the following way

$$x_j(t_1, t_2, \ldots, t_m) = \bigvee_{i=1}^{m} x_{j-1}(t_1, t_2, \ldots, t_{i-1}, t_i - p_j, t_{i+1}, \ldots, t_m).$$

(Of course, $x_j(t_1, t_2, \ldots, t_m) = false$, for any $t_i < 0$).

For $j = 0, 1, \ldots, n$, compute $x_j(t_1, t_2, \ldots, t_m)$ for $t_i = 0, 1, \ldots, C: i = 1, 2, \ldots, m$, where C denotes an upper bound on the optimal schedule length C^*_{max}. Then C^*_{max} is determined as

$$C^*_{max} = \min\{\max\{t_1, t_2, \ldots, t_m\} : x_n(t_1, t_2, \ldots, t_m) = true\}.$$

The above procedure solves problem $P \| C_{max}$ in $O(nC^m)$ time, thus, for fixed m it is a pseudopolynomial-time algorithm. Hence, for small values of m and C the algorithm can be used even in computer applications. In general, however, its application is limited (see [102] for a survey of other enumerative approaches).

4. Scheduling on parallel processors

4.1. Introduction

A general scheme for the analysis of scheduling problems has been presented in Subsection 2.3 and illustrated by an exemplary problem $P \| C_{max}$ in Section 3. This pattern, however, cannot be fully repeated in this section, for other scheduling problems. First of all, there are many open questions in that area concerning especially the worst-case and probabilistic analysis of approximation algorithms for NP-hard problems. On the other hand, even the existing results form a large collection and we are not able to describe them all. Thus, we had to choose only some. Our choice was based on the relative significance of the results and on the inspiring role for further research they have had. We focused our attention on polynomial-time optimization algorithms and NP-hardness results, mentioning also some important results concerning worst-case analysis of approximation algorithms.

At this point we would like to comment briefly on the existing books and surveys concerning the area of scheduling problems which allow the reader to study the subject more deeply and give him references to many source papers. Two classical books [44, 7] give a good introduction to the theory of scheduling. The same audience is addressed by [48]. Other books [36, 106, 123] survey the state of the modern scheduling theory (i.e. taking into account most of the issues discussed in Sections 2.3 and 3) around the mid-seventies. Up-to-date collections of results can be found in the following survey-papers [70, 82, 102, 99, 110]. There are also some other surveys but they deal with certain areas of deterministic scheduling theory only. Let us also mention here an interesting approach to automatic generation of problem complexity listings containing the hardest easy problems, the easiest open ones, the hardest open ones, and the easiest hard ones [94, 95]. This generation program explores known complexity results and uses simple polynomial transformations as shown in Fig. 4.1. For each graph in the figure, the presented problems differ only by one parameter and the arrows indicate the direction of the polynomial transformation. These simple transformations are very useful in many practical situations when analyzing new scheduling problems. Thus, many of the results presented in that study can be immediately extended to cover a broader class of scheduling problems.

As we mentioned, in this section we will present some basic results concerned with scheduling on parallel processors. The presentation is divided into three subsections concerning the minimization of schedule length, mean flow time and maximum lateness, respectively. This is because these are the most important criteria and, moreover, they require quite different approaches.

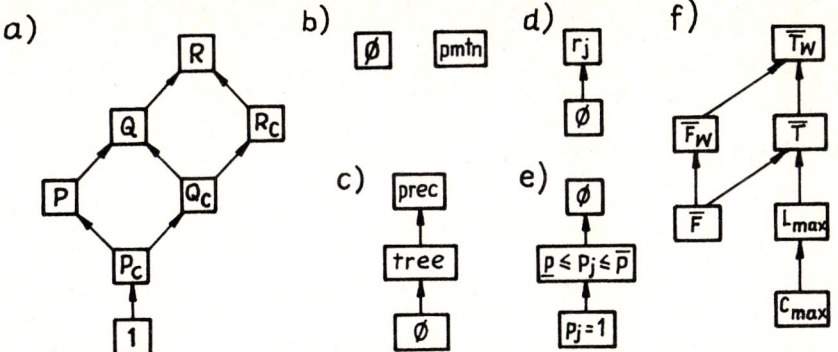

Fig. 4.1. Simple polynomial transformations among scheduling problems that differ by: type and number of processors (a), mode of processing (b), type of precedence constraints (c), ready times (d), processing times (e) and criterion (f).

4.2. Minimizing schedule length

A. Identical processors

The first problem to be considered is that of scheduling independent tasks on identical processors. However, it was discussed in detail in Section 3. We can only add that the interested reader may find several other approximation algorithms for the problem in question in the survey paper [42].

Let us pass now to the case of dependent tasks. At first tasks are assumed to be scheduled nonpreemptively. It is obvious that there is no hope of finding a polynomial-time optimization algorithm for scheduling tasks of arbitrary length since already $P \| C_{max}$ is NP-hard. However, one can try to find such an algorithm for unit-processing times of all the tasks. The first algorithm has been given for sceduling *forests*, consisting of either *in-trees* or *out-trees* [77]. We will first present Hu's algorithm for the case of an *in-tree*, i.e. a graph in which each task (node) has at most one immediate successor and there is only one task (root) with no successors (cf. Fig. 4.2). The algorithm is based on the notion of a *task level* in an in-tree which is defined as the number of tasks in the path to the root of the graph. (Thus, its name is also a *level algorithm*). The algortithm is as follows.

Algorithm 4.1. (Problem $P\,|\,\text{in-tree}, p_j = 1\,|\,C_{max}$)
1. Calculate the levels of tasks.
2. At each time unit, if the number of tasks without predecessors is no greater than m, assign them to processors and remove these tasks from the graph. Otherwise, assign to processors m non-assigned tasks with the highest levels and also remove them from the graph.

This algorithm can be implemented to run in $O(n)$ time. An example of its

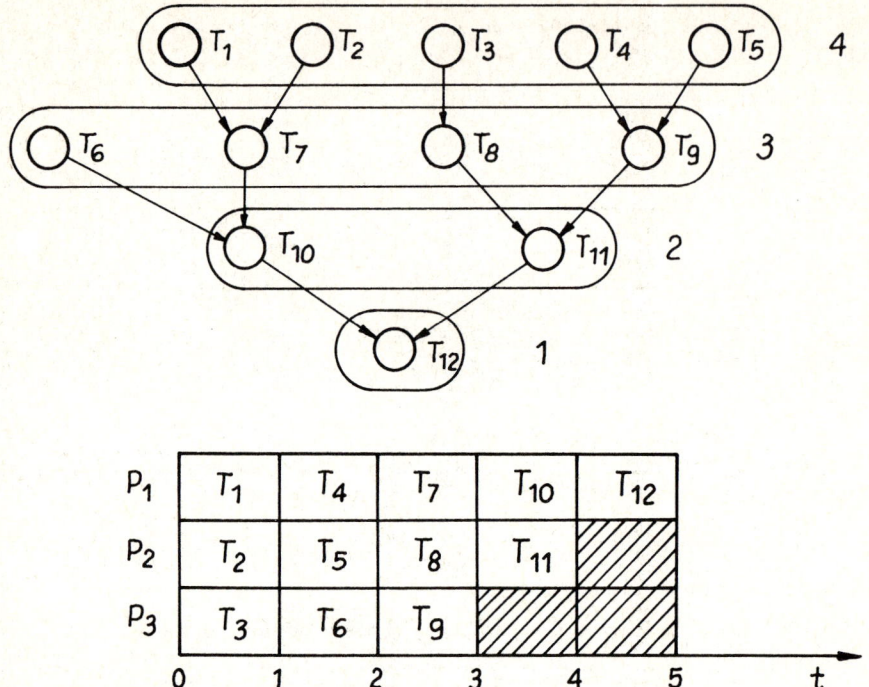

Fig. 4.2. An example of the application of Algorithm 4.1 for three processors.

application is shown in Fig. 4.2.

A forest consisting of in-trees can be scheduled by adding a dummy task that is an immediate successor of only the roots of in-trees, and then applying Algorithm 4.1. A schedule for an *out-tree*, i.e. a graph in which each task has at most one immediate predecessor and where there is only one task with no predecessor, can be constructed by changing the orientation of arcs, applying Algorithm 4.1 to the obtained in-tree and then reading the schedule backward (i.e., from right to left). It is interesting to note that the problem of scheduling opposing forests (that is, combinations of in-trees and out-trees) on an arbitrary number of processors is NP-hard [60]. However, when then number of processors is limited to 2, the problem is easily solvable even for arbitrary precedence graphs [40, 51, 52]. We present the algorithm given in [40] since it can be further extended to cover the preemptive case. The algorithm uses labels assigned to tasks, which take into account the level of the tasks and the numbers of their immediate successors. The following algorithm assigns labels and then uses them to find the shortest schedule ($A(T)$ denotes the set of all immediate successors of T).

Algorithm 4.2. (Problem $P2 \mid \text{prec}, p_j = 1 \mid C_{\max}$).
1. Assign label 1 to any T_0 for which $A(T_0) = \phi$.
2. Let labels $1, 2, \ldots, j-1$ be assigned. Let S be the set of unlabelled tasks with

no unlabelled successors. For each $T \in S$ let $1(T)$ denote a list of labels of tasks belonging to $A(T)$ ordered in decreasing order of their values. Let T^* be an element of S such that for all T in S, $1(T^*)$ is lexicographically smaller than $1(T)$. Assign label j to T^*. Repeat step 2 until all tasks are labelled.
3. Assign the tasks to processors in the way described in Step 2 of Algorithm 4.1, using labels instead of levels.

A careful analysis shows that the above algorithm can be implemented to run in time which is almost linear in n, plus the number of arcs in the precedence graph [131], (thus, $O(n^2)$), if the graph has no transitive arcs. Otherwise, they can be deleted in $O(n^{2.8})$ time [4]. An example of the application of Algorithm 4.2 is given in Fig. 4.3.

It must be stressed that the question concerning the complexity of problem with a fixed number of processors, unit processing time tasks, and arbitrary, precedence graphs is still open despite the fact that many papers have been devoted to solving various subcases (see [110]). Hence, several papers have dealt with approximation algorithms for these and more complicated problems. We quote some of the most interesting results. The application of level (critical path) algorithm (Algorithm 4.1) to solve $P \,|\, \text{prec}, p_j = 1 \,|\, C_{\max}$ has been analyzed in [34, 92]. The following bound has been proved.

$$R_{\text{level}} = \begin{cases} 4/3 & \text{for } m = 2 \\ 2 - \dfrac{1}{m-1} & \text{for } m \geqslant 3. \end{cases}$$

Slightly better is Algorithm 4.2 [96], for which we have

$$R = 2 - \frac{2}{m} \quad (m \geqslant 2).$$

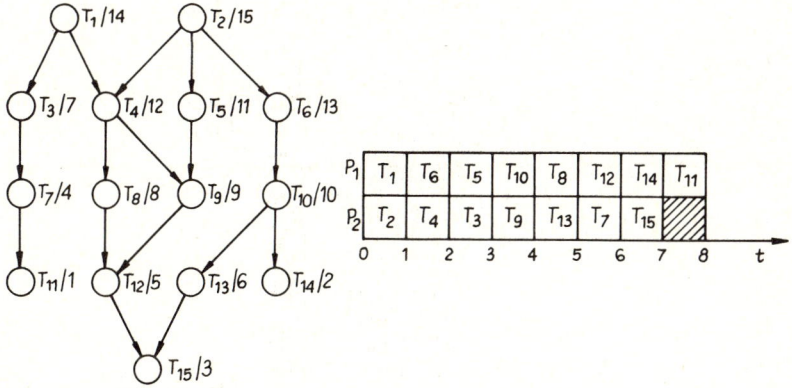

Fig. 4.3. An example of the application of Algorithm 4.2 (nodes are denoted by task/label).

In this context one should not forget the results presented in Section 3, where the list scheduling anomalies have been analyzed.

The analysis carried out in Section 3 showed also that preemptions can be profitable from the viewpoint of two factors. Firstly, they can make problems easier to solve, and secondly, they can shorten the schedule. In fact, in the case of dependent tasks scheduled on processors in order to minimize schedule length, these two factors are valid and one can construct an optimal preemptive schedule for tasks of arbitrary length and with other parameters the same as in Algorithm 4.1 and 4.2 [120, 121]. The approach again uses the notion of the level of task T_j in a precedence graph, by which is now understood the sum of processing times (including p_j) of tasks along the longest path between T_j and a terminal task (a task with no successors). Let us note that the level of a task which is being executed is decreasing. We have the following algorithm.

Algorithm 4.3. (Problems $P2\,|\,$pmtn, prec$\,|\,C_{max}$ and $P\,|\,$pmtn, forest$\,|\,C_{max}$)
1. Compute the level of all the tasks in the precedence graph.
2. Assign tasks at the highest level (say g tasks) to available processors (say h processors) as follows. If $g > h$, assign $\beta = h/g$ processors to each of the g tasks, thus obtaining a *processor-shared schedule*. Otherwise assign one processor to each task. If there are any processors left, consider the tasks at the next highest level and so on.
3. Process the tasks assigned in step 2 until one of the following occurs:
   — a task is finished,
   — a point is reached at which continuing with the present assignment means that a task at a lower level will be executed at a faster rate than a task at a higher one. In both cases go to step 2.
4. Between each pair of successive reassignment points (obtained in step 3), the tasks are rescheduled by means of Algorithm 3.1.

The above algorithm can be implemented to run in $O(n^2)$ time. An example of its application to an instance of problem $P2\,|\,$pmtn, prec$\,|\,C_{max}$ is shown in Fig. 4.4.

At this point let us also mention another structure of the precedence graph which enables one to solve a scheduling problem in polynomial time. To do this we have to present precedence constraints in the form of an activity network (task on arc precedence graph). Now, let S_I denote the set of all the tasks which may be performed between the occurrence of event (node) I and $I+1$. Such sets will be called *main sets*. Let us number from 1 to K the *processor feasible sets*, i.e. those main sets and those subsets of the main sets whose cardinalities are not greater than m. Now, let Q_j denote the set of indices of processor feasible sets in which task T_j may be performed, and let x_i denote the duration of set i. Now, the linear programming problem may be formulated in the following way [152, 19] (another LP formulation is presented for unrelated processors).

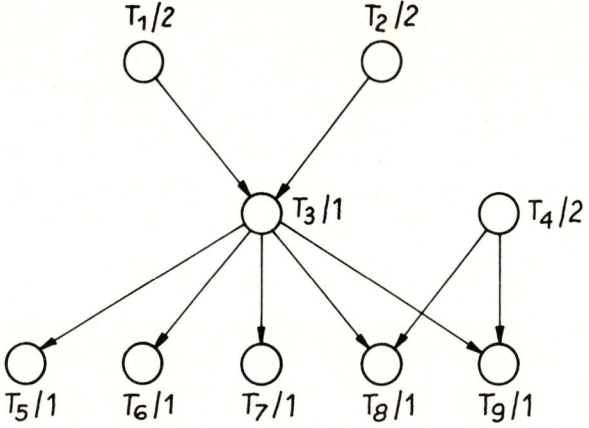

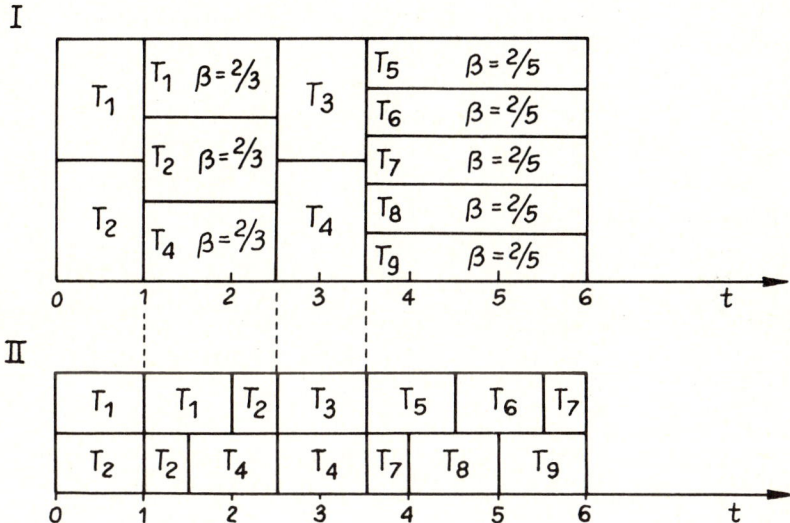

Fig. 4.4. An example of the application of Algorithm 4.3 for $m = 2$ (I-processor-shared schedule, II-preemptive one).

$$\text{minimize} \quad C_{\max} = \sum_{i=1}^{K} x_i \quad (4.1)$$

$$\text{subject to} \quad \sum_{i \in Q_j} x_i = p_j, \quad j = 1, 2, \ldots, n. \quad (4.2)$$

It is clear that the solution obtained depends on the ordering of the nodes of the activity network, hence an optimal solution is found when this topological ordering in unique. Such a situation takes place for a *uniconnected activity network* (uan), i.e. one in which any two nodes are connected by a directed path in one direction only. An example of a uniconnected activity network (and the corresponding precedence graph) is shown in Fig. 4.5. On the other hand, the number of variables in the above LP problem depends polynomially on the input length, when the number of processors m is fixed. We may now use Khachijan's procedure [89] which solves an LP problem in time which is polynomial in the number of variables and constraints. Hence, we may conclude that the above procedure solves problem $Pm\,|\,\text{pmtn, uan}\,|\,C_{\max}$ in polynomial time.

The general precedence graph, however, results in NP-hardness of the scheduling problem [143]. The worst-case behavior of Algorithm 4.3 in the case of problem $P\,|\,\text{pmtn, prec}\,|\,C_{\max}$ has been analysed in [96]:

$$R_{\text{Alg}\,4.3} = 2 - \frac{2}{m} \qquad (m \geqslant 2).$$

B. Uniform processors

Let us start with an analysis of independent tasks and nonpreemptive scheduling. Since the problem with arbitrary processing times is already NP-hard for identical processors, all we can hope to find is a polynomial-time optimization algorithm for tasks with unit processing times only. Such an algorithm has been given in [70], where a transportation network approach han been presented to solve problem $Q\,|\,p_j = 1\,|\,C_{\max}$. We describe it briefly below.

Let there be n sources j, $j = 1, 2, \ldots, n$ and mn sinks (i, k), $i = 1, 2, \ldots, m$, $k = 1, 2, \ldots, n$. Sources correspond to tasks and sinks to processors and positions on them. Set the cost of arc $(j, (i, k))$ equal to $c_{ijk} = k/b_i$; it corresponds to the completion time of task T_j processed on P_i in the k-th position. The arc flow x_{ijk} has the following interpretation:

$$x_{ijk} = \begin{cases} 1 & \text{if } T_j \text{ is processed on } P_i \text{ in the } k\text{-th position,} \\ 0 & \text{otherwise.} \end{cases}$$

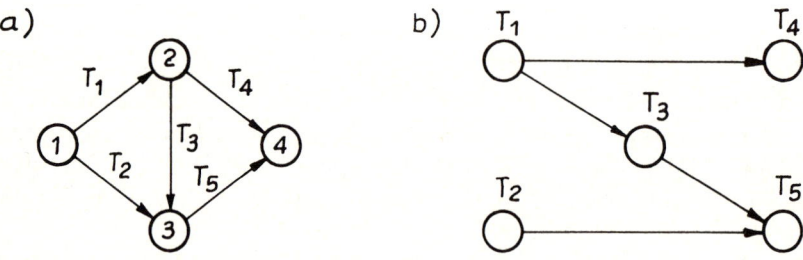

Fig. 4.5. An example of a simple uniconnected activity network (a) and the corresponding precedence graph (b): $S_1 = \{T_1, T_2\}$, $S_2 = \{T_2, T_3, T_4\}$, $S_3 = \{T_4, T_5\}$.

The min-max transportation problem can be now formulated as follows:

$$\text{minimize} \quad \max_{i,j,k} \{c_{ijk} x_{ijk}\} \tag{4.3}$$

$$\text{subject to} \quad \sum_{i=1}^{m} \sum_{k=1}^{n} x_{ijk} = 1 \quad \text{for all } j \tag{4.4}$$

$$\sum_{j=1}^{n} x_{ijk} \leq 1 \quad \text{for all } i, k \tag{4.5}$$

$$x_{ijk} \geq 0 \quad \text{for all } i, j, k. \tag{4.6}$$

The above problem can be solved in $O(n^3)$ time [70].

Since other problems of nonpreemptive scheduling of independent tasks are NP-hard, one may be interested in applying some heuristics. One of them has been presented in [113]. This is a list scheduling algorithm. Tasks are ordered on the list in nonincreasing order of their processing times and processors are ordered in nonincreasing order of their processing speeds. Now, whenever a processor becomes free the first nonassigned task on the list is scheduled on it (if there are more free processors, the fastest is chosen). The worst-case behavior of the algorithm has been evaluated for the case of $m+1$ processors in the system, m of which have processing speed factors equal to 1 and the remaining processor has the processing speed factor equal to b. The bound is as follows

$$R = \begin{cases} \dfrac{2(m+b)}{b+2} & \text{for } b \leq 2 \\ \dfrac{m+b}{2} & \text{for } b > 2. \end{cases}$$

It is clear that the algorithm does better when b and m decrease. Other algorithms have been analyzed in [114, 115, 62].

By allowing preemptions one can find optimal schedules in polynomial time. That is, problem $Q \mid \text{pmtn} \mid C_{\max}$ can be solved in polynomial time. We will present the algorithm given in [76], despite the fact that there is a more efficient one [65]. This is because the first algorithm also covers more general precedence constraints than the second, and it generalizes the ideas presented in Algorithm 4.3. It is based on two concepts: the task level (defined as previously as the processing requirement of the unexecuted portion of a task) and *processor sharing* (i.e. the possibility of assigning a task a part β ($0 \leq \beta \leq \max_i \{b_i\}$) of processing capacity). Let us assume that tasks are given in order of nonincreasing p_j's and processors in order of nonincreasing b_i's. It is quite clear that the minimum schedule length is

$$C_{\max}^* \geq C = \max \left\{ \max_{1 \leq k \leq m} \left\{ \frac{X_k}{B_k} \right\}, \frac{X_n}{B_m} \right\} \tag{4.7}$$

where X_k is the sum of processing requirements (standard processing times p_j) of the first k tasks and B_k the collective processing capacity (the sum of processing speed factors b_i) of the first k processors. The algorithm that constructs a schedule with the length equal to C, may be presented as follows.

Algorithm 4.4. (Problem $Q\,|\,\text{pmtn}\,|\,C_{\max}$)
1. Let h be the number of free (available) processors and g be the number of tasks at the highest level. If $g \leqslant h$, assign g tasks to be executed at the same rate on the fastest g processors. Otherwise, assign the g tasks onto the h processors. If there are any processors left, consider the tasks at the next highest level.
2. Preempt the processing and repeat step 1 whenever one of the following occurs:
   (a) a task is finished,
   (b) a point is reached at which continuing with the present assignment means that a task at a lower level is being executed at a faster rate than a task at a higher level.
3. To construct a preemptive schedule from the obtained shared one, reassign a portion of the schedule between every pair of events (denote the length of this interval by y).
   (a) If g tasks have shared g processors, assign each task to each processor for y/g time units.
   (b) Otherwise, i.e. if the number of tasks has been greater than the number of processors ($g > h$), let each of the g tasks have a processing requirement of p in the interval. If $p/b < y$, where b is the processing speed factor of the slowest processor, then the tasks can be assigned as in Algorithm 3.1, ignoring the different processor speeds. Otherwise, divide the interval into g equal subintervals. Assign the g tasks so that each task occurs in exactly h intervals, each time on a different processor.

The complexity of Algorithm 4.4 is $O(mn^2)$. An example of its application is shown in Fig. 4.6.

When considering dependent tasks, only preemptive optimization algorithms exist. Algorithm 4.4 also solves problem $Q2\,|\,\text{pmtn, prec}\,|\,C_{\max}$. Now, the level of a task is understood as in Algorithm 4.3, assuming standard processing times for all the tasks. When considering this problem one should also take into account the possibility of solving it for unconnected activity networks via a slightly modified linear programming approach (4.1) - (4.2) or another LP formulation described in the next subsection.

C. Unrelated processors

The case of unrelated processors is the most difficult. (For example it makes no sense to speak about unit-length tasks). Hence, no polynomial-time optimiza-

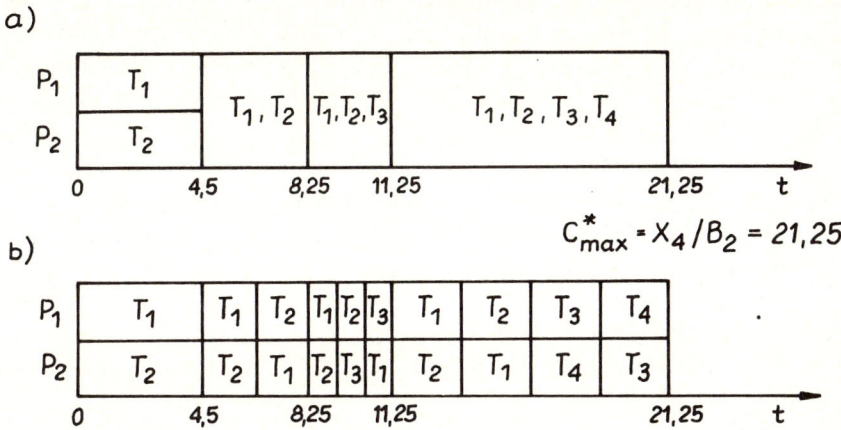

Fig. 4.6. An example of the application of Algorithm 4.4: $n = 4$, $m = 2$, $\overline{p} = [35, 26, 14, 10]$, $\overline{b} = [3,1]$; a processor shared schedule (a) and an optimal one (b).

tion algorithms are known for problems other than preemptive ones. Also, very little is known about approximation algorithms for this case. Some results have appeared in [79], but the obtained bounds are not very encouraging. Thus, we will pass to a preemptive scheduling model.

Problem $R \mid \text{pmtn} \mid C_{\max}$ can be solved by a two-phase method. The first phase consists in solving a linear programming problem formulated independently in [18, 20] and in [101]. The second phase uses the solution of the above LP problem and produces an optimal preemptive schedule [101, 141].

Let $x_{ij} \in [0, 1]$ denote a part of T_j processed on P_i. The LP formulation is as follows:

$$\text{minimize} \quad C_{\max} \tag{4.8}$$

subject to

$$C_{\max} - \sum_{j=1}^{n} p_{ij} x_{ij} \geq 0, \quad i = 1, 2, \ldots, m \tag{4.9}$$

$$C_{\max} - \sum_{i=1}^{m} p_{ij} x_{ij} \geq 0, \quad j = 1, 2, \ldots, n \tag{4.10}$$

$$\sum_{i=1}^{m} x_{ij} = 1 \quad j = 1, 2, \ldots, n. \tag{4.11}$$

Solving the above problem, we get $C_{\max} = C_{\max}^*$ and optimal values x_{ij}^*. However, we do not know the schedule i.e. the assignment of these parts to processors in time. It may be constructed in the following way. Let $t_{ij}^* = p_{ij} x_{ij}^*$, $i = 1, 2, \ldots, m$;

$j = 1, 2, \ldots, n$. Let $T = [t_{ij}^*]$ be an $m \times n$ matrix. The j-th column of T corresponding to task T_j will be called *critical* if $\sum_{i=1}^{m} t_{ij}^* = C_{max}^*$. By Y we denote an $m \times m$ diagonal matrix whose element y_{kk} is the total idle time on processor k, i.e. $y_{kk} = C_{max}^* - \sum_{i=1}^{n} t_{kj}^*$. Columns of Y correspond to dummy tasks. Let $V = [T, Y]$ be an $m \times (n+m)$ matrix. Now a set U containing m positive elements of matrix V can be defined as having exactly one element from each critical column and at most one element from other columns, and having exactly one element from each row. We see that U corresponds to a task set which may be processed in parallel in an optimal schedule. Thus, it may be used to construct a partial schedule of length $\delta > 0$. An optimal schedule is then produced as the union of the partial schedules. This procedure is summarized in Algorithm 4.5.

Algorithm 4.5.
1. Find set U.
2. Calculate the length of a partial schedule

$$\delta = \begin{cases} v_{min} & \text{if } C_{max} - v_{min} \geq v_{max} \\ C_{max} - v_{max} & \text{otherwise} \end{cases}$$

where

$$v_{min} = \min_{v_{ij} \in U} \{v_{ij}\}, \quad v_{max} = \max_{v_{ij} \in U} \{v_{ij}\}.$$

3. Decrease C_{max} and $v_{ij} \in U$ by δ. If $C_{max} = 0$, an optimal schedule has been constructed. Otherwise go to step 1.

Now we only need an algorithm that finds set U for a given matrix V. One of the possible algorithms is based on the network flow approach. A corresponding network has m nodes corresponding to processors (rows of V) and $n + m$ nodes corresponding tasks (columns of V) (c.f. Fig. 4.7). A node i from the first group is connected by an arc to a node j of the second group if and only if $V_{ij} > 0$. Arc flows are constrained by b from below and by $c = 1$ from above. The value of b is equal to 1 for arcs joining the source with processor-nodes and critical task nodes with the sink, and to 0 for other arcs. We see that finding a feasible flow in this network is equivalent to finding set U.

The overall complexity of the above approach is bounded from above by a polynomial in the input length. This is because the LP problem may be solved in polynomial time using Khachijan's algorithm [89]; the loop in Algorithm 4.5 is repeated at most mn times and solving the network flow problem requires $O(z^3)$ time, where z is the number of network nodes [87].

When dependent tasks are considered, linear programming problems similar to (4.1) - (4.2) or to (4.8) - (4.11) can again be formulated taking into account an activity network presentation. For example, in the latter formulation one defines x_{ijk} as a part of task T_j processed on processor P_i in the main set S_k.

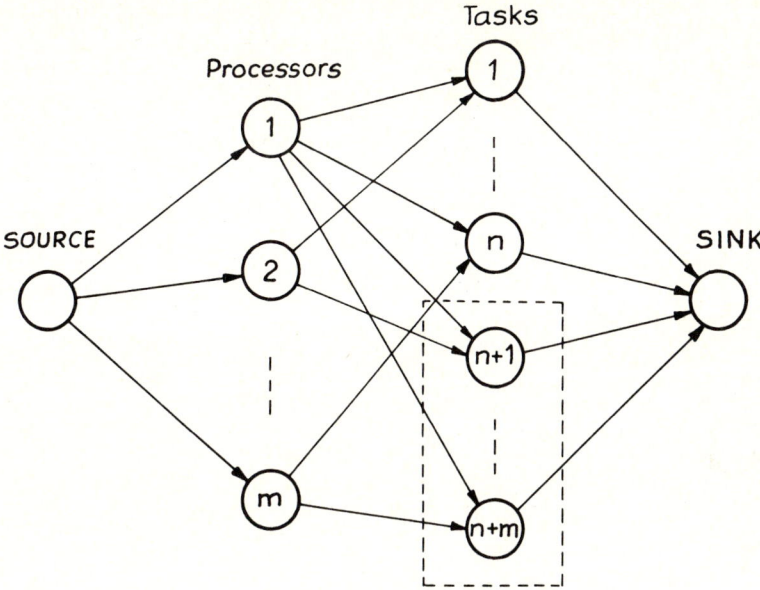

Fig. 4.7. Finding set U by the network flow approach.

Solving the LP problem for x_{ijk}, one then applies Algorithm 4.5 for each main set. If the activity network is uniconnected, an optimal schedule is constructed in this way.

We can complete this subsection by remarking that introducing different ready times into the considered problems is equivalent to minimizing maximum lateness. We will consider such problems in Section 4.4.

4.3. Minimizing mean flow time

A. Identical processors

In the case of identical processors preemptions are not profitable (when equal ready times are assumed) from the viewpoint of the value of the mean flow time [116]. Thus, we can limit ourselves to considering nonpreemptive schedules only.

When analyzing the nature of the criterion, one may expect that by assigning tasks in nondecreasing order of their processing times the mean flow time will be minimized. In fact, this simple SPT rule (shortest processing time) minimizes the criterion for a one-processor case [114]. A proper generalization will also produce an optimization algorithm for $P \| \Sigma C_j$ [44]. It is as follows:

Algorithm 4.6. (Problem $P \| \Sigma C_j$)
1. Order tasks in nondecreasing order of their processing times.
2. Take consecutive m-tuples from the list and for each m-tuple assign the tasks

to m different processors.
3. Process tasks assigned to each processor in SPT order.

The complexity of the algorithm is obviously $O(n \log n)$.

In this context let us also mention that introducing different ready times makes the problem strongly NP-hard even for the case of one processor [112]. On the other hand, problem $1 \| \Sigma\, w_j\, C_j$, i.e. with non-equal weights of the tasks, can be solved by the weighted-SPT rule, that is, the tasks are scheduled in nondecreasing order of ratios p_j/w_j [142]. However, unlike the mean flow time criterion, an extension to the multiprocessor case does not exist here, since problem $P2\|\Sigma\, w_j\, C_j$ is already NP-hard [30].

Let us pass now to dependent tasks and first consider the case of one processor. It appears that the WSPT rule can be extended to cover also the case of tree-like precedence constraints [73, 3, 133]. The algorithm is based on the observation that two tasks T_i, T_j can be treated as one with processing time $p_i + p_j$ and weight $w_i + w_j$, if $T_i < T_j$, $p_i/w_i > p_j/w_j$ and all other tasks either precede T_i, succeed T_j, or are incomparable with either. In order to describe the algorithm we need the notion of a feasible successor set of task T_i. This set Z_i, $i = 1, 2, \ldots, n$, is defined as one containing tasks that may be processed after T_i has been finished without the completion of any tasks not belonging to Z_i. Thus, this set has the following properties:
- $T_i \in Z_i$,
- if $T_k \in Z_i$ and $T_i \neq T_k$, then $T_i < T_k$,
- if $T_k \in Z_i$ and $T_j < T_k$, then either $T_j \in Z_i$ or $T_j < T_i$.

Below we present an algorithm for out-trees.

Algorithm 4.7. (Problem $1 \mid$ out-tree $\mid \Sigma\, w_j\, C_j$)
1. For each task calculate

$$f_i = \min \left\{ \left(\sum_{T_k \in Z_i} p_k\right) \bigg/ \left(\sum_{T_k \in Z_i} w_k\right) \right\},$$

where the minimum is taken over all feasible successor sets of task T_i.
2. Process tasks in non-decreasing order of f_i, observing precedence constraints.

The above algorithm can be implemented to run in $O(n \log n)$ time. The case of in-trees can be solved by a similar modification as in the case of Algorithm 4.1. Let us note that a more general case involving a *series-parallel* precedence graph can also be solved in $O(n \log n)$ time [98], but general precedence constraints result in NP-hardness of the problem even for equal weights or processing times [98, 107].

When the multiple processor case in considered, then $P \mid$ out-tree, $p_j = 1 \mid \Sigma\, C_j$ is solved by Algorithm 4.1 adapted to the out-tree case [128], and $P2 \mid$ prec, $p_j =$

$= 1|\Sigma C_j$ is strongly NP-hard [107], as are almost all problems with arbitrary processing times since already problems $P2|\text{in-tree}|\Sigma C_j$ and $P2|\text{out-tree}|\Sigma C_j$ are NP-hard [132]. Unfortunately, no approximation algorithms for these problems are evaluated from the point of view of their worst-case behavior.

B. Uniform and unrelated processors

The results of Subsection A also indicate that scheduling dependent tasks on uniform (or unrelated) processors is in general an NP-hard problem. No heuristics have been investigated either. Thus, we will not consider this subject below.

On the other hand, in this case preemptions may be worthwhile, thus we have to treat non-preemptive and preemptive scheduling separately. Let us start with uniform processors and independent tasks to be scheduled nonpreemptively. In this case the flow time of a task is given as $F_{i[k]} = (1/b_i) \Sigma_{j=1}^{k} p_{i[j]}$, where $i[k]$ is the index of a task which is processed in the k-th position on P_i. Let us denote by n_i the number of tasks processed on processor P_i. Thus, $n = \Sigma_{i=1}^{m} n_i$. The mean flow time is now given by the formula

$$F = \frac{\sum_{i=1}^{m} \frac{1}{b_i} \sum_{k=1}^{n_i} (n_i - k + 1) p_{i[k]}}{n}. \tag{4.12}$$

It is easy to see that the numerator in the above formula is the sum of n terms each of which is the product of a processing time and one of the following coefficients:

$$\frac{1}{b_1} n_1, \frac{1}{b_1}(n_1 - 1), \ldots, \frac{1}{b_1}, \frac{1}{b_2} n_2,$$

$$\frac{1}{b_2}(n_2 - 1), \ldots, \frac{1}{b_2}, \ldots, \frac{1}{b_m} n_m, \frac{1}{b_m}(n_m - 1), \ldots, \frac{1}{b_m}.$$

It is known that such a sum is minimized by matching n smallest coefficients in nondecreasing order with processing times in nonincreasing order [44]. An $O(n \log n)$ implementation of this rule has been given in [75].

In the case of preemptive scheduling, it is possible to show that there exists an optimal schedule for $Q|\text{pmtn}|\Sigma C_j$ in which $c_j \leq c_k$ if $p_j < p_k$ [101]. On the basis of this observation, the following algorithm may be proposed [61] (assume that processors are ordered in nonincreasing order of their processing speed factors).

Algorithm 4.8. (Problem $Q|\text{pmtn}|\Sigma C_j$)
1. Place the tasks on the list in the SPT order.
2. Having scheduled tasks $T_1, T_2, \ldots, T_j$, schedule task T_{j+1} to be completed as

early as possible (preempting when necessary).

The complexity of the algorithm is $O(n \log n + mn)$. An example of its application is given in Fig. 4.8.

Let us now consider unrelated processors and problem $R \| \Sigma C_j$. An approach to its solution is based on the observation that T_j processed on processor P_i as the last task contributes its processing time p_{ij} to $\overline{F}$. The same task processed in the last but one position contributes $2p_{ij}$, and so on [30]. This reasoning allows one to construct a matrix Q presenting contributions of particular tasks processed in different positions on different processors, to the value of $\overline{F}$.

$$Q = \begin{bmatrix} [p_{ij}] \\ 2\,[p_{ij}] \\ \vdots \\ n\,[p_{ij}] \end{bmatrix}.$$

The problem is now to choose n elements from Q such that:
— there is exactly one element from each column,
— there is at most one element from each row,
— the sum of elements is minimum.

The above problem may be solved in a natural way via the transportation problem. The corresponding transportation network is shown in Fig. 4.9.

Careful analysis of the problem allows its being solved in $O(n^3)$ [30].

To end this subsection, let us mention that the complexity of problem $R \,|\, \text{pmtn} \,|\, \Sigma C_j$ is still an open question.

4.4. Minimizing maximum lateness

A. Identical processors

It seems to be quite natural that in the case of this criterion the general rule

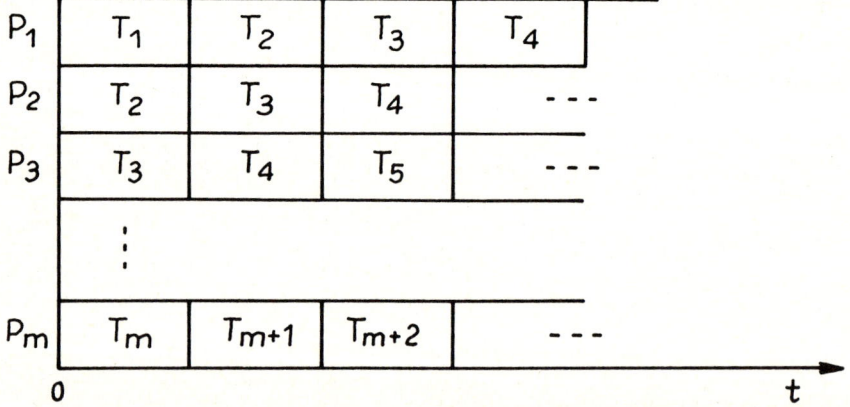

Fig. 4.8. An example of the application of Algorithm 4.8.

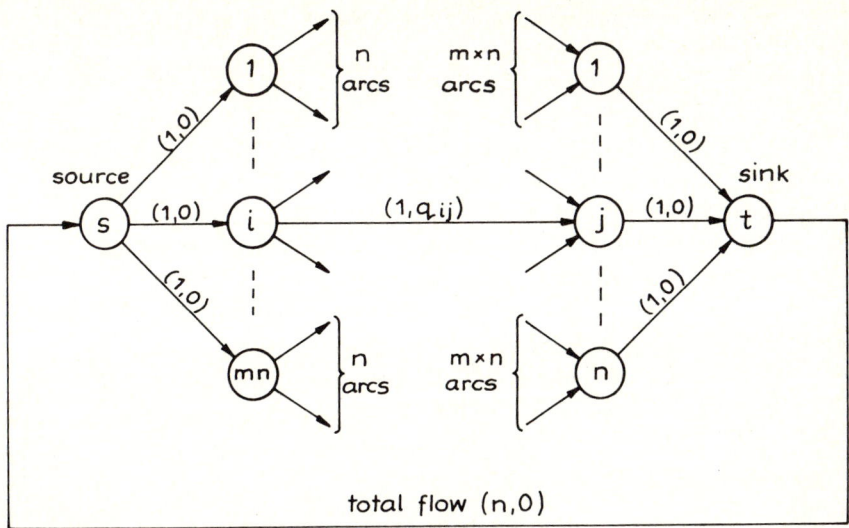

Fig. 4.9. The transportation network for problem $R\|\Sigma C_j$: arcs are denoted by (c, y), where c is a capacity and y the cost of a unit flow.

should be to schedule tasks according to their earliest due-dates (EDD-rule). However, this simple rule of Jackson [80] produces optimal schedules in only a few cases. In most cases more complicated algorithms are necessary or the problems are NP-hard. Let us start with nonpreemptive scheduling and independent tasks.

Problem $1\|L_{\max}$ may be solved by the EDD rule in $O(n \log n)$ time, but introducing different ready times makes the problem NP-hard in the strong sense [112]. Assuming, however, equal processing times, the problem of finding a feasible schedule with respect to given ready times and deadlines, denoted by $1|p_j = p, r_j, d_j|-$, can be solved in $O(n \log n)$ time [59] by slightly modifying the EDD rule. A bisection search over the possible $L_{\max}$ values leads to a polynomial algorithm for $1|r_j, p_j = p|L_{\max}$.

Of course taking into account Fig. 4.1 and the relation between the $C_{\max}$ and $L_{\max}$ criteria, we see that all the problems that are NP-hard with the $C_{\max}$ criterion remain NP-hard with the $L_{\max}$ one as well. For example $P2\|L_{\max}$ is NP-hard. On the other hand, unit processing times of tasks make the problem easy and $P|p_j = 1, r_j|L_{\max}$ can be solved by an obvious application of the EDD rule [12]. Moreover, problem $P|p_j = p, r_j|L_{\max}$ can also be solved in polynomial time by an extension of the above-mentioned single processor algorithm [59]. Unfortunately, very little is known about the worst-case behavior of approximation algorithms for the NP-hard problem in question.

The preemptive mode of processing makes the solution of the scheduling problem much easier. The fundamental approach in that area is testing for feasibility via the network flow approach [74] (problem $P\,|\,\text{pmtn}, r_j, d_j\,|\,-$). Let the ready times and the deadlines in $P\,|\,\text{pmtn}, r_j, d_j\,|\,-$ be ordered on a list in such a way that: $e_0 < e_1 < \ldots < e_k$, $k \leqslant 2n$. A corresponding network has two sets of nodes. (Fig. 4.10). The first set corresponds to time intervals in a schedule, i.e. node W_i, $i = 1, 2, \ldots, k$, corresponds to interval $[e_{i-1}, e_i]$. The second set corresponds to the task set. The capacity of the arc joining the source of the network with node W_i is equal to $m(e_i - e_{i-1})$ and thus corresponds to the total processing of m processors in this interval. If task T_j can be processed in interval $[e_{i-1}, e_i]$ (because of its ready time and deadline) then W_i is joined with T_j by an arc of capacity $e_i - e_{i-1}$. Node T_j is joined with the sink of the network by an arc with a capacity equal to p_j and a lower bound also equal to p_j. We see that finding a feasible flow pattern corresponds to constructing a feasible schedule and this test can be made in $O(n^3)$ time. Then a binary search can be conducted on the optimal value of $L_{\max}$, with each trial value of $L_{\max}$ inducing deadlines which are checked for feasibility by means of the network computation. This procedure can be implemented to solve problem $P\,|\,\text{pmtn}, r_j\,|\,L_{\max}$ in $O(n^3 \min\{n^2, \log n + \log \max\{p_j\}\})$ [93].

Now let us pass to dependent tasks. A general approach in this case consists in assigning modified due dates to tasks, depending on the number and due dates of their successors. Of course, the way in which modified due dates are calculated depends on the parameters of the problem in question. Let us start with nonpreemptive scheduling and with one processor. Problem $1\,|\,\text{prec}\,|\,L_{\max}$ may be solved in $O(n^2)$ time by assigning the following due dates

$$d_j^* = \min\{d_j, \min\{d_i : T_j < T_i\}\}, \quad j = 1, 2, \ldots, n \tag{4.13}$$

and then scheduling tasks in nondecreasing order of these due dates, observing precedence constraints [104]. Let us note that this algorithm can be generalized

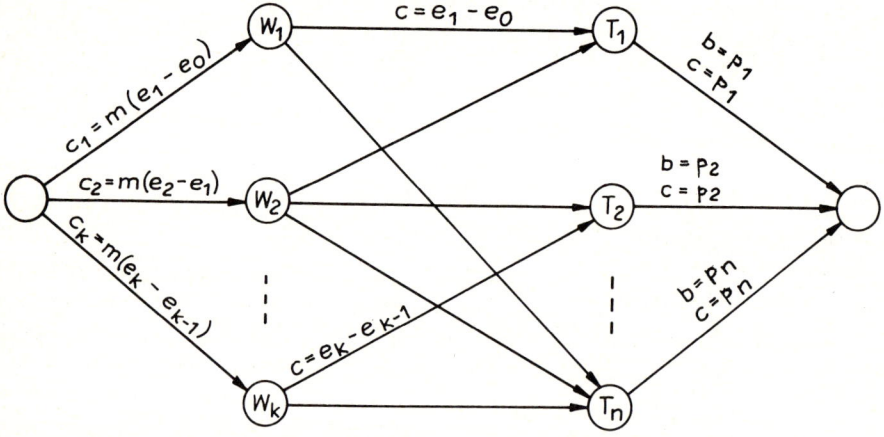

Fig. 4.10. The network corresponding to problem $P\,|\,\text{pmtn}, r_j\, d_j\,|\,-$.

to cover the case of an arbitrary nondecreasing cost function [97]. When scheduling on a multiple processor system only unit processing times can result in polynomial-time scheduling algorithms. Let us start with in-tree precedence constraints and assume that if $T_i < T_j$ then $i > j$. The following algorithm minimizes L_{max} [28] ($is(j)$ denotes an immediate successor of T_j).

Algorithm 4.9. (Problem $P |$ in-tree, $p_j = 1 | L_{max}$)
1. Set $d_1^* := 1 - d_1$.
2. For T_k, $k = 2, 3, \ldots, n$, compute a modified due date according to the formula

$$d_k^* = \max \{1 + d_{is(k)}^*, 1 - d_k\}.$$

3. Schedule tasks in nonincreasing order of their modified due dates subject to precedence constraints.

This algorithm can be implemented to run in $O(n \log n)$ time. Surprisingly out-tree precedence constraints result in the NP-hardness of the problem [29]. However, when we limit ourselves to two processors, a different way of computing modified due dates can be proposed which allows one to solve the problem in $O(n^2)$ time [56].

Algorithm 4.10. (Problem $P2 |$ prec, $p_j = 1 | L_{max}$)
1. Choose T_k which is not assigned a modified due date and all of whose successors have been assigned these due dates.
2. Define set $A(T_k)$ consisting of all successors of T_k having modified due dates not greater than d_i^*. Define a modified due date of T_k as:

$$d_k^* := \min \{d_k, \min \{(d_i^* - \lceil 1/2 g(k, d_i^*) \rceil) : T_i \in A(T_k)\}\} \quad (7).$$

If there is any task with no modified due date assigned, go to step 1.
3. Schedule tasks in nondecreasing order of their modified due-dates subject to precedence constraints.

The above algorithm may be generalized to cover the case of different ready times too but the running time is then $O(n^3)$ [57] and this is as much as we can get assuming nonpreemptive scheduling. Preemptions allow one to solve problems with different processing times but, in general, other parameters remain the same as in the nonpreemptive case.

When tasks are to be scheduled on one processor, problem $1 |$ pmtn, prec, $r_j | L_{max}$ can be solved in $O(n^2)$ time by defining modified due dates according to formula (4.13) and then scheduling available tasks in nondecreasing order of these due dates, subject to precedence constraints [11]. Let us note that this approach can

---

(7) $\lceil x \rceil$ denotes the smallest integer not smaller than x.

be generalized again to cover the case of an arbitrary nondecreasing cost function [8].

When considering the multiple processor case it is interesting to note that there are preemptive counterparts of all of the nonpreemptive scheduling algorithms for unit-length tasks, presented above. In [100] algorithms have been presented that are preemptive counterparts of Algorithms 4.9, 4.10 and the one presented in [57]. Hence problems $P|\text{pmtn, in-tree}|L_{max}$, $P2|\text{pmtn, prec}|L_{max}$ and $P2|\text{pmtn, prec}, r_j|L_{max}$ are solvable in polynomial-time. These preemptive scheduling algorithms employ essentially the same techniques for dealing with precedence constraints as the corresponding algorithms for unit-length tasks. However, these algorithms are more complex and their lengthy description prevents us from presenting them here.

B. Uniform and unrelated processors

From the considerations of Subsection 4.2 and 4.4 A, we see that nonpreemptive scheduling to minimize L_{max} is in general a hard problem, and practically the only known polynomial-time optimization algorithm is that for problem $Q|p_j=1|L_{max}$. This can be solved via transportation problem formulation (4.3) - (4.6), but now $c_{ijk} = k/b_i - d_j$. Thus, below we can concentrate on preemptive scheduling.

We will consider uniform processors first. One of the most interesting algorithms in that area has been presented for problem $Q|\text{pmtn}, r_j|L_{max}$ [47]. It is a generalization of the network flow approach to the feasibility testing of problem $P|\text{pmtn}, r_j, d_j|-$ described above.

In the case, of precedence constraints, $Q2|\text{pmtn, prec}|L_{max}$ and $Q2|\text{pmtn, prec}, r_j|L_{max}$ can be solved in $O(n^2)$ and $O(n^6)$ time, respectively, by the algorithms already mentioned [100].

As far as unrelated processors are concerned, problem $R|\text{pmtn}|L_{max}$ can be solved by LP formulation very similar to (4.8) - (4.11) [101], but now x_{ij}^k denotes the amount of T_j which is processed on P_i in time interval $[d_{k-1} + L_{max}, d_k + L_{max}]$ (due dates are ordered: $d_1 \leqslant d_2 \leqslant \ldots \leqslant d_n$). Moreover, Algorithm 4.5 is now applied to each matrix $T^{(k)} = [t_{ij}^{k*}]$, $k = 1, 2, \ldots, n$. It this context let us also mention that the case when precedence constraints form a uniconnected activity network, can also be solved via the same modification of the LP problem as described for the C_{max} criterion [138].

5. Scheduling on dedicated processors

5.1. Introduction

In this section we will consider scheduling problem in which a task requires processing on more than one machine and hence may be represented as a set of operations. Unfortunately, in this case most of the scheduling problems are NP-hard, and this is especially true for criteria other than C_{max}. Below we will concentrate on polynomial-time algorithms and because of that only open-shop and flow-shop scheduling problems are considered since the job shop scheduling problem is NP-hard for almost all the cases (see [102] as the survey).

5.2. Open shop scheduling

Let us consider nonpreemptive scheduling first. Problem $O2 \| C_{max}$ can be solved in $O(n)$ time [63]. We give here a simplified description of the algorithm presented in [102]. For convenience let us denote $a_j = p_{1j}$, $b_j = p_{2j}$, $A = \{T_j : a_j \geqslant b_j\}$, $B = \{T_j : a_j < b_j\}$ and $J_1 = \sum_{j=1}^{n} a_j$, and $J_2 = \sum_{j=1}^{n} b_j$.

Algorithm 5.1. (Problem $O2 \| C_{max}$)
1. Choose any two tasks T_k and T_l for which we have

$$a_k \geqslant \max_{T_j \in A} \{b_j\}, \qquad b_l \geqslant \max_{T_j \in B} \{a_j\}.$$

Let $A' = A - \{T_k, T_l\}$ and $B' = B - \{T_k, T_l\}$.
2. Construct separate schedules for $B' \cup \{T_l\}$ and $A' \cup \{T_k\}$, as depicted in Fig. 5.1. Other tasks from A' and B' are scheduled arbitrarily.
3. Suppose that $J_1 - a_l \geqslant J_2 - b_k$ ($J_1 - a_l < J_2 - b_k$ is symmetric). Join both schedules in the way shown in Fig. 5.2. Move tasks from $B' \cup \{T_l\}$ processed on P_2 to the right.
4. Change the order of processing on P_2 in such a way that O_{2k} is processed as the first one on this processor.

The above problem becomes NP-hard as the number of processors increases to 3. But again preemptions result in a polynomial time algorithm. That is, problem $O \mid pmtn \mid C_{max}$ can be solved [63] by taking

$$C^*_{max} = \max\left\{\max_{j}\left\{\sum_{i=1}^{m} p_{ij}\right\}, \max_{i}\left\{\sum_{j=1}^{n} p_{ij}\right\}\right\}$$

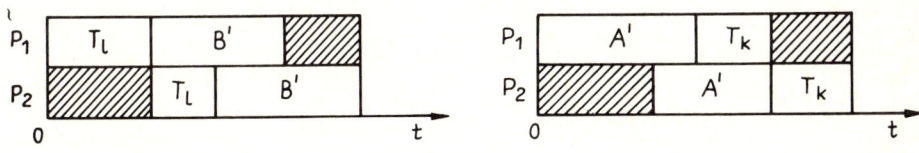

Fig. 5.1.

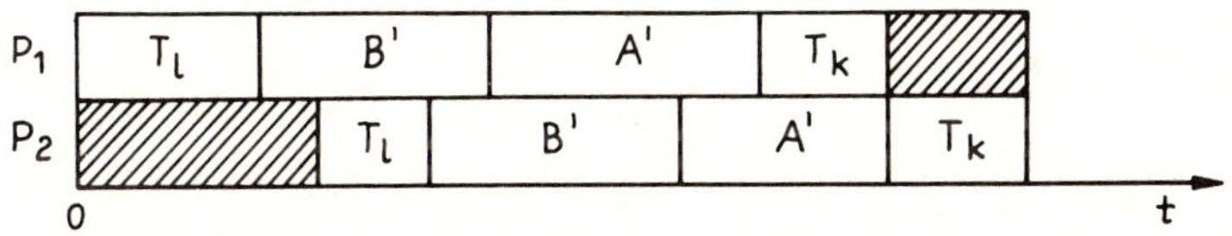

Fig. 5.2.

and then applying Algorithm 4.5.

Let us also mention here that problems $O2\,\|\,\Sigma\,C_j$ and $O2\,\|\,L_{max}$ are NP-hard [1] and [103], and problem $O\,|\,pmtn, r_j\,|\,L_{max}$ is solvable via the linear programming approach [35]. As far as heuristics are concerned, an arbitrary list scheduling and SPT algorithms have been evaluated for $O\,\|\,\Sigma\,C_j$ [1]. Their asymptotic performance ratios are as follows

$$R_L^\infty = n, \qquad R_{SPT}^\infty = m.$$

Since the number of tasks is usually much larger than the number of processors, the bounds indicate the advantage of SPT schedules over arbitrary ones.

5.3. Flow shop scheduling

One of the most classical algorithms in this area is that for problem $F2\,\|\,C_{max}$ due to S. Johnson [81]. It is as follows.

Algorithm 5.2. (Problem $F2\,\|\,C_{max}$)
1. Choose tasks such that $p_{1j} \leqslant p_{2j}$. Schedule them in nondecreasing order of their p_{1j}'s.
2. Schedule other tasks in nonincreasing order of their p_{2j}'s.

It is clear that this algorithm requires $O(n \log n)$ time. It can be extended to cover also the special case of three-processor scheduling in which $\min_j \{p_{1j}\} \geqslant \max_j \{p_{2j}\}$ or $\min_j \{p_{3j}\} \geqslant \max_j \{p_{2j}\}$. In this case the processing time on the second processor is not important and an optimal schedule can be obtained by applying Algorithm 5.2 to processing times $(p_{1j} + p_{2j}, p_{2j} + p_{3j})$. However, more complicated assumptions concerning problem parameters (i.e. precedence constraints, more processors involved or other criteria) make the problem strongly NP-hard. Thus, one is interested in evaluating some heuristic algorithms, but not much work has been done in this area. In fact, the only results have appeared in [64], [9] and [127]. In the first paper, the worst-case behavior of an algorithm H based on Algorithm 5.2 has been proposed for problem $F\,\|\,C_{max}$. Its worst-case behavior is proved to be

$$R_H = \left\lceil \frac{m}{2} \right\rceil.$$

In the second paper, a quite complicated heuristic A has been proposed whose absolute performance ratio does not depend on n and is proved to be

$$R_A = (m-1)(3m-1)p_{max}/2.$$

In the last paper, heuristics are proposed that replace m machines by two machines, and the task processing times are defined as the sums of the appropriate original processing times. Absolute performance ratios are proportional to m.

As far as the preemptive scheduling is concerned, the situation is the same as described above, i.e. $F2\,|\,\text{pmtn}\,|\,C_{\max}$ is solved by Algorithm 5.2 and other problems are strongly NP-hard. The only exception is $F2\,|\,\text{pmtn}\,|\,\Sigma\,C_j$, which is open.

6. New directions in scheduling theory

6.1. Introduction

In this section we will present some of the new directions in determinstic scheduling theory. We have chosen two of them: scheduling under resource constraints when each task, besides a processor, may also require certain additional scarce resources, and scheduling in microprocessor systems, when each task may require more than one processor at a time. Our choice has been primarily motivated by the relevance of the chosen topics to computer scheduling. Thus, we have not been able to include some other very interesting areas, as for example:

— scheduling with a different model of task processing, e.g. relating the processing speed of a task to the amounts of resources granted (see e.g. [147, 148, 149, 151] and [150] as a survey);

— multicriteria analysis of scheduling problems (see e.g. [78, 139, 140, 144, 145]);

— vehicle routing and scheduling problems (see [108] for an interesting survey of the results obtained).

Also, we have not been able to discuss the connections between scheduling and inventory theory, despite the fact that an integration of these two areas seems necessary and may lead to better understanding of the general production planning problem (cf. e.g. [111]).

6.2. Scheduling under resource constraints

The model considered here is very similar to the one defined in Section 2. The only difference is the existence of s types of additional resources $R_1, R_2, \ldots, R_s$, available in $m_1, m_2, \ldots, m_s$ units, respectively. Each task T_j requires for its processing one processor and certain amounts of additional resources specified by resource requirement vector $\bar{r}(T_j) = [r_1(T_j), r_2(T_j), \ldots, r_s(T_j)]$, where $r_l(T_j)$ $(0 \leq r_l(T_j) \leq m_l)$, $l = 1, 2, \ldots, s$, denotes the number of units of R_l required for the processing of T_j. (In the case of job shop scheduling each operation is characterized by a resource requirement vector). We will assume here that all required resources are granted to a task before its processing begins or resumes (in the case of preemptive scheduling), and they are returned by the task after its completion or in the case of its preemption. These assumptions define the simplest rule of preventing system deadlocks (see e.g. [37]) which is often used in practice, despite the fact that it may lead to a not very efficient use of additional resources. (For other methods which tend to maximize resource usage at the expense of tedious overhead involved see e.g. [71, 31, 32]). On the other hand, this method allows the construction of optimization algorithms for a wide class of scheduling problems.

It is not hard to see that the above model is especially justified in computer systems, where additional resources can stand for primary memory, mass storage, channels, i/o devices, etc. But one should not forget about other applications of this model in which tasks besides machines, can also require other limited resources for their processing, such as manpower, tools, space, etc.

At this point we would also like to present possible transformations among scheduling problems that differ only in their resource requirements (see Fig. 6.1). In this figure six basic resource requirements are presented. All but two of these transformations are quite obvious. A transformation Π (res...) $\propto \Pi$ (res1 ..) has been proved for the case of saturation of processors and additional resources [55] and we will not present it here. The second Π (res1 ..) $\propto \Pi$ (res.11) has been proved quite recently [17] and we will quote this proof below. For a given instance of the first problem we construct a corresponding instance of the second problem by assuming the parameters all the same, except resource contraints. Then for each pair T_i, T_j such that $r_1(T_i) + r_1(T_j) > m_1$ (in the first problem), resource R_{ij} available in the amount of one unit is defined. Tasks T_i, T_j require a unit of R_{ij}. Other tasks do not require this resource. It follows that $r_1(T_i) + r_1(T_j) \leq m_1$ in the first problem if and only if for each resource R_k, $r_k(T_i) + r_k(T_j) \leq 1$ in the second problem.

We will now pass to the description of basic results obtained in the area of resource constrained scheduling. Space limitations prohibit us even from only quoting all the results available in that area. Thus, we have chosen as an example the problem of scheduling tasks on parallel identical processors to minimize schedule length. Other cases have been considered in the following papers:

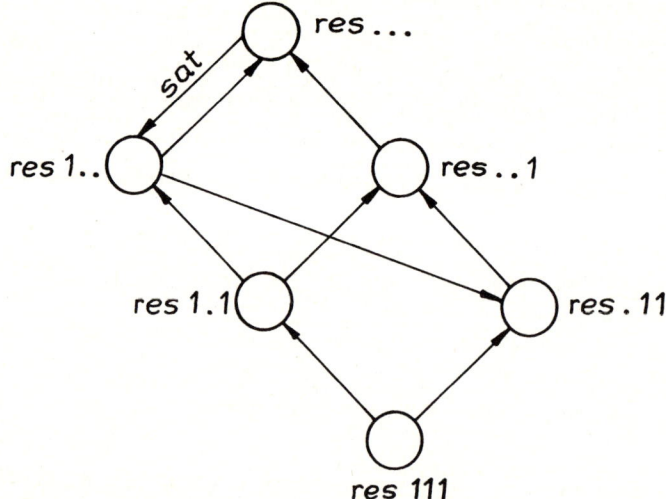

Fig. 6.1. Polynomial transformations among resource-constrained scheduling problems.

scheduling on uniform and unrelated processors [27, 137, 138, 153] minimizing mean flow time in [14, 25, 27], minimizing maximum lateness in [15, 16, 17] and scheduling in job shop systems in [27, 125, 26].

Let us first consider independent task and nonpreemptive scheduling. The problem of scheduling unit-length tasks on two processors with arbitrary resource constraints and requirements may be solved by the following algorithm [55].

Algorithm 6.1. (Problem $P2 \mid \text{res.} \ldots, p_j = 1 \mid C_{\max}$).
1. Construct an n-node (undirected) graph G with each node labelled as a distinct task and with an edge joining T_i to T_j if and only if

$$r_k(T_i) + r_k(T_j) \leqslant m_k, \qquad k = 1, 2, \ldots, s.$$

2. Find a maximal matching F of graph G. Put the minimal value of schedule length $C^*_{\max} = n - |F|$.
3. Process in parallel the pairs of tasks joined by the edges comprising set F. Process other tasks individually.

Notice that the key idea here is the correspondence between a maximum matching in a graph displaying resource constraints and the minimum-length schedule. The complexity of the above algorithm clearly depends on the complexity of the algorithm determining the maximum matching. There are several algorithms for finding it, the complexity of the most efficient being $O(n^{2.5})$ [83].

We can do better if we restrict ourselves to the one-resource case. It is not hard to see that in this case an optimal schedule will be produced by ordering tasks in nonincreasing order of their resource requirements and assigning tasks in that order to the first free processor on which a given task can be processed because of resource constraints. Thus, problem $P2 \mid \text{res}1 \ldots, p_j = 1 \mid C_{\max}$ can be solved in $O(n \log n)$ time.

If in the last problem tasks are allowed only for 0-1 resource requirements, the problem can be solved in $O(n)$ time even for arbitrary ready times and an arbitrary number of processors, by first assigning tasks with unit resource requirements up to m_1 in each slot, and then filling these slots with tasks with zero resource requirements [13].

When we fix resource limits and resource requirements, problem $P \mid \text{res spr}, p_j = 1 \mid C_{\max}$ is still solvable in linear time, even for an arbitrary number of processors [24]. We describe this approach below, since it has more general application. Let us note that we can divide all the tasks into k (which is fixed) classes depending on their resource requirements $(r_1(T_i), r_2(T_i), \ldots, r_s(T_i)) \in \{0, 1, \ldots, r\}^s$. For each possible resource requirement we define one such class. The correspondence between the resource requirements and the classes will be described by a 1-1 function $f: \{0, 1, \ldots, r\}^s \rightarrow \{1, 2, \ldots, k\}$, where k is the number of different resource requirements, i.e. $k = (r+1)^s$. For a given instance, we denote by n_i the number of tasks belonging to the i-th class, $i = 1, 2, \ldots, k$, thus having their resource requirements equal to $f^{-1}(i)$. Let us observe that most of the input information describing any instance of problem $P\,|\,\text{res spr}, p_j = 1\,|\,C_{max}$ is given by the n resource requirements (we bypass for the moment the number of processors m, the number of additional resources s and resource limits p). We may now replace this input by another vector, making use of the classes introduced above. That is, we introduce a new vector $\bar{v} = (v_1, v_2, \ldots, v_k) \in N_0^k$ (⁸), where v_i is the number of tasks having resource requirements equal to $f^{-1}(i)$, $i = 1, 2, \ldots, k$. Of course, the sum of the components of this vector is equal to the number of tasks, i.e. $\Sigma_{i=1}^k v_i = n$.

We now introduce some definitions useful in the following discussion. An *elementary instance* of $P\,|\,\text{res spr}, p_j = 1\,|\,C_{max}$ is defined as a sequence $\bar{r}(T_1)$, $\bar{r}(T_2), \ldots, \bar{r}(T_l)$, where each $\bar{r}(T_i) \in \{0, 1, \ldots, r\}^s - (0, 0, \ldots, 0)$, with properties $l \leq m$ and $\Sigma_{i=1}^l r(T_i) \leq (p, p, \ldots, p)$. Note that the minimal schedule length for an elementary instance is always equal to 1. An *elemenaty vector* is a vector $\bar{v} \in N_0^k$ which corresponds to an elementary instance. If we calculate the number K of different elementary instances, we will see that it cannot be greater than $(p+1)^{(r+1)^s-1}$. However, this number is obviously much less than this upper bound.

Let us denote the elementary vectors (in any order) by $\bar{b}_1, \bar{b}_2, \ldots, \bar{b}_K$. Let us now observe two facts. Firstly, any input $\bar{r}(T_1), \bar{r}(T_2), \ldots, \bar{r}(T_n)$ can be considered as a union of elementary instances. This is because any input consisting of one task is elementary. Secondly, each schedule is also constructed from elementary instances, since all the tasks which are executed at the same time form an elementary instance.

Now, taking into account the fact that the minimal length of a schedule for any elementary instance is equal to one, we may formulate our problem as that of finding a decomposition of a given instance into the minimal number of elementary instances. One may easily notice that this is equivalent to finding a decomposition of the vector $\bar{v} = (v_1, v_2, \ldots, v_k) \in N_0^k$ into a linear combination of elementary vectors $\bar{b}_1, \bar{b}_2, \ldots, \bar{b}_K$, for which the sum of coefficients is minimal: Find $e_1, e_2, \ldots, e_K \in N_0$ such that $\Sigma_{i=1}^K e_i \bar{b}_i = \bar{v}$ and $\Sigma_{i=1}^K e_i$ is minimal.

Thus, we have obtained a linear integer programming problem, which in the general case, would be NP-hard. Fortunately, in our case the number of variables K is fixed. It follows that we can apply a result due to H.W. Lenstra [105], which states that the linear integer programming problem with a fixed number of variables can be solved in polynomial time in the number of constraints and $\log a$, but not in the number of variables, where a is the maximum of all the coefficients in the linear integer programming problem. Thus, the complexity of the problem is $O(2^{K^2}(k \log a)^{cK})$, for some constant c. In our case the complexity of that

---

(⁸) N_0 denotes the set of nonnegative integers.

algorithm is $O(2^{K^2}(k \log n)^{cK}) < O(n)$. Since the time needed to construct the data for this integer programming problem is $O(2^s(K + \log n)) = O(\log n)$, we have proved that the problem $P|\text{res spr}, p_j = 1|C_{\max}$ may be solved in linear time.

It follows that when we consider the nonpreemptive case of scheduling of unit length tasks we have four polynomial time algorithms and this is probably as much as we can get in this area, since other problems of nonpreemptive scheduling under resource constraints have been proved to be NP-hard. Let us mention the parameters that have an influence on the hardness of the problem. Firstly, different ready times cause the strong NP-hardness of the problem even for two processors and very simple reource requirements, i.e. problem $P2|\text{res1}\ldots,r_j,p_j = 1|C_{\max}$ is already strongly NP-hard [17]. (From Fig. 6.1 we see that problem $P2|\text{res}.11,r_j,p_j = 1|C_{\max}$ is strongly NP-hard as well). Secondly, an increase in the number of processors from 2 to 3 results in the strong NP-hardness of the problem. That is, problem $P3|\text{res1}\ldots,p_j = 1|C_{\max}$ is strongly NP-hard [55]. (Note that this is the famous 3-partition problem, the first strongly NP-hard problem). Again from Fig. 6.1 we conclude that problem $P3|\text{res}.11,p_j = 1|C_{\max}$ is NP-hard in the strong sense. Finally, even the simplest precedence constraints result in the NP-hardness of the scheduling problem, that is, the $P2|\text{res}111,\text{chain},p_j = 1|C_{\max}$ is NP-hard in the strong sense [27].

Because of many NP-hard problems, there is a need to work out heuristics with the guaranteed worst-case behavior. We quote some of the results. All of the approximation algorithms considered here are list scheduling algorithms which differ from each other in the ordering of tasks on the list. We mention three approximation algorithms analyzed for the problem (9).

1. *First fit* (FF). Each task is assigned to the earliest time slot in such a way that no resource (and processor) limits are violated.
2. *First fit decreasing* (FFD). A variant of the first algorithm applied to a list ordered in nonincreasing order of $r_{\max}(T_j)$, where $r_{\max}(T_j) = \max\{r_l(T_j)/m_l : 1 \leq l \leq s\}$.
3. *Iterated lowest fit decreasing* (ILFD - applied for $s = 1$ and $p_j = 1$ only). Order tasks as in the FFD algorithm. Put C as a lower bound on $C^*_{\max}$. Place T_1 in the first time slot and proceed through the list of tasks, placing T_j in a time slot for which the total resource requirement of tasks already assigned is minimum. If we ever reach a point where T_j cannot be assigned to any of C slots, we halt the iteration, increase C by 1 and start over. □

---

(9) Let us note that the resource constrained scheduling for the unit processing times of the tasks is equivalent to the variant of a bin packing problem in which the number of items per bin is restricted to m. On the other hand, several other approximation algorithms have been analyzed for the general bin packing problem and the interested reader is referred to [42] for an excellent survey of the results obtained in that area.

Below, we will present the main known bounds for the case $m < n$. In [90] several bounds have been established. Let us start with the problem $P\,|\,\text{res}\,1\ldots$, $p_j = 1\,|\,C_{\max}$ for which the three above-mentioned algorithms have the following bounds:

$$\frac{27}{10} - \left\lceil \frac{37}{10m} \right\rceil < R_{FF}^\infty < \frac{27}{10} - \frac{24}{10m}$$

$$R_{FFD}^\infty = 2 - \frac{2}{m}$$

$$R_{ILFD}^\infty \leq 2.$$

We see that the use of an ordered list improves the bound by about 30%. Let us pass now to the arbitrary processing times. Here some other bounds have been established. For problem $P\,|\,\text{res}\ldots|\,C_{\max}$ the first fit algorithm has been analyzed [54]:

$$R_{FF}^\infty = \min\left\{\frac{m+1}{2}, s + 2 - \frac{2s+1}{m}\right\}.$$

Finally, when dependent tasks are considered, the first fit algorithm has been evaluated for problem $P\,|\,\text{res}\ldots,\text{prec}\,|\,C_{\max}$ [54]:

$$R_{FF}^\infty = m.$$

Unfortunately, no results are reported on the probabilistic analysis of approximation algorithms for resource constrained scheduling. Now let us pass to preemptive scheduling. Problem $P\,|\,\text{pmtn},\text{res}\,1.1\,|\,C_{\max}$ can be solved via a modification of Mc Naughton's rule (Algorithm 3.1), by taking as the minimum schedule length

$$C_{\max}^* = \max\left\{\max_j \{p_j\}, \sum_{j=1}^n p_j/m, \sum_{T_j \in Z_R} p_j/m_1\right\}.$$

where Z_R is the set of tasks for which $r_1(T_j) = 1$. The tasks are scheduled as in Algorithm 3.1, the tasks from Z_R being scheduled first. The complexity of the algorithm is obviously $O(n)$.

Let us consider now the problem $P2\,|\,\text{pmtn},\text{res}\ldots|\,C_{\max}$. This can be solved via a transformation into the transportation problem [27].

Without loss of generality we may assume that task T_j, $j = 1, 2, \ldots, n$ spends exactly $p_j/2$ time units on each of the two processors. Let (T_j, T_r), $j \neq r$ denote a resource feasible task pair, i.e. a pair for which $r_k(T_j) + r_k(T_r) \leq m_k$, $k = 1, 2, \ldots, s$. Let Z be the set of all resource feasible pairs of tasks. Z also includes all pairs of the type (T_j, T_{n+1}) $j = 1, 2, \ldots, n$, where T_{n+1} is an idle time (dummy) task. Now we may construct a transportation network. Let $n + 1$ sender nodes correspond to the $n+1$ tasks (including an idle time task) which are processed on processor P_1 and let $n+1$ receiver nodes correspond to the $n+1$ tasks processed on processor P_2. Stocks and requirements of nodes corresponding to T_j, $j = 1, 2, \ldots, n$, are equal to $p_j/2$, since the amount of time each task spends on each processor is equal to $p_j/2$. The stock and requirement of two nodes corresponding to T_{n+1} are equal to $\Sigma_{j=1}^{n} p_j/2$, since these are the maximum amounts of time each processor may be idle. Then we draw directed arcs (T_j, T_r) and (T_r, T_j) if and only if $(T_j, T_r) \in Z$, to express the possibility of processing in parallel tasks T_j and T_r on processors P_1 and P_2. In addition we draw an arc (T_{n+1}, T_{n+1}). Then we assign for each pair $(T_j, T_r) \in Z$ a cost associated with arcs (T_j, T_r) and (T_r, T_j) equal to 1, and a cost associated with the arc (T_{n+1}, T_{n+1}) equal to 0. (This is because an interval with idle times on both processors does not lengthen the schedule). Now it is quite clear that the solution of the corresponding transportation problem, i.e. the set x_{jr}^*, is simply the set of the numbers of time units during which corresponding pairs of tasks are processed (T_j being processed on P_1 and T_r on P_2).

The complexity of the above algorithm is $O(n^4 \log \Sigma p_j)$ since this is the complexity of finding a minimum cost flow in a network (the number of vertices in the transportation network is $O(n)$).

Now let us pass to the problem $Pm \mid \text{pmtn, res} \ldots \mid C_{max}$. It can still be solved in polynomial time via the linear programming approach (4.1)-(4.2) [157], but now, instead of the processor feasible set, the notion of a *resource feasible set* is used. By the latter we mean the set of tasks which can be simultaneously processed because of resource limits (including processor limit). At this point let us also mention that problem $P \mid \text{pmtn, res} \ldots 1 \mid C_{max}$ can be solved by the generalization of another linear programming approach (4.8)-(4.11) [137, 138, 153] ([10]). Let us also add that the latter approach can handle different ready times and the L_{max} criterion. On the other hand, both approaches can be adapted to cover the case of the uniconnected activity network in the same way as that described in Subsection 4.2.

Finally, when analyzing preemptive scheduling let us present bounds on the worst-cass behavior of certain approximation algorithms [90] (problem $P \mid \text{pmtn, res} 1 \ldots \mid C_{max}$):

$$R_{FF}^{\infty} = 3 - \frac{3}{m} \qquad R_{FFD}^{\infty} = 3 - \frac{3}{m}.$$

Surprisingly, using an ordered list does not improve the bound.

([10]) Strictly speaking, in this approach resource constraints slightly different from res..1 can be handled. Details can be found in [153].

6.3. Scheduling in microprocessor systems

One of the assumptions imposed in Section 2 was that each task is processed on at most one processor at a time. However, in recent years with the rapid development of microprocessor and especially multi-microprocessor systems, the above assumption has ceased to be justified in some important applications. There are, for example, self-testing multi-microprocessor systems in which one processor is used to test others or diagnostic systems in which testing signals stimulate the tested elements and their corresponding outputs are simultaneously analyzed [5, 46]. When formulating scheduling problems in such systems, one must take into account the fact that some tasks have to be processed on more than one processor at a time. These problems create a new direction in processor scheduling theory.

We will set up the subject more precisely [22, 23]. Tasks are to be processed on a set of identical processors denoted as previously $P_1, P_2, \ldots, P_m$. The set of tasks is divided into k (or less) subsets $\mathcal{T}^1 = \{T_1, T_2, \ldots, T_{n_1}\}, \mathcal{T}^2 = \{T_1^2, T_2^2, \ldots, T_{n_2}^2\}, \ldots, \mathcal{T}^k = \{T_1^k, T_2^k, \ldots, T_{n_k}^k\}$, where $n = n_1 + n_2 + \ldots + n_k$. Each task T_i^1, $i = 1, 2, \ldots, n_1$, requires one arbitrary processor for its processing and its processing time is equal to t_i^1. On the other hand, each task T_i^k when $k > 1$, requires k arbitrary processors simultaneously for its processing during a period of time whose length is equal to t_i^k. We will call tasks from $\mathcal{T}^k$ *width-k tasks* or *T^k-tasks*. All the tasks considered here are assumed to be *independent*, i.e. there are no precedence constraints among them. A schedule will be called *feasible* if, besides the usual conditions, each T^1-task is processed by one processor and each T^k-task is processed by k processors at a time. The schedule length is taken as a criterion.

Let us start with the nonpreemptive scheduling. It is clear that the general problem is NP-hard (cf. section 3), thus, we may concentrate on unit-length tasks. Let us start with the problem of scheduling tasks which belong to two sets only: $\mathcal{T}^1$ and $\mathcal{T}^k$, for arbitrary k. This problem can be solved optimally by the following algorithm [23].

Algorithm 6.2

1. Calculate the length of an optimal schedule according to the formula

$$C^*_{\max} = \max\left\{ \left\lceil \frac{n_1 + kn_k}{m} \right\rceil, \left\lceil \frac{n_k}{\left\lfloor \frac{m}{k} \right\rfloor} \right\rceil \right\} \quad (11). \tag{6.1}$$

2. Schedule the width-k tasks first in time interval $[0, C^*_{\max}]$. Then assign unit

---

[11] $\lfloor x \rfloor$ denotes the greatest integer not greater than x.

width tasks to the remaining free processors.

It should be clear that (6.1) gives a lower bound on the schedule length of an optimal schedule and this bound is always met by a schedule constructed by Algorithm 6.2.

Let us consider now the case of scheduling tasks belonging to sets $\mathcal{T}^1, \mathcal{T}^2, \mathcal{T}^3, \ldots, \mathcal{T}^k$, where k is a fixed integer. The approach used to solve the problem is similar to that for the problem of nonpreemptive scheduling of unit processing time tasks under fixed resource constraints [24]. We have described that approach in Subsection 6.2.

Now, we will pass to preemptive scheduling. Firstly, let us consider the problem of scheduling tasks from sets $\mathcal{T}^1$ and $\mathcal{T}^k$ in order to minimize the schedule length. In [22, 23] it has been proved that among minimum-length schedules for the problem there exists a feasible A-schedule, i.e. one in which first all T^k-tasks are assigned in time interval $[0, C^*_{\max}]$ using McNaughton's rule, and then all T^1-tasks are assigned, using the same rule, in the remaining part of the schedule (cf. Fig. 6.2).

Following the above result, we will concentrate on finding an optimal schedule among A-schedules. Now, we give a lower bound on a schedule length for our problem. Define

$$X = \sum_{i=1}^{n_1} t_i^1, \qquad Y = \sum_{i=1}^{n_k} t_i^k, \qquad Z = X + kY$$

$$t_{\max}^1 = \max\{t_i^1 : T_i^1 \in \mathcal{T}^1\}, \qquad t_{\max}^k = \max\{t_i^k : T_i^k \in \mathcal{T}^k\}.$$

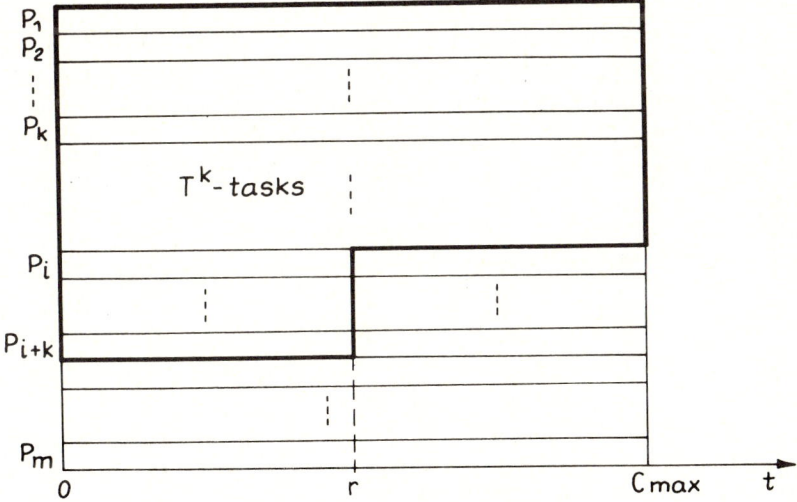

Fig. 6.2. An example A-schedule.

A bound on C_{max} is obtained as follows:

$$C_{max} \geq C = \max\{Z/m, Y/\lfloor m/k \rfloor, t^1_{max}, t^k_{max}\}. \tag{6.2}$$

It is clear that no feasible schedule can be shorter that the maximum of the above factors, i.e. mean processing requirement on one processor, mean processing requirement of T^k-tasks on k processors, the maximum processing time of a T^1-task and the maximum processing time of a T^k-task. If $m \cdot C > Z$, there will be an idle time in any schedule and its minimum length $IT = m \cdot C - Z$.

On the basis of bound (6.2) and the reasoning preceding it one can try to construct a preemptive schedule of minimum length equal to C. However, this is not always possible. Thus, one has to lengthen the schedule. Below we present the reasoning that allows one to find the optimal schedule length. Let $p = \lfloor Y/C \rfloor$. It is quite clear that the optimal schedule length C^*_{max} must obey the following inequality

$$C \leq C^*_{max} \leq \frac{Y}{p}.$$

We know that there exists an optimal A-schedule with $k \cdot p$ processors devoted entirely to T^k-tasks, k processors devoted to T^k-tasks in time interval $[0, r]$ and T^1-tasks scheduled in the remaining time (cf. Fig. 6.2). Let the number of processors that can process T^1-tasks in time interval $[0, r]$ be

$$m_1 = m - (p+1) \cdot k.$$

An A-schedule which completes all tasks by time D, where $C \leq D \leq Y/p$, will have $r = Y - Dp$. Thus, the optimum value C^*_{max} will be the smallest value of D ($D \geq C$) such that the T^1-tasks can be scheduled on $m_1 + k$ processors available in the interval $[0, D]$. Below we give necessary and sufficient conditions for the unit width tasks to be scheduled. To do this, let us assume that these tasks are ordered in such a way that $t^1_1 \geq t^1_2 \geq \ldots \geq t^1_{n_1}$. For a given pair D, r ($r = Y - Dp$), let $t^1_1, t^1_2, \ldots, t^1_j$ be the only processing times greater than $D - r$. Then the T^1-tasks can be scheduled if and only if

$$\sum_{i=1}^{j} [t^1_i - (D-r)] \leq m_1 r. \tag{6.3}$$

To prove that the above condition really is necessary and sufficient, let us first observe that if (6.3) is violated the T^1-tasks cannot be scheduled. Suppose now that (6.3) holds. Then one should schedule the excess (exceeding $D-r$) of long tasks $T_1, T_2, \ldots, T_j$ and (if (6.3) holds without equality) some other tasks on m_1 processors in time interval $[0, r]$ using McNaughton's rule. After this operation the interval is completely filled with unit width tasks on m_1 processors.

Now we describe how the optimum value of the schedule length (C^*_{max}) can be found. Let $P_j = \Sigma_{i=1}^{j} t_i^1$. Inequality (6.3) may be rewritten as

$$P_j - j(D-r) \leq m_1(Y - Dp).$$

Solving it for D we get

$$D \geq \frac{(j-m_1)Y + P_j}{(j-m_1)p + j}.$$

Define

$$C_j = \frac{(j-m_1)Y + P_j}{(j-m_1)p + j}.$$ Thus, we may write

$$C^*_{max} = \max\{C, C_1, C_2, \ldots, C_{n_1}\}.$$

Finding the above maximum can clearly be done in $O(n_1 \log n_1)$ time by sorting the unit-width tasks by t_i. But one can do better by taking into account the following facts.

1. $C_i \leq C$ for $i \leq m_1$ and $i \geq m_1 + k$.
2. C_i has no local maxima for $i = m_1 + 1, \ldots, m_1 + k - 1$.

Thus, to find a maximum over $C_{m_1+1}, \ldots, C_{m_1+k-1}$ and C we only need to apply a linear time median finding algorithm [4] and binary search. This will result in an $O(n_1)$ algorithm that calculates C^*_{max}. (Finding the medians takes $O(n_1)$ the first time, $O(n_1/2)$ the second time, $O(n_1/4)$ the third time, ... Thus the total time to find the medians is $O(n_1)$).

Now we can give an optimization algorithm for the considered case of scheduling unit-width and width-k tasks

Algorithm 6.3
1. Calculate the minimum length C^*_{max}.
2. Schedule all the T^k-tasks (in any order) in interval $[0, C^*_{max}]$ using Mc Naughton's rule.
3. Assign the excess of the long tasks (that exceed $C^*_{max} - r$) and possibly some other tasks to m_1 processors in interval $[0, r]$. Schedule the remaining processing requirement in interval $[r, C^*_{max}]$ using Mc Naughton's rule.

The optimality of the above algorithm follows from the discussion preceding it. Its complexity is $O(n_1 + n_k)$, thus we get $O(n)$.

To this end let us consider the general case of preemptively scheduling tasks form sets $\mathcal{T}^1, \mathcal{T}^2, \ldots, \mathcal{T}^k$, k fixed. Once more, we can use the very useful linear programming approach (4.1)-(4.2) to solve this problem in polynomial time.

Now we will comment briefly on the possible extensions and refinements.

Firstly, one can consider additional resources besides central processors, thus combining the models considered in Subsections 6.2 and 6.3. This subject is partially covered by a forthcomong paper [21]. Then, one can consider slightly different models, for example one assuming a kind of dedicated tests, where each test (each task) is to be carried out by an *a priori* specified set of processors. Preliminary results in that area have been obtained in [91, 119].

Appendix

To distinguish scheduling problems in a short way the three-field notation $\alpha|\beta|\gamma$ has been proposed in [70, 102, 27].

The first field $\alpha = \alpha_1 \alpha_2$ describes the processor environment. Parameter $\alpha_1 \in \{\phi, P, Q, R, O, F, J\}$ characterizes the type of processor used:

$\alpha_1 = \phi$: one processor [12]; $\alpha_1 = P$: identical processors;
$\alpha_1 = Q$: uniform processor; $\alpha_1 = R$: unrelated processors;
$\alpha_1 = 0$: dedicated processors: open shop system;
$\alpha_1 = F$: dedicated processors: flow shop system; $\alpha_1 = J$: dedicated processors, job shop system.

Parameter $\alpha_2 \in \{\phi, k\}$ denotes the number of processors in the problem:

$\alpha_2 = \phi$: the number of processors is assumed to be variable;
$\alpha_2 = k$: the number of processors is equal to k (k is a positive integer).

The second field $\beta = \beta_1, \beta_2, \beta_3, \beta_4, \beta_5$ describes task and resource characteristics. Parameter $\beta_1 \in \{\text{pmtn}, \phi\}$ indicates the possibility of task preemption:

$\beta_1 = \text{pmtn}$: preemptions are allowed;
$\beta_1 = \phi$: no preemption is allowed.

Parameter $\beta_2 \in \{\phi, \text{res } \lambda\sigma\rho\}$ characterizes additional resources:

$\beta_2 = \phi$: no additional resources exist;
$\beta_2 = \text{res } \lambda\sigma\rho$: there are specified resource constraints;
$\lambda, \sigma, \rho \in \{\cdot, k\}$: denote the number of additional resource types, resource limits and resource requirements, respectively;
if
$\lambda, \sigma, \rho = \cdot$ then the number of additional resource types, resource limits and resource requirements are respectively arbitrary,
and if
$\lambda, \sigma, \rho = k,$ then respectively: the number of additional resource types is equal to k, each resource is

---

[12] In this notation ϕ denotes an empty symbol, which will be omitted in presenting problems.

available in the system in the amount of k units and the resource requirements of each task are at most equal to k units.

Parameter $\beta_3 \in \{\phi, \text{prec}, \text{uan}, \text{tree}, \text{chain}\}$ reflects the precedence contraints:

$\beta_3 = \phi$, prec, uan, tree, chain: denotes respectively independent tasks, general precedence constraints, uniconnected activity networks, precedence constraints forming a tree or a chain.

Parameter $\beta_4 \in \{r_j, \phi\}$ describes ready times:

$\beta_4 = r_j$: ready times differ per task,

$\beta_4 = \phi$: all ready times are equal to zero.

Parameter $\beta_5 \in \{p_j = p, \underline{p} \leq p_j \leq \bar{p}, \phi\}$ describes task processing times:

$\beta_5 = (p_j = p)$: all tasks have processing times equal to p units,

$\beta_5 = (\underline{p} \leq p_j \leq \bar{p})$: no p_j is less than $\underline{p}$ and no greater than $\bar{p}$,

$\beta_5 = \phi$: tasks have arbitrary processing times.

The third field γ denotes an optimality criterion (performance measure) i.e. $\gamma \in \{C_{max}, \Sigma C_j, \Sigma w_j C_j, L_{max}, \Sigma T_j, \Sigma w_j T_j, \Sigma U_j, \Sigma w_j U_j\}$, where $\Sigma C_j = \bar{F}$, $\Sigma w_j C_j = \bar{F}_w$, $\Sigma T_j = \bar{T}$, $\Sigma w_j T_j = \bar{T}_w$, $\Sigma U_j = \bar{U}$ and $\Sigma w_j U_j = \bar{U}_w$.

Acknowledgment

I would like to thank Gerd Finke, Matteo Fischetti, Silvano Martello, Alexander Rinnooy Kan and Jan Weglarz for valuable comments concerning the first draft of this paper.

References

[1] J.O. Achugbue and F.Y. Chin, «Scheduling the open shop to minimize mean flow time», *SIAM Journal on Computing* 11, 709 - 720, 1982.

[2] ACM, *Record of the project MAC conference on concurrent system and parallel computation*, Wood's Hole, Mass., 1970.

[3] D. Adolphson and T.C. Hu, «Optimal linear ordering», *SIAM Journal on Applied Mathematics* 25, 403 - 423, 1973.

[4] A.V. Aho, J.E. Hopcroft and J.D. Ullman, *The Design and Analysis of Computer Algorithms*, Addison-Wesley, Reading, Mass., 1974.

[5] A. Avizienis, «Fault tolerance: the survival attribute of digital systems», *Proceedings of the IEEE* 66, No. 10, 1109 - 1125, 1978.

[6] J.L. Baer, «A survey of some theoretical aspects of multiprocessing», *Computing Surveys* 5, No. 1, 1973.

[7] K.R. Baker, *Introduction to Sequencing and Scheduling*, J. Wiley & Sons, New York, 1974.

[8] K.R. Baker, E.L. Lawler, J.K. Lenstra and A.H.G. Rinnooy Kan, «Preemptive scheduling of a single machine to minimize maximum cost subject to release dates and precedence constraints», *Operations Research 31*, 381 - 386, 1983.

[9] I. Barany, «A vector-sum theorem and its application to improving flow guarantees», *Mathematics of Operations Research*, 6, 445 - 452, 1981.

[10] A.J. Bernstein, «Analysis of programs for parallel programming», *IEEE Transactions on Computers EC-15,* No. 5, 1966.

[11] J. Blazewicz, «Scheduling dependent tasks with different arrival times to meet deadlines», in E. Gelenbe and H.Beilner (eds.), *Modelling and Performance Evaluation of Computer Systems*, North Holland, Amsterdam, 57 - 65, 1976.
[12] J. Blazewicz, «Simple algorithms for multiprocessor scheduling to meet deadlines», *Information Processing Letters* 6, 162 - 164, 1977.
[13] J. Blazewicz, «Complexity of computer scheduling algorithms under resource constraints», *Proceedings of I Meeting AFCET-SMF on Applied Mathematics*, Palaiseau, 169 - 178, 1978.
[14] J. Blazewicz, «Scheduling tasks on parallel processors under resource constraints to minimize mean finishing time», *Methods of Operations Research*, No. 35, 67 - 72, 1979.
[15] J. Blazewicz, «Deadline scheduling of tasks with ready times and resource constraints», *Information Processing Letters* 8, 60 - 63, 1979.
[16] J. Blazewicz, «Solving the resource constrained deadline scheduling problem via reduction to the network flow problem», *European Journal of Operational Research*, 6, 75 - 79, 1981.
[17] J. Blazewicz, J. Barcelo, W. Kubiak and H. Röck, «Scheduling tasks on two processors with deadlines and additional resources» *European Journal of Operational Research* (to appear).
[18] J. Blazewicz, W. Cellary, R. Slowinski and J. Weglarz, «Deterministyczne problemy szeregowania zadan na rownoleglych procesorach», *Cz. I. Zbiory zadan niezaleznych, Podstawy Sterowania* 6, 155 - 178, 1976.
[19] J. Blazewicz, W. Cellary, R. Slowinski and J. Weglarz, «Deterministyczne problemy szeregowania zadan na rownoleglych processorach», *Cz. II. Zbiory zadan zaleznych, Podstawy Sterowania* 6, 297 - 320, 1976.
[20] J. Blazewicz, W. Cellary and J. Weglarz, «A strategy for scheduling splittable tasks to reduce schedule length», *Acta Cybernatica* 3, 99 - 106, 1977.
[21] J. Blazewicz, M. Drabowski, K. Ecker and J. Weglarz, «Scheduling multiprocessor tasks with resource constraints», (to appear).
[22] J. Blazewicz, M. Drabowski and J. Weglarz, «Scheduling independent 2-processor tasks to minimize schedule length», *Information Processing Letters* 18, 267 - 273, 1984.
[23] J. Blazewicz, M.Drabowski and J. Weglarz, «Scheduling multiprocessor tasks to minimize schedule length», *IEEE Transactions on Computers* (to appear).
[24] J. Blazewicz and K. Ecker, «A linear time algorithm for restricted bin packing and scheduling problems», *Operations Research Letters* 2, 80 - 83, 1983.
[25] J. Blazewicz, W. Kubiak, H. Röck and J. Szwarcfiter, «Minimizing mean flow time under resource constraints on parallel processors», *Report Technical University of Poznan*, December 1984.
[26] J. Blazewicz, W. Kubiak, H. Röck and J. Szwarcfiter, «Scheduling unit-time jobs on flow-shops under resource constraints» (to appear).
[27] J. Blazewicz, J.K. Lenstra and A.H.G. Rinnooy Kan, «Scheduling subject to resource constraints: classification and complexity», *Discrete Applied Mathematics* 5, 11 - 21, 1983.
[28] P.J. Brucker, «Sequencing unit-time jobs with treelike precedence on m machines to minimize maximum lateness», *Proceedings IX International Symposium on Mathematical Programming*, Budapest, 1976.
[29] P. Brucker, MR. Garey and D.S. Johnson, «Scheduling equal-length tasks under treelike precedence constraints to minimize maximum lateness», *Mathematics of Operations Research* 2, 275 - 284, 1977.
[30] J. Bruno, E.G. Coffman, Jr. and R. Sethi, «Scheduling independent tasks to reduce mean finishing time», *Communications of the ACM* 17, 382 - 387, 1974.
[31] W. Cellary, «Task scheduling in systems with nonpreemptible resources» in E. Gelenbe and H. Beilner, eds., *Modelling, Measuring and Peformance Evaluation of Computer Systems*, North Holland, Amsterdam, 1978.
[33] W. Cellary, «Scheduling dependent tasks from an infinite stream in systems with nonpreemptible resource», in G. Bracchi and P.G. Lockemann, eds., *Lecture Notes in Computer Science*, vol 65, Springer Verlag, Berlin, 536 - 547, 1979.
[34] N.-F. Chen and C.L. Liu, «On a class of scheduling algorithms for multiprocessors computing systems», in T.–Y. Feng, ed., *Parallel Processing, Lecture Notes in Computer Science* 24, Springer Verlag, Berlin, 1-16, 1975.
[35] Y. Cho and S. Sahni «Preemptive scheduling of independent jobs with release and due times on open, flow and job shops», *Operations Research* 29, 511 - 522, 1981.
[36] E.G. Coffman, Jr., ed., *Computer & Job/Shop Scheduling*, J. Wiley, New York, 1976.
[37] E.G. Coffman, Jr., P.J. Denning, *Operating Systems Theory*, Prentice-Hall, Englewood Cliffs, N.J., 1973.
[38] E.G. Coffman, Jr., G.N. Grederickson and G.S. Lueker, «A note on expected makespans for largest--first sequences of independent tasks on two processors», *Mathematics of Operations Research* 9, 260 - 266, 1984.

[39]   E.G. Coffman, Jr., G.N. Frederickson and G.S. Lueker, «Probabilistic analysis of the LPT processor scheduling heuristic», (unpublished paper), 1983.
[40]   E.G. Coffman, Jr. and R.L. Graham, «Optimal scheduling for two processor systems», *Acta Informatica* 1, 200 - 213, 1972.
[41]   E.G. Coffman, Jr., M.R. Garey and D.S. Johnson, «An application of bin-packing to multiprocessor scheduling», *SIAM Journal on Computing* 7, 1 - 17, 1978.
[42]   E.G. Coffman, Jr., M.R. Garey and D.S. Johnson, «Approximation algorithms for bin-packing, an updated survey», in G. Ausiello, M. Luccertini, P. Serafini, eds., *Algorithm Design for Computer System Design*, Springer Verlag, Wien, 49 - 106, 1984.
[43]   E.G. Coffman, Jr. and R. Sethi, «A generalized bound on LPT sequencing», *RAIRO-Informatique* 10, 17 - 25, 1976.
[44]   R.W. Conway, W.L. Maxwell and L.W. Miller, *Theory of Scheduling*, Addison-Wesley, Reading, Mass., 1967.
[45]   S.A. Cook, «The complexity of theorem proving procedures», *Proceedings ACM Symposium on Theory of Computing*, 151 - 158, 1971.
[46]   M. Dal Cin and E. Dilger, «On the diagnosability of self-testing multimicroprocessor systems», *Microprocessing and microprogramming* 7, 177 - 184, 1981.
[47]   A. Federgruen and H. Groenevelt (to appear).
[48]   S. French, *Sequencing and Scheduling: an Introduction to the Mathematics of the Job-Shop*, Horwood, Chichester, 1982.
[49]   J.B.G. Frenk and A.H.G. Rinnooy Kan, «The asymptotic optimality of the LPT scheduling heuristic», *Report Erasmus University*, Rotterdam, 1984.
[50]   J.B.G. Frenk and A.H.G. Rinnooy Kan, «The rate of convergence to optimality of the LPT heuristic», *Report Erasmus University*, Rotterdam, 1984.
[51]   M. Fujii, T. Kasami and K. Ninomiya, «Optimal sequencing of two equivalent processors», *SIAM Journal on Applied Mathematics* 17, 784 - 789, 1969, Err: 20, 141, 1971.
[52]   H.N. Gabow, «An almost-linear algorithm for two-processor scheduling», *Journal of the ACM* 29, 766 - 780, 1982.
[53]   M.R. Garey, Unpublished result.
[54]   M.R. Garey and R.L. Graham, «Bounds for multiprocessor scheduling with resource constraints», *SIAM Journal on Computing* 4, 187 - 200, 1975.
[55]   M.R. Garey and D.S. Johnson, «Complexity results for multiprocessor scheduling under resource constraints», *SIAM Journal on Computing* 4, 397 - 411, 1975.
[56]   M.R. Garey and D.S. Johnson, «Scheduling tasks with nonuniform deadlines on two processors», *Journal of the ACM* 23, 461 - 467, 1976.
[57]   M.R. Garey and D.S. Johnson, «Two-processor scheduling with start-times and deadlines», *SIAM Journal on Computing* 6, 416 - 426, 1977.
[58]   M.R. Garey and D.S. Johnson, *Computers and Intractability: A Guide to the Theory of NP – Completeness*, W.H. Freeman, San Francisco, 1979.
[59]   M.R. Garey, D.S. Johnson, B.B. Simons and R.E. Tarjan, «Scheduling unit time tasks with arbitrary release times and deadlines», *SIAM Journal on Computing* 10, 256 - 269, 1981.
[60]   M.R. Garey, D.S. Johnson, R.E. Tarjan and M. Yannakakis, «Scheduling opposing forests», *SIAM Journal on Algebraic and Discrete Mathematics* 4, 72 - 93, 1983.
[61]   T. Gonzalez, «Optimal mean finish time preemptive schedules», *Technical Report 220, Computer Science Departement*, Pennsylvania State Univ. 1977.
[62]   T. Gonzalez, O.H. Ibarra and S. Sahni, «Bounds for LPT schedules on uniform processors», *SIAM Journal on Computing* 6, 155 - 166, 1977.
[63]   T. Gonzales and S. Sahni, «Open shop scheduling to minimize finish time», *Journal of the ACM* 23, 665 - 679, 1976.
[64]   T. Gonzalez and S. Sahni, «Flowshop and jobshop schedules: complexity and approximation», *Operations Research* 20, 36 -52, 1978.
[65]   T. Gonzales and S. Sahni, «Preemptive scheduling for uniform processor systems, *Journal of the ACM* 25, 81 - 101, 1978.
[66]   J.A. Gosden, «Explicit parallel processing description and control in programs for multi- and uni--processor computers», *AFIPS Conference Proceedings*, vol. 29, Fall Joint Computer Conference 651 - 660' 1966.
[67]   R.L. Graham, «Bounds for certain multiprocessing anomalies», *Bell System Technical Journal* 25, 1563 - 1581, 1966.
[68]   R.L. Graham, «Bounds on multiprocessing timing anomalies», *SIAM Journal on Applied Mathematics* 17, 263 - 269, 1969.

[69] R.L. Graham, «Bounds on performance of scheduling algorithms», Chapter 5 in [36].
[70] R.L. Graham, E.L. Lawler, J.K. Lenstra and A.H.G. Rinnooy Kan, «Optimization and approximation in deterministic sequencing and scheduling theory: a survey», *Annals of Discrete Mathematics* 5, 287 - 326, 1979.
[71] A.N. Habermann, «Prevention of system deadlocks», *Communications of the ACM* 12, 373-377, 1969.
[72] D.S. Hochbaum and D.B. Shmoys, «Using dual approximation algorithms in scheduling», *Journal of the ACM* (to appear).
[73] W.A. Horn, «Single-machine job sequencing with treelike precedence ordering and linear delay penalties», *SIAM Journal on Applied Mathematics* 23, 189 - 202, 1972.
[74] W.A. Horn, «Some simple scheduling algorithms», *Naval Research Logistics Quarterly* 21, 177 - 185, 1974.
[75] E. Horowitz and S. Sahni, «Exact and approximate algorithms for scheduling non-identical processors», *Journal of the ACM* 23, 317 - 327, 1976.
[76] E.G. Horvath, S. Lam and R. Sethi, «A level algorithm for preemptive scheduling», *Journal of the ACM* 24, 32 - 43, 1977.
[77] T.C. Hu, «Parallel sequencing and assembly line problems», *Operations Research* 9, 841 - 848, 1961.
[78] K. Huckert, R. Rhode and R. Weber, «On the interactive solution to a multicriteria scheduling problem», *Zeitschrift für Operations Research* 24, 47 - 60, 1980.
[79] O.H. Ibarra and C.E. Kim, «Heuristic algorithms for scheduling independent tasks on nonidentical processors», *Journal of the ACM* 24, 280 - 289, 1977.
[80] J.R. Jackson, «Scheduling a production line to minimize maximum tardiness», *Research Report* 43, *Management Research Project, Univ. of California*, Los Angeles, 1955.
[81] S.M. Johnson, «Optimal two-and-three-stage production schedules», *Naval Research Logistics Quarterly* 1, 61 - 68, 1954.
[82] D.S. Johnson, «The NP-completeness column: an ongoing guide», *Journal of Algorithms* 4, 189 - 203, 1983.
[83] O. Kariv and S. Even, «An $O(n^{2.5})$ algorithm for maximum matching in general graphs», *16 - th Annual Symposium on Foundations of Computer Science IEEE*, 100 - 112, 1975.
[84] N. Karmarkar and R.M. Karp, «The differencing method of set partitioning», *Mathematics of Operations Research* (to appear).
[85] R.M. Karp, «Reducibility among combinatorial problems», in R.E. Miller and J.W. Thatcher, eds., *Complexity of Computer Computation*, Plenum Press, New York, 85 - 101, 1972.
[86] R.M. Karp, J.K. Lenstra, C.J.H. McDiarmid and A.H.G. Rinnooy Kan, «Probabilistic analysis of combinatorial algorithms: an annotated bibliography», in M.O'h Eigearthaigh, J.K. Lenstra and A.H.G. Rinnoy Kan, eds., *Combinatorial Optimization: Annotated Bibliographies*, J. Wiley, Chichester, 1984.
[87] A.W. Karzanov, «Determining the maximal flow in a network by the method of preflows», *Soviet Math. Dokl.* 15, 434 - 437, 1974.
[88] S.K. Kedia, A job scheduling problem with parallel machines, *Unpublished report, Dept. of Ind. Eng. University of Michigan*, Ann Arbor, 1970.
[89] L.G. Khachiyan, «A polynomial algorithm for linear programming» (in Russian), *Doklady Akademii Nauk USSR*, 244, 1093 - 1096, 1979.
[90] K.L. Krause, V.Y. Shen and H.D. Schwetman, «Analysis of several task-scheduling algorithms for a model of multiprogramming computer systems», *Journal of the ACM* 22, 522 - 550, 1975: Err. 24, 527, 1977.
[91] H. Krawczyk and M. Kubale, «An approximation algorithm for diagnostic test scheduling in multicomputer systems», *IEEE Transactions on Computers* (to appear).
[92] M. Kunde, «Bests Schranke beim LP-Scheduling», *Bericht 7630, Institut für Informatik and Praktische Mathematik, Universität Kiel*, 1976.
[93] J. Labetoulle, E.L. Lawler, J.K. Lenstra and A.H.G. Rinnooy Kan, «Preemptive scheduling of uniform processors subject to release dates», *Report BW99, Mathematisch Centrum Amsterdam* 1979.
[94] B.J. Lageweg, E.L. Lawler, J.K. Lenstra and A.H.G. Rinnooy Kan, «Computer-aided complexity classification of deterministic scheduling problems» *Report BW 138, Centre for Mathematics and Computer Science*, Amsterdam, 1981.
[95] B.J. Lageweg, J.K. Lenstra, E.L. Lawler and A.H.G. Rinnooy Kan, «Computer aided complexity classification of combinatorial problems», *Communications of the ACM* 25, 817 - 822, 1982.
[96] S. Lam and R. Sethi, «Worst case analysis of two scheduling algorithms», *SIAM Journal Computing*, 518 - 536, 1977.
[97] E.L. Lawler, «Optimal sequencing of a single machine subject to precedence constraints», *Management Science* 19, 544 - 546, 1973.

[98]   E.L. Lawler, «Sequencing jobs to minimize total weighted completion time subject to precedence constraints», *Annals of Discrete Mathematics* 2, 75 - 90, 1978.
[99]   E.L. Lawler, «Recent results in the theory of machine scheduling», in A. Bachem, M. Grötschel and B. Korte, eds., *Mathematical Programming: The State of Art-Bonn 1982*, Springer Verlag, Berlin, 202 - 234, 1982.
[100] E.L. Lawler, «Preemptive scheduling of precedence-constrained jobs on parallel machines», in M.A.H. Dempster, J.K. Lenstra and A.H.G. Rinnooy Kan, eds., *Deterministic and Stochastic Scheduling*, Reidel, Dovdrecht, 101 - 123, 1982.
[101] E.L. Lawler and J. Labetoulle, «Preemptive scheduling of unrelated parallel processors by linear programming», *Journal of the ACM* 25, 612 - 619, 1978.
[102] E.L. Lawler, J.K. Lenstra and A.H.G. Rinnooy Kan, «Recent developments in deterministic sequencing and scheduling: a survey» , in M.A.H. Dempster, J.K. Lenstra and A.H.G. Rinnooy Kan, eds., *Deterministic and Stochastic Scheduling*, Reidel, Dordrecht, 35 - 73, 1982.
[103] E.L. Lawler, J.K. Lenstra and A.H.G. Rinnooy Kan, «Minimizing maximum lateness in two machine open shop», *Mathematics of Operations Research* 6, 153 - 158, 1981.
[104] E.L. Lawler, J.M. Moore, «A functional equation and its application to resource allocation and scheduling problems», *Management Science* 16, 77 - 84, 1969.
[105] H.W. Lenstra, Jr., «Integer programming with a fixed number of variables», *Report, University of Amsterdam*, 1981.
[106] J.K. Lenstra, «Sequencing by Enumerative Methods», *Mathematical Centre Tracte 69, Mathematisch Centrum*, Amsterdam, 1981.
[107] J.K. Lenstra and A.H.G. Rinnooy Kan, «Complexity of scheduling under precedence constraints», *Operations Research* 26, 22 - 35, 1978.
[108] J.K. Lenstra and A.H.G. Rinnooy Kan, «Complexity of vehicle routing and scheduling problem», *Report BW 111, Mathematisch Centrum*, Amsterdam, 1979.
[109] J.K. Lenstra and A.H.G. Rinnooy Kan, «An introduction to multiprocessor scheduling», *Report BW 121, Mathematisch Centrum*, Amsterdam, 1980.
[110] J.K. Lenstra and A.H.G. Rinnooy Kan, «Scheduling Theory since 1981: an annotated bibliography», in M.O. hEigeartaigh, J.K. Lenstra and A.H.G. Rinnooy Kan, eds., *Combinatorial Optimization: Annotated Bibliographies*, J. Wiley, Chichester, 1984.
[111] J.K. Lenstra and A.H.G. Rinnooy Kan, «New directions in scheduling theory», *Operations Research Letters* 2, 255 - 259, 1984.
[112] J.K. Lenstra, A.H.G. Rinnooy Kan and P. Brucker, «Complexity of machine scheduling problems», *Annals of Discrete Mathematics* 1, 343 - 362, 1977.
[113] J.W.S. Liu and C.L. Liu, «Bounds on scheduling algorithms for heterogeneous computing systems», *Proceedings IFIPS* 74, North Holland, Amsterdam 349 - 353, 1974.
[114] J.W.S. Liu and C.L. Liu, «Performance analysis of heterogeneous multiprocessor computing systems», in E. Gelenbe and R. Mahl, eds., *Computer Architecture and Networks*, North Holland, Amsterdam, 331 - 343, 1974.
[115] J.W.S. Liu and C.L. Liu, «Bounds on scheduling algorithms for heterogeneous computing systems», *Technical Report UIUCDC-R-74-632, Dept of Computer Science, University of Illinois at Urbana Champaign*, 1974.
[116] R. McNaughton, «Scheduling with deadlines and loss functions», *Management Science* 12, 1 - 12, 1959.
[117] R.H. Möhring, F.J. Radermacher and G. Weiss, «Stochastic scheduling problems I: general strategies», *Zeitschrift für Operations Research* 28, 193 - 260, 1984.
[118] R.H. Möhring, F.J. Radermacher and G. Weiss, «Stochastic scheduling problems II: set strategies», *Zeitschrift für Operations Research* (to appear).
[119] C.L. Monma, «A scheduling problem with simultaneous machine requirement», *TIMS XXVI, Copenhagen*, 1984.
[120] R.R. Muntz and E.G. Coffman, Jr., «Optimal preemptive scheduling on two-processor systems», *IEEE Transactions on Computers. C-18*, 1014 - 1029, 1969.
[121] R.R. Muntz and E.G. Coffman, Jr., «Preemptive scheduling of real time tasks on multiprocessor systems», *Journal of the ACM* 17, 324 - 338, 1970.
[122] C.V. Ramamoorthy and M.J. Gonzales, «A survey of techniques for recognizing parallel processable streams in computer programs», *AFIPS Conference Proceedings, Fall Joint Computer Conference*, 1 - 15, 1969.
[123] A.H.G. Rinnooy Kan, *Machine Scheduling Problems: Classification, Complexity and Computations*, Nijhoff, The Hague, 1978.
[124] A.H.G. Rinnooy Kan, «Probabilistic analysis of algorithms», this volume.

[125] H. Röck, «Some new results in no-wait flow shop scheduling», *Zeitschrift für Operations Research* 28, 1 - 16, 1984.
[126] H. Röck, «The three-machine no-wait flow shop problem is NP-complete», *Journal of the ACM* 51, 336 - 345, 1984.
[127] H. Röck and G. Schmidt, «Machine aggregation heuristics in shop scheduling», *Bericht 82 - 11, Fachbereich 20 Informatik, Technische Universität Berlin*, 1982.
[128] L. Rosenfeld, unpublished result.
[129] M.H. Rothkopf, «Scheduling independent tasks on parallel processors», *Management Science* 12, 347 - 447, 1966.
[130] E.C. Russel, «Automatic program analysis», *Ph. D. Thesis, Dept. of Eng. University of California*, Los Angeles, 1969.
[131] R. Sethi, «Algorithms for minimal-length schedules», Chapter 2 in [36].
[132] R. Sethi, «On the complexity of mean flow time scheduling», *Mathematics of Operations Research*, 320 - 330, 1977.
[133] J.B. Sidney, «Decomposition algorithms for single-machine sequencing with precedence relations and deferral costs», *Operations Research* 23, 283 - 298, 1975.
[134] E.A. Silver, R.V. Vidal and D. de Werra, «A tutorial on heuristic methods», *European Journal of Operational Research* 5, 153 - 162, 1980.
[135] L. Slominski, «Probabilistic analysis of combinatorial algorithms: a bibliography with selected annotation», *Computing* 28, 257 - 267, 1982.
[136] R. Slowinski, «Scheduling preemptible tasks on unrelated processors with additional resources to minimize schedule length», in G. Bracchi and R.C. Lockemann, eds., *Lecture Notes in Computer Science, vol. 65*, Springer Verlag, Berlin, 536 - 547, 1978.
[137] R. Slowinski, «Two approaches to problems of resource allocation among project activities - a comparative study», *Journal of the Operational Research Society* 31, 711 - 723, 1980.
[138] R. Slowinski, «L'ordonnancement des tâches préemptives sur les processeurs indépendants en présence de ressources supplémentaires», *RAIRO Informatique* 15, 155 - 166, 1981.
[139] R. Slowinski, «Multiobjective network scheduling with efficient use of renewable and nonrenewable resources», *European Journal of Operational Research* 7, 265 - 273, 1981.
[140] R. Slowinski, «Modelling and solving multicriteria project scheduling problems», in A. Straszak, ed., *Large Scale Systems Theory and Applications*, Pergamon Press, Oxford, 469 - 574, 1984.
[141] R. Slowinski and J. Weglarz, «Minimalno-czasowy model sieciowy z roznymi sposobami wykonywania czynnosci», *Przegląd Statystyczny* 24, 409 - 416, 1977.
[142] W.E. Smith, «Various optimizers for single-stage production», *Naval Research Logistics Quarterly* 3, 59 - 66, 1956.
[143] J.D. Ullman «Complexity of Sequencing Problems», Chapter 4 in [36].
[144] L.N. Van Wassenhove and K.R. Baker, «A bicriterion approach to time/cost trade offs in sequencing», *European Journal of Operational Research* 11, 48 - 54, 1982.
[145] L.N. Van Wassenhove, and L.F. Gelders, «Solving a bicriterion scheduling problem», *European Journal of Operational Research* 4, 42 - 48, 1980.
[146] S. Volansky, «Graph model analysis and implementation of computational sequences», *Ph.D. thesis, Report No UCLA-ENG-7048, School Engineering and Applied Sciences, University of California*, Los Angeles, 1970.
[147] J. Weglarz, «Project scheduling with discrete and continuous resources», *IEEE Trans. on Syst. Man and Cybernet*, SMC-9, 644 - 650, 1979.
[148] J. Weglarz, «Multiprocessor scheduling with memory allocation - a deterministic approach», *IEEE Transactions on Computers* C-29, 703 - 1053, 1981.
[149] J. Weglarz, «Project scheduling with continuously-divisible doubly constrained resources», *Management Science* 27, 1040 - 1053, 1980.
[150] J. Weglarz, «Modelling and control of dynamic resource allocation project scheduling systems», in S.G. Tzafestas, ed., *Optimization and Control of Dynamic Operational Research Models*, North Holland, Amsterdam, 1982.
[151] J. Weglarz, «Deadline scheduling of independent jobs under continuous processing speed-resources amount functions», *Report, Technical University of Poznan*, June 1984.
[152] J. Weglarz, J. Blazewicz, W. Cellary and R. Slowinski, «An automatic revised simplex method for constrained resource network scheduling», *ACM Transactions on Mathematical Software* 3, 295 - 300, 1977.
[153] D. de Werra, «Preemptive scheduling linear programming and network flows», *SIAM Journal Algebraic and Discrete Mathematics* 5, 11,- 20, 1984.

Jacek Błażewicz
Instytut Automatyki
Politechnika Poznanska
Poznan
Poland

QUADRATIC ASSIGNMENT PROBLEMS

Gerd FINKE*, Rainer E. BURKARD** and Franz RENDL**

1. Introduction

In general, the benefit or cost resulting from an economic activity at some location is also depending on the locations of the other facilities. In order to model such location problems, Koopmans and Beckmann [28] introduced «Quadratic Assignment Problems» (QAP). Mathematically, QAPs can be described as follows.

Given a set $N = \{1, 2, \ldots, n\}$ and three $(n \times n)$ matrices $A = (a_{ik})$, $B = (b_{j\ell})$, $C = (c_{ij})$, find a permutation π of the set N which minimizes

$$\sum_{i=1}^{n} c_{i\pi(i)} + \sum_{i=1}^{n} \sum_{k=1}^{n} a_{ik} b_{\pi(i)\pi(k)}. \tag{1}$$

In the framework of location problems the set N describes the set of sites on which plants are to be built. The $(n \times n)$ matrix $A = (a_{ik})$ is the distance matrix between the sites, whereas the connection matrix $B = (b_{j\ell})$ describes the interdependance or the flow between the plants j and ℓ. The linear term, described by $C = (c_{ij})$, can be interpreted as building and running cost for a plant j that is situated in location i. Thus the objective (1) tries to assign the plants to the possible sites such that the total cost of building and operating the plants becomes minimal. For a discussion of QAPs within the framework of locational decisions see Francis and White [17]. There are many other application areas which can be modelled by QAPs. We shall summarize them in the next section.

It is the aim of this article to survey the theory and the solution procedures for QAPs. Formulations equivalent to (1) will be given. Most of them are integer programming formulations. There is, however, a further interesting way of expressing (1) which uses the trace of the underlying matrices. This formulation has basically been suggested by Lawler [31] who has expressed (1) as a dot

---

*This research was supported in part under NSERC grant A4117.
**This work was supported by the Austrian «Fonds zur Förderung der wissenschaftlichen Forschung, Projekt S 32/01».

product. The final trace form was then introduced by Edwards [13].

Subsequently, the various exact and approximate solution methods for QAPs will be outlined. Presently, there is only one successful exact solution technique available which is based on matrix reductions and on the Gilmore-Lawler bounds. It will be demonstrated that the trace form is providing a very convenient tool to derive the theory of this method.

There is also a second use of the trace formulation. This approach will yield completely new competitive bounds for the symmetric case based on the eigenvalues of the underlying matrices. The eigenvalue related bounds give access to an optimal reduction procedure and also help to characterize QAPs which are almost linear. Some preliminary numerical comparisons with the classical lower bounds are reported.

2. Applications

In the last twenty-five years this model turned out to be of great importance for quite a variety of situations. Steinberg [47] used QAPs for minimizing backboard wiring on electronical circuits. QAPs are a basic ingredient for placement algorithms, for example in VLSI design. Another application concerns the design of typewriter keyboards and control panels (Burkard and Offerman [7]; McCormick [35]). QAPs have extensively been used in facility lay-out problems, in particular in hospital planning (Krarup and Pruzan [29]; Elshafei [15]). Dickey and Hopkins [12] modelled the location of new buildings on a campus as a QAP. Heffley [24] points out that assigning runners to a relay team leads to QAPs. These problems occur moreover in information retrieval, where an optimal ordering of interrelated data on a disk is planned, as well as in scheduling theory, for example if parallel production lines with changeover costs are to be scheduled (Geoffrion and Graves [21]). Further applications involve the ranking of archaeological data (Krarup and Pruzan [29]), and the analysis of chemical reactions (Ugi et al. [48]).

3. Problem Formulations

3.1. Integer Programs

We get formulations of QAPs as 0-1 integer programs by introducing an $(n \times n)$ matrix $X = (x_{ij})$ of Boolean variables. A permutation π can be expressed as a permutation matrix $X = (x_{ij})$ which satisfies

$$\sum_{j=1}^{n} x_{ij} = 1 \qquad i = 1, 2, \ldots, n \qquad (2)$$

$$\sum_{i=1}^{n} x_{ij} = 1 \qquad j = 1, 2, \ldots, n$$

$$x_{ij} \in \{0, 1\} \qquad i, j = 1, \ldots, n. \tag{2}$$

Therefore, we can write (1) in the form

$$\sum_{i=1}^{n} \sum_{j=1}^{n} c_{ij} x_{ij} + \sum_{i=1}^{n} \sum_{j=1}^{n} \sum_{k=1}^{n} \sum_{\ell=1}^{n} a_{ik} b_{j\ell} x_{ij} x_{k\ell} \tag{3}$$

subject to (2). The objective function (3) contains quadratic terms in the unknowns x_{ij} and usual assignment contraints (2). This motivates the name «Quadratic Assignment Problems». Lawler [31] generalized QAPs by introducing n^4 coefficients $d_{ijk\ell}$ and considering the objective

$$\sum_{i=1}^{n} \sum_{j=1}^{n} \sum_{k=1}^{n} \sum_{\ell=1}^{n} d_{ijk\ell} x_{ij} x_{k\ell}. \tag{4}$$

Problem (3) can be transformed to (4) by

$$d_{ijij} := c_{ij} + a_{ii} b_{jj}$$

$$d_{ijk\ell} := a_{ik} b_{j\ell} \qquad (i, j) \neq (k, \ell).$$

Throughout this paper we shall treat however QAPs in «Koopmans-Beckmann form» as formulated in (1) or (3).

There have been many attempts to get rid of the quadratic terms in the objective function (3). Lawler [31], for instance, introduces n^4 new variables $y_{ijk\ell}$ by setting

$$y_{ijk\ell} := x_{ij} \cdot x_{k\ell}$$

and shows that the minimization of (4) subject to (2) is equivalent to the following integer program in 0 - 1 variables:

$$\min \sum_{i=1}^{n} \sum_{j=1}^{n} \sum_{k=1}^{n} \sum_{\ell=1}^{n} d_{ijk\ell} y_{ijk\ell}$$

s.t. (2) and

$$\sum_{i=1}^{n} \sum_{j=1}^{n} \sum_{k=1}^{n} \sum_{\ell=1}^{n} y_{ijk\ell} = n^2$$

$$x_{ij} + x_{k\ell} - 2 y_{ijk\ell} \geq 0 \qquad (i, j, k, \ell = 1, 2, \ldots, n)$$

$$y_{ijk\ell} \in \{0, 1\} \qquad (i, j, k, \ell = 1, 2, \ldots, n).$$

This Boolean integer program has $O(n^4)$ variables and constraints. Further linearizations are due to Bazaraa and Sherali [2], who showed the equivalence of a QAP with a mixed integer program in n^2 binary and $n^2(n-1)^2/2$ real variables and $2n^2$ linear constraints. The smallest linearization known (with respect to additionally introduced variables and constraints) is due to Kaufman and Broeckx [26] who used Glover's linearization technique in order to transform a QAP into a mixed integer program of the form

$$\min \sum_{i=1}^{n} \sum_{j=1}^{n} w_{ij}$$

s.t.(2) and

$$c'_{ij} x_{ij} + \sum_{k=1}^{n} \sum_{\ell=1}^{n} a_{ik} b_{j\ell} x_{k\ell} - w_{ij} \leq c'_{ij} \quad \text{for} \quad i,j = 1, 2, \ldots, n$$

$$w_{ij} \geq 0 \quad \text{for} \quad i,j = 1, 2, \ldots, n$$

where

$$c'_{ij} := \sum_{k=1}^{n} \sum_{\ell=1}^{n} a_{ik} b_{j\ell}.$$

Further linearizations of QAPs have been given by Balas and Mazzola [1], Christofides, Mingozzi and Toth [10], Burkard and Bönniger [5], and Frieze and Yadegar [8].

3.2. Trace Form

Another formulation of QAPs is its trace form which was used for the first time by Edwards [13] and [14]. Since we shall make extensive use of it in the subsequent sections, we describe it here in more detail. Let M by any real $(n \times n)$ matrix. We denote by M^t the transpose and by tr M the trace of the matrix M. Let us recall the properties

$$\text{tr}(M+N) = \text{tr } M + \text{tr } N, \quad \text{tr } M^t = \text{tr } M, \quad \text{tr } MN = \text{tr } NM. \tag{5}$$

Moreover, we define for any permutation π the matrix πM, which is obtained by rearranging the columns of M according to π, i.e. the column j of M becomes the column $\pi(j)$ of πM for all $j = 1, 2, \ldots, n$. Writing π as permutation matrix X yields the identities $\pi M = MX$ and $\pi^{-1} M = MX^t$. One may verify the equations $\Sigma_i c_{i\pi(i)} = \text{tr } \pi^{-1} C = \text{tr } CX^t$.

Hence the classical linear assignment problem AP with matrix C possesses the following equivalent forms

$$\min_{\pi} \text{tr } \pi^{-1} C \quad \text{and} \quad \min_{X \in \Pi} \text{tr } CX^t \tag{6}$$

denoting by Π the set of all $(n \times n)$ permutation matrices.

Theorem 1. *Equivalent formulations for the pure quadratic assignment problem are given by*

$$\min_{\pi} \operatorname{tr} \pi A \pi^{-1} B^t \quad \text{and} \quad \min_{X \in \Pi} \operatorname{tr} A X B^t X^t. \tag{7}$$

Proof. Consider the diagonal element d_{ii} of the matrix $\pi A \pi^{-1} B^t$. By definition, one has $d_{ii} = \Sigma_j a_{i\pi^{-1}(j)} b_{\pi(i)j} = \Sigma_k a_{ik} b_{\pi(i)\pi(k)}$ setting $k = \pi^{-1}(j)$. Hence $\operatorname{tr} \pi A \pi^{-1} B^t = \Sigma_i d_{ii} = \Sigma_i \Sigma_k a_{ik} b_{\pi(i)\pi(k)}$. □

Combining (6) and (7) yields the equivalent trace form of the general QAP in (1):

$$\min_{X \in \Pi} \operatorname{tr} (C + A X B^t) X^t. \tag{8}$$

In most practical applications, at least one of the two matrices, the distance matrix A or the flow matrix B is symmetric. In addition, the matrices have usually zero diagonal elements. It should be noted that problems with one symmetric matrix are equivalent to symmetric QAPs with both matrices symmetric. Suppose that $A = A^t$ and B is arbitrary. Then, for any permutation matrix $X \in \Pi$,

$$\operatorname{tr} A X B^t X^t = \operatorname{tr} A X D^t X^t \quad \text{with} \quad D = \frac{1}{2} (B + B^t).$$

This transformation is well-known and follows by means of the trace properties (5) from the identities

$$\operatorname{tr} A X B^t X^t = \operatorname{tr} X B X^t A^t = \operatorname{tr} A^t X B X^t = \operatorname{tr} A X B X^t.$$

For a square matrix M of order n with eigenvalues $\lambda_1, \lambda_2, \ldots, \lambda_n$ one has the identities

$$\operatorname{tr} M = \sum_{i=1}^{n} \lambda_i \quad \text{and} \quad \det M = \prod_{i=1}^{n} \lambda_i. \tag{9}$$

Consequently, the QAP seeks to minimize the sum of the eigenvalues of the matrices $(C + AXB^t)X^t$. The connection between QAPs and eigenvalues will be explored later on in detail and will lead to completely new families of lower bounds.

4. Solution Methods

4.1. Complexity and Exact Algorithms

QAPs are known to be *NP*-hard. Even the problem of finding an ϵ-approxima-

tion of the solution for all problem instances is *NP*-hard (Sahni and Gonzalez [46]). This explains why only implicit enumeration algorithms are known to solve QAPs exactly. There are basically three different types of enumeration procedures applied to QAPs:
- single assignment algorithms
- pair assignment algorithms and
- cutting plane methods.

Single assignment algorithms are branch and bound methods of the following type: In every step just one index is assigned to another or such a «single assignment» is revoked. One of the first single assignment algorithms is due to Gilmore [22]. Refinements were made by Lawler [31], Burkard [4] and many others. The computationally most successful exact algorithms belong to this class. QAPs up to a problem size of about $n = 15$ can be solved by such approaches.

Pair assignment algorithms are branch and bound methods in which always pairs of facilities are assigned to pairs of sites or this «pair assignment» is disallowed. Pair assignment algorithms were developed among others by Land [30], Gavett and Plyter [20], Nugent et al. [40] and others. Though investigated by quite a number of researchers, this approach did not prove to be very successful in computational tests.

For surveys on branch and bound algorithms for QAPs see Pierce and Crowston [41], Hanan and Kurtzberg [23] as well as Mirchandani and Obata [36].

Integer programming formulations suggest *cutting plane methods* for solving QAPs. Such methods were developed e.g. by Kaufman and Broeckx [26], Balas and Mazzola [1], and Bazaraa and Sherali [2]. This approach is intriguing from a mathematical point of view, but was not very successful in computational tests. The largest problem solved exactly by a cutting plane method had a size of $n = 8$. Cutting plane methods may however yield good heuristics, see e.g. Bazaraa and Sherali [2] or Burkard and Bönniger [5].

4.2. Heuristics

Because of the difficulty to find exact solutions to QAPs of even small problem sizes, there is obviously a need for good heuristics. Four main classes of heuristic approaches are in use, namely
- limited enumeration methods
- construction methods
- improvement methods and
- simulation approaches.

Limited enumeration methods rely on the experience that in a branch and bound method an optimal or good suboptimal solution is often found at an early stage in the branching process. Several techniques can be used to reduce the size of the enumeration tree, such as time limits, modified upper bounds subject to decreasing demands and others. The suboptimal cutting plane methods mentioned above belong also to this class of heuristics.

Construction methods build up a suboptimal solution step by step, e.g. by enlarging the set of assigned indices systematically by one further assignment at a time. Gilmore's heuristic and the method of increasing degree of freedom (Müller-Merbach [38]) belong, for instance, to this category. Usually, construction methods are easy to implement. They have a short running time and can be used even on small (micro-)computers. The quality of their solutions is, however, often very unsatisfactory.

Improvement methods start with a feasible solution and try to improve it by interchanges of single assignments. Such «pair exchanges» have been proposed among others by Buffa, Armour and Vollmann [3] as well as by Heider [25].

Recently a *simulation approach* with a motivation from thermodynamics has been introduced by Kirkpatrick, Gellati and Vecchi [27]. They called their method «Simulated Annealing». The same principle has been applied to QAPs by Burkard and Rendl [8] with very promising results. The main idea is to accept a solution not only if it has an improved value, but also if the value of the objective function is not improved. These inferior solutions are accepted with probabilities depending on the change of the objective function and a parameter that decreases during the simulation procedure. This enables the algorithm to move out of local minima.

5. Single Assignment Algorithm

5.1. Reduction

We would like to focus on the most successful exact solution method in the present state-of-the art. As mentioned earlier, this is a single assignment algorithm which, more exactly, is based on a reduction mechanism and the Gilmore-Lawler bound.

The *reduction* of a square matrix $M = (m_{ij})$ is simply a transformation to a matrix of the form $\bar{M} = (\bar{m}_{ij}) = (m_{ij} - u_i - v_j)$.

The Hungarian method to solve assignment problems is a systematic procedure to generate reduced matrices. Reductions have also been applied to other combinatorial problems, for instance to the traveling salesman problem by Little et al. [32]. The aim for QAPs is obvious: one would like to reduce the magnitude of the quadratic term and augment the influence of the linear term. Such an approach to QAPs has first been used by Conrad [11] and then extended by Burkard [4], Roucairol [43, 44], Edwards [13, 14], and Frieze and Yadegar [18].

The trace form seems to be the appropriate framework to establish the two reduction rules $(Rd1)$ and $(Rd2)$.

$(Rd1)$ Set $A = \bar{A} + E + F$ and $B = \bar{B} + G + H$ where E, G are matrices with constant rows and F, H are matrices with constant columns. Then, for every $X \in \Pi$,

$$\operatorname{tr}(AXB^t + C)X^t = \operatorname{tr}(\bar{A}X\bar{B}^t + \bar{C})X^t$$

with $\overline{C} = \overline{A}G^t + \overline{A}^tH + EB^t + F^tB + C$.

Proof. $\operatorname{tr} \overline{A}X\overline{B}^tX^t = \operatorname{tr}(A - E - F)X(B^t - G^t - H^t)X^t$. We use the identities $XG^t = G^t$, $H^tX^t = H^t$, $EX = E$, $X^tF = F$ to simplify. One also has $\operatorname{tr}(A - E - F)XH^tX^t = \operatorname{tr} \overline{A}XH^t = \operatorname{tr} XH^t\overline{A} = \operatorname{tr} \overline{A}^tHX^t$ and $\operatorname{tr} FXB^tX^t = \operatorname{tr} XB^tX^tF = \operatorname{tr} XB^tF = \operatorname{tr} F^tBX^t$. A straightforward computation will now yield the given identity. □

The second reduction rule allows to eliminate the non-zero diagonal elements from the matrices of the quadratic term.

(Rd 2)   Set $A = \overline{A} + R$ and $B = \overline{B} + S$ with diagonal matrices $R = \operatorname{diag}(r_1, r_2, \ldots, r_n)$ and $S = \operatorname{diag}(s_1, s_2, \ldots, s_n)$.
Then, for every $X \in \Pi$,

$$\operatorname{tr}(AXB^t + C)X^t = \operatorname{tr}(\overline{A}X\overline{B}^t + \overline{C})X^t$$

with $\overline{C} = (\overline{c}_{ij}) = (c_{ij} + \overline{a}_{ii}s_j + r_i\overline{b}_{jj} + r_is_j)$.

Proof. $\operatorname{tr} AXB^tX^t = \operatorname{tr} \overline{A}X\overline{B}^tX^t + \operatorname{tr} RX\overline{B}^tX^t + \operatorname{tr} \overline{A}XSX^t + \operatorname{tr} RXSX^t$. QAPs with a diagonal matrix can be solved as assignment problems: $\operatorname{tr} RX\overline{B}^tX^t = \operatorname{tr}(r_i\overline{b}_{jj})X^t$, $\operatorname{tr} \overline{A}XSX^t = \Sigma \overline{a}_{ii}s_{\pi(i)} = \operatorname{tr}(\overline{a}_{ii}s_j)X^t$, and $\operatorname{tr} RXSX^t = \operatorname{tr}(r_is_j)X^t$. □

Usually, both matrices are reduced by the same scheme. The main choices for the entries of the reduction matrices are listed below. We may skip the diagonal elements altogether because of (Rd 2).

(a) Burkard [4]
    Subtract from each column the minimal element.

    Roucairol [43]
(b) Use the full opening phase of the Hungarian method. Subtract from each row the minimum, then do the same for the columns.

(c) Decrease, at each step, as much as possible the maximal entry of the current (partially reduced) matrix.

(d) Edwards [13]
    Without giving the details, it is reduced with the mean values of the rows and columns. The target is to obtain a final reduced matrix with zero main diagonal and row and column sums equal to zero.

Edwards seems to be the only author who uses reduced matrices that are not restricted in sign. We shall establish a similar reduction procedure which originates, however, from a different viewpoint.
One further possible reduction method should be mentioned. One may apply

the complete Hungarian method and use the final reduced matrix. This matrix has the following intriguing theoretical property in the set of all reduced matrices: all entries are nonnegative and their total sum is minimal. However, some (limited) empirical experimentations with this scheme did not indicate an improvement which made the additional computation time worthwhile.

The effectiveness of the procedures (a) - (d) can only be judged in connection with an appropriate lower bounding method. As pointed out by Frieze and Yadegar [18], probably none of these reductions is the best possible one. This can also be illustrated by the following simple calculation. Consider an arbitrary symmetric 3×3 matrix A. One may verify that in general A cannot be reduced to the zero matrix $\bar{A} = 0$ by the methods described above. Now let us write A in the form

$$A = D + \sum_{i=1}^{3} \alpha_i M_i \quad \text{with a diagonal matrix } D \text{ and}$$

$$M_1 = \begin{pmatrix} 0 & 1 & 0 \\ 1 & 0 & 0 \\ 0 & 0 & 0 \end{pmatrix}, \quad M_2 = \begin{pmatrix} 0 & 0 & 1 \\ 0 & 0 & 0 \\ 1 & 0 & 0 \end{pmatrix} \quad \text{and } M_3 = \begin{pmatrix} 0 & 0 & 0 \\ 0 & 0 & 1 \\ 0 & 1 & 0 \end{pmatrix}.$$

The generating matrices M_i can be brought into diagonal form

$$M_1 + \frac{1}{2} \begin{pmatrix} -1 & -1 & -1 \\ -1 & -1 & -1 \\ 1 & 1 & 1 \end{pmatrix} + \frac{1}{2} \begin{pmatrix} -1 & -1 & 1 \\ -1 & -1 & 1 \\ -1 & -1 & 1 \end{pmatrix} = \begin{pmatrix} -1 & 0 & 0 \\ 0 & -1 & 0 \\ 0 & 0 & 1 \end{pmatrix}$$

$$M_2 + \frac{1}{2} \begin{pmatrix} -1 & -1 & -1 \\ 1 & 1 & 1 \\ -1 & -1 & -1 \end{pmatrix} + \frac{1}{2} \begin{pmatrix} -1 & 1 & -1 \\ -1 & 1 & -1 \\ -1 & 1 & -1 \end{pmatrix} = \begin{pmatrix} -1 & 0 & 0 \\ 0 & 1 & 0 \\ 0 & 0 & -1 \end{pmatrix}$$

$$M_3 + \frac{1}{2} \begin{pmatrix} 1 & 1 & 1 \\ -1 & -1 & -1 \\ -1 & -1 & -1 \end{pmatrix} + \frac{1}{2} \begin{pmatrix} 1 & -1 & -1 \\ 1 & -1 & -1 \\ 1 & -1 & -1 \end{pmatrix} = \begin{pmatrix} 1 & 0 & 0 \\ 0 & -1 & 0 \\ 0 & 0 & -1 \end{pmatrix}.$$

Hence A can be reduced to diagonal form with $(Rd\,1)$. Then $(Rd\,2)$ may be used to eliminate the quadratic term altogether.

5.2. The Gilmore-Lawler Bound

Let $\langle x, y \rangle$ denote the scalar product of the vectors x and y. Having arranged (and renumbered for simplicity) the components so that $x_1 \leq x_2 \leq \ldots \leq x_n$ and $y_1 \geq y_2 \geq \ldots \geq y_n$, yields the minimal scalar product

$$\langle x, y \rangle_- = \min_{X \in \Pi} \langle x, Xy \rangle = \sum_{i=1}^{n} x_i y_i \qquad (10a)$$

and the maximal scalar product

$$\langle x, y \rangle_+ = \max_{X \in \Pi} \langle x, Xy \rangle = \sum_{i=1}^{n} x_i y_{n+1-i}. \qquad (10b)$$

The Gilmore-Lawler bound GLB is based on minimal scalar products. Consider a QAP in the form (8). Let a_i be row i of A and a'_i be the vector with the diagonal element a_{ii} omitted.

Define similarly b'_j for the j-th column b_j of B^t. Introduce a kind of matrix product for A and B^t as follows:

$$A * B^t = (\gamma_{ij}) \quad \text{with} \quad \gamma_{ij} = a_{ii} b_{jj} + \langle a'_i, b'_j \rangle_-. \qquad (11)$$

The inequality

$$\text{tr}\,(A * B^t) X^t \leqslant \text{tr}\, AXB^t X^t$$

is valid for all permutation matrices $X \in \Pi$. In fact, the inequality holds for each diagonal element. The matching of a_{ii} with b_{jj} is thereby implied by the fact that the mapping $XB^t X^t$ preserves the elements of the main diagonal. Thus we obtain the lower bound

$$\text{GLB} = \min_{X \in \Pi} \text{tr}\,(C + A * B^t) X^t \qquad (12)$$

which is according to (6) an assignment problem with respect to the matrix $C + A * B^t$.

The matrices A and B are first reduced to $\bar{A}$ and $\bar{B}$ with suitable matrices E, F, G, H and then the bound GLB = GLB (E, F, G, H) is determined. Frieze and Yadegar [18] proved the following redundancy.

Theorem 2. *The row reductions with matrices E and G are redundant for the Gilmore-Lawler bounds, i.e.*

$$\text{GLB}\,(E, F, G, H) = \text{GLB}\,(0, F, 0, H).$$

Proof. Let $A = \bar{A} + E$ and $B = \bar{B} + G$ where E and G have constant rows. According to $(Rd\,1)$, we have

$$\text{tr}\,(AXB^t + C) X^t = \text{tr}\,(\bar{A}X\bar{B}^t + \bar{A}G^t + EB^t + C) X^t.$$

Consider the operation (11) for the products $\bar{A} * \bar{B}^t$ and $A * B^t = (\bar{A} + E) * * (\bar{B} + G)^t$. Since E and G have constant rows, the minimal scalar products for both cases are formed with respect to the same orderings. Therefore, one obtains the matrix identity $A * B^t = \bar{A} * \bar{B}^t + \bar{A} G^t + E B^t$ which implies the given redundancy. □

For the reduction method (a), this redundancy is already implemented. In procedure (b), first the row minima are subtracted (which is redundant). However, then the column minima of the row-reduced matrix are taken. Therefore, this reduction lies «somewhere» between no reduction and the full column reduction (a). Even with Theorem 2, methods of the type (c) and (d) cannot be abandoned. It may well be that the best column reductions F and H can only be found by means of some simultaneous row and column reduction scheme.

The comparisons of the Gilmore-Lawler bounds with different reductions by Frieze and Yadegar [18] fail to establish a clear trend. Only the procedure (c) was practically always dominated. We carried out a computer run on the data given by Nugent et al. [40]. Also, some random data were tested with elements of A being uniform in $[0, 100]$ and those of B uniform in $[0, 10]$. The results are summarized in Table 1 (the bounds in the last three columns are explained later). GLB_1 refers to the unreduced data, GLB_2 uses reduction (a), and GLB_3 uses (d) in the version of Frieze and Yadegar [18]. Our results are similar to the ones by Frieze. However, really surprising in the small effect of the reduction step altogether.

5.3. Enumeration Schemes and Computer Codes

Computer codes for some of the suboptimal algorithms can be found in Burkard and Derigs [6], Burkard and Bönniger [5], and West [49] [here the complete code is available from the ACM Algorithms Distribution Service].

Complete listings of FORTRAN routines for the single assignment algorithm are given in the book of Burkard and Derigs [6] and by Mautor and Savoye [34] using the method of Roucairol [43]. The computation times to solve optimally QAPs by this method grow exponentially. The limit of solvable problems is practically reached at $n = 15$. A CPU time of almost 50 min. on a CDC CYBER 76 has been reported in Burkard and Derigs [6] to solve an example by Nugent et al. [40] of this size.

The single assignment algorithm is a branch and bound method which requires the computation of a Gilmore-Lawler bound at each node of the enumeration tree. Suppose that, at some intermediate node, the indices of a subset N_1 of N are already fixed and those of N_2 are still unassigned. Thus we are looking for the best extension of a partial permutation $\pi(i)$, $i \in N_1$. The objective function (1) may be written as follows:

$$\sum_{i \in N_1} \sum_{k \in N_1} a_{ik} b_{\pi(i)\pi(k)}$$

$$+ \sum_{k \in N_2} \sum_{i \in N_1} (a_{ik} b_{\pi(i)\pi(k)} + a_{ki} b_{\pi(k)\pi(i)})$$

$$+ \sum_{i \in N_2} \sum_{k \in N_2} a_{ik} b_{\pi(i)\pi(k)}.$$

Setting

$$c_{k\ell} := \sum_{i \in N_1} (a_{ik} b_{\pi(i)\ell} + a_{ki} b_{\ell\pi(i)})$$

for all $k \in N_2$, $\ell \notin \pi(N_1)$, one obtains a QAP of the form

$$\text{const.} + \text{tr}\, (C + AXB^t) X^t$$

with matrices reduced to the order $|N_2|$. Hence the Gilmore-Lawler bounding technique applies to each node of the branching tree.

At each branching step, a certain element i is assigned to j or i is not assigned to j. Since the bound produces the final matrix of an assignment problem, the selection of the branching pair (i,j) is based on the principle of «alternative costs» by Little et al. [32]. Branching procedures with $k > 2$ successors have recently been investigated by Roucairol [45]. This approach could lead to a parallel enumeration scheme which may accelerate the search.

There is, however, the following main difficulty with QAPs. Take from Table 1, for instance, the best bound GLB that has been discussed so far, i.e. GLB = $= \max\{\text{GLB}_1, \text{GLB}_2, \text{GLB}_3\}$. Table II is listing the relative error of GLB with respect to the optimal solution SOL, which is given by (SOL − GLB)/SOL. Unfortunately one can observe a deterioration of the bounds with increasing n. As well known from branch and bound methods for other combinatorial problems, the quality of the lower bounds is by far more decisive for the performance of the algorithm than any branching scheme or strategy.

This deterioration of the bounds with n seems to be the main source for the computational difficulties. A significant extension of the sizes of solvable QAPs will probably require an advancement of the lower bound computations. We will show that the trace formulation gives access to completely new families of bounds which, without claiming any breakthrough, are advantageous for certain categories of problems.

6. Eigenvalue Approach to Symmetric QAPs

6.1. Eigenvalue Related Bounds

We concentrate on symmetric QAPs which form the most important class for applications. With $A = A^t$ and $B = B^t$, all eigenvalues are real. Let $\lambda_1, \lambda_2, \ldots, \lambda_n$ and $\mu_1, \mu_2, \ldots, \mu_n$ be the eigenvalues of A and B, respectively.

We can assume the ordering $\lambda_1 \leq \lambda_2 \leq \ldots \leq \lambda_n$ and $\mu_1 \geq \mu_2 \geq \ldots \geq \mu_n$ and set $\lambda^t = (\lambda_1, \lambda_2, \ldots, \lambda_n)$ and $\mu^t = (\mu_1, \mu_2, \ldots, \mu_n)$. The matrices A and B possess diagonalizations of the form $A = P_1 \Lambda_1 P_1^t$ and $B = P_2 \Lambda_2 P_2^t$ with orthogonal matrices P_1, P_2 and diagonal matrices $\Lambda_1 = \text{diag}(\lambda_1, \lambda_2, \ldots, \lambda_n)$, $\Lambda_2 = \text{diag}(\mu_1, \mu_2, \ldots, \mu_n)$. Let $x_1, x_2, \ldots, x_n$ denote the columns of P_1 and $y_1, y_2, \ldots, y_n$ the columns of P_2.

Lemma 1.

(i) $\quad \text{tr } AB = \lambda^t S \mu$ with the doubly stochastic matrix

$$S = (\langle x_i, y_j \rangle^2);$$

(ii) $\quad \langle \lambda, \mu \rangle_- \leq \text{tr } AB \leq \langle \lambda, \mu \rangle_+.$

Proof. See Gaffke and Krafft [19]. Their proof is stated for positive semidefinite matrices but is, in fact, valid for all symmetric matrices. One may verify the presentation

$$\text{tr } AB = \text{tr}\left(\sum_{i=1}^{n} \sum_{j=1}^{n} \lambda_i \mu_j x_i x_i^t y_j y_j^t\right) = \sum_{i=1}^{n} \sum_{j=1}^{n} \lambda_i \mu_j \langle x_i, y_j \rangle^2 = \lambda^t S \mu$$

where S is doubly stochastic since $x_1, \ldots, x_n$ and $y_1, \ldots, y_n$ are orthonormal bases. Using Birkhoff's theorem (e.g. in Marcus and Minc [33]), there exists a convex linear combination of the form

$$S = \sum_{X \in \Pi} \alpha_x X \quad \text{with} \quad \alpha_x \geq 0 \quad \text{and} \quad \Sigma \alpha_x = 1.$$

Thus, we obtain

$$\text{tr } AB = \lambda^t S \mu = \sum_{X \in \Pi} \alpha_x \langle \lambda, X\mu \rangle$$

which implies (ii). $\qquad \square$

Theorem 3. *Let A and B be symmetric. Then, for all $X \in \Pi$,*

1) $\quad \text{tr } AXBX^t = \lambda^t S(X) \mu$

   where $S(X) = (\langle x_i, Xy_j \rangle^2)$ is a doubly stochastic matrix;

2) $\quad \langle \lambda, \mu \rangle_- \leq \text{tr } AXBX^t \leq \langle \lambda, \mu \rangle_+.$

Proof. The mapping XBX^t is a similarity transformation. Hence B and XBX^t have the same eigenvalues μ. Thus 2) is implied by (ii). We have the diagonalization $XBX^t = (XP_2) \Lambda_2 (XP_2)^t$ with the orthogonal matrix $XP_2 = (Xy_j)$. Therefore 1) follows from (i). □

As illustration, consider the following matrices and their diagonalizations

$$A = \begin{pmatrix} 17 & -1 & 4 \\ -1 & 17 & 4 \\ 4 & 4 & 20 \end{pmatrix} \text{ and } B = \begin{pmatrix} 5 & 0 & 1 \\ 0 & 2 & 0 \\ 1 & 0 & 5 \end{pmatrix}$$

$$A = P_1 \Lambda_1 P_1^t = \begin{pmatrix} -1/\sqrt{3} & 1/\sqrt{2} & 1/\sqrt{6} \\ -1/\sqrt{3} & -1/\sqrt{2} & 1/\sqrt{6} \\ 1/\sqrt{3} & 0 & 2/\sqrt{6} \end{pmatrix} \begin{pmatrix} 12 & 0 & 0 \\ 0 & 18 & 0 \\ 0 & 0 & 24 \end{pmatrix} \begin{pmatrix} -1/\sqrt{3} & -1/\sqrt{3} & 1/\sqrt{3} \\ 1/\sqrt{2} & -1/\sqrt{2} & 0 \\ 1/\sqrt{6} & 1/\sqrt{6} & 2/\sqrt{6} \end{pmatrix}$$

$$B = P_2 \Lambda_2 P_2^t = \begin{pmatrix} 1/\sqrt{2} & 1/\sqrt{2} & 0 \\ 0 & 0 & 1 \\ 1/\sqrt{2} & -1/\sqrt{2} & 0 \end{pmatrix} \begin{pmatrix} 6 & 0 & 0 \\ 0 & 4 & 0 \\ 0 & 0 & 2 \end{pmatrix} \begin{pmatrix} 1/\sqrt{2} & 0 & 1/\sqrt{2} \\ 1/\sqrt{2} & 0 & -1/\sqrt{2} \\ 0 & 1 & 0 \end{pmatrix}.$$

One obtains the optimal permutation $\pi = \begin{pmatrix} 1 & 2 & 3 \\ 3 & 1 & 2 \end{pmatrix}$ or $X = \begin{pmatrix} 0 & 0 & 1 \\ 1 & 0 & 0 \\ 0 & 1 & 0 \end{pmatrix}$.

The corresponding eigenvalue representation of Theorem 3 is as follows:

$$\text{tr } AXBX^t = \lambda^t S(X) \mu = (12, 18, 24) \begin{pmatrix} 2/3 & 0 & 1/3 \\ 0 & 1 & 0 \\ 1/3 & 0 & 2/3 \end{pmatrix} \begin{pmatrix} 6 \\ 4 \\ 2 \end{pmatrix} = 208.$$

Theorem 3 yields the eigenvalue bound

$$\text{EVB} = \sum_i \lambda_i \mu_i = \langle \lambda, \mu \rangle_- \tag{13}$$

which works out to EVB = $12(6) + 18(4) + 24(2) = 192$ for the example.

The eigenvalue bound EVB is optimal for certain pairs of matrices which one may characterize as follows. Let $A = P_1 \Lambda_1 P_1^t$ be given. Consider the set $M(A, X_*)$ of matrices $B = P_2 \Lambda_2 P_2^t$ with arbitrary diagonal matrix $\Lambda_2 = \text{diag}(\mu_1, \mu_2, \ldots, \mu_n)$, $\mu_1 \geq \mu_2 \geq \ldots \geq \mu_n$, and orthogonal matrix P_2 satisfying $X_* P_2 = P_1$ for some permutation matrix $X_* \in \Pi$.

Theorem 4. *The QAP for symmetric matrices A and B with $B \in M(A, X_*)$ is solved by the permutation X_* and has optimal value $\langle \lambda, \mu \rangle_- = \Sigma_i \lambda_i \mu_i$.*

Proof. It is $X_* P_2 = P_1$, i.e. $X_* y_j = x_j$. One has, therefore,

$$\operatorname{tr} AX_*BX_*^t = \lambda^t(\langle x_i, X_*y_j\rangle^2)\mu = \lambda^t(\delta_{ij})\mu = \sum_i \lambda_i\mu_i = \langle\lambda,\mu\rangle_-$$

which is minimal (Theorem 3). □

Suppose A is positive definite. Define $A^p = P_1\Lambda_1^p P_1^t$ with $\Lambda_1^p = \operatorname{diag}(\lambda_1^p, \lambda_2^p, \ldots, \lambda_n^p)$ for any real p. Since by definition $0 < \lambda_1 \leq \lambda_2 \leq \ldots \leq \lambda_n$, we get $\lambda_1^{-p} \geq \lambda_2^{-p} \geq \ldots \geq \lambda_n^{-p} > 0$ for all $p > 0$. Hence $A^{-p} \in M(A, I)$ for $p > 0$ and the identical permutation I solves the QAP for the pair A and A^{-p}. In particular, the case $p = 1$ yields the following peculiar characterization of a determinant.

Corollary 1. *Let A be positive definite. Consider the mapping*

$$a_{ij} \to A_{k\ell}; \quad k = \pi(i), \quad \ell = \pi(j),$$

where $A_{k\ell}$ is the cofactor and π a permutation. Then

$$\det A = \frac{1}{n}\sum_i\sum_j a_{ij}A_{ij} = \frac{1}{n}\min_\pi \sum_i\sum_j a_{ij}A_{\pi(i)\pi(j)}.$$

Proof. The identity solves the QAP for the matrices A and A^{-1}. The assertion follows since $A^{-1} = (A_{ij}/\det A)$. □

In the case of positive definite matrices, a different lower bound, based on determinants, can be given.

Theorem 5. *Let A and B be positive definite matrices. Then*

$$\min_{X\in\Pi} \operatorname{tr} AXBX^t \geq n(\det A \det B)^{1/n}.$$

Proof. AB is in general not symmetric. However, one obtains a symmetric matrix with equal trace as follows: $A^{1/2}$ is defined and $A^{1/2}A^{1/2} = A$. Hence $\operatorname{tr} AB = \operatorname{tr} A^{1/2}A^{1/2}B = \operatorname{tr} A^{1/2}BA^{1/2}$. The matrix $A^{1/2}BA^{1/2}$ is symmetric and also positive definite since $x^t A^{1/2}BA^{1/2}x = y^t By > 0$ for all x, setting $y = A^{1/2}x$. Let $\lambda_1, \lambda_2, \ldots, \lambda_n$ be the positive eigenvalues of $A^{1/2}BA^{1/2}$. Using the arithmetic-geometric mean inequality, we obtain

$$\frac{1}{n}\operatorname{tr} A^{1/2}BA^{1/2} = \frac{1}{n}\sum_{i=1}^n \lambda_i \geq \left(\prod_{i=1}^n \lambda_i\right)^{1/n} = (\det A^{1/2}BA^{1/2})^{1/n} = (\det A \det B)^{1/n}.$$

Therefore $\operatorname{tr} AB \geq n(\det A \det B)^{1/n}$ and the Theorem follows since $\det XBX^t = \det B$. □

The numerical value of the lower bound in Theorem 5 for the example is

$$n(\det A \det B)^{1/n} = 188.69 < EVB.$$

This bound applies also to general symmetric matrices A and B since we can add a multiple of the unit matrix according to reduction rule $(Rd\,2)$. Our numerical tests showed, however, that this bound is usually dominated by the eigenvalue bound, as in our example.

6.2. Optimal Reduction

According to the reduction rules $(Rd\,1)$ and $(Rd\,2)$, we may reduce a symmetric matrix A in the form $A = \bar{A} + E + E^t + R$ in order to maintain the symmetry for $\bar{A}$. Theorem 3 shows that the reduced quadratic term $\operatorname{tr} \bar{A} X \bar{B} X^t$ has values ranging from $\langle \bar{\lambda}, \bar{\mu} \rangle_-$ to $\langle \bar{\lambda}, \bar{\mu} \rangle_+$, denoting by $\bar{\lambda}$ and $\bar{\mu}$ the eigenvalues of $\bar{A}$ and $\bar{B}$. The best way to minimize the length of this interval seems to be to reduce the eigenvalue fluctuation of both matrices as much as possible. This suggests the following approach. Let A be a symmetric matrix with eigenvalues $\lambda_1, \lambda_2, \ldots, \lambda_n$ $(n \geq 3)$. Define the *spread*

$$s(A) = \max_{i,j} |\lambda_i - \lambda_j|. \tag{14}$$

Since there are no simple formula for the spread, Finke et al. [16] use the following upper bound $m(A)$ for $s(A)$ by Mirsky [37]:

$$s(A) \leq m(A) = \left[2 \sum_i \sum_j a_{ij}^2 - \frac{2}{n} (\operatorname{tr} A)^2 \right]^{1/2}. \tag{15}$$

The minimal value of $s(\bar{A})$ may now be approximated by minimizing $m(\bar{A})$ or equivalently

$$f(e,r) = \sum_{\substack{i,j \\ i \neq j}} (a_{ij} - e_i - e_j)^2 + \sum_i (a_{ii} - 2e_i - r_i)^2 - \frac{1}{n} \left[\sum_i (a_{ii} - 2e_i - r_i) \right]^2.$$

The conditions $\partial f / \partial e_k = 0$ and $\partial f / \partial r_k = 0$ yield the following system of equations:

$$(n-2)e_k + \sum_i e_i = \sum_j a_{kj} - a_{kk}$$

$$n(a_{kk} - 2e_k - r_k) = \sum_i a_{ii} - 2 \sum_i e_i - \sum_i r_i$$

for all k. These equations can be solved explicitly. Note that one value r_k may be assigned arbitrarily since shifting the diagonal elements to $r_k + r$ for all k leaves the spread unchanged. Setting $r_n = a_{nn} - 2e_n$ gives a zero diagonal. Defining $z = \Sigma_i e_i$, results in the following computational scheme:

$$z := \frac{1}{2(n-1)} \left(\sum_i \sum_j a_{ij} - \operatorname{tr} A \right)$$

$$e_k := \frac{1}{n-2} \left(\sum_j a_{kj} - a_{kk} - z \right) \quad \text{for } k = 1, \ldots, n \qquad (16)$$

$$r_k := a_{kk} - 2e_k \quad \text{for } k = 1, \ldots, n.$$

One may verify that the reduced matrices have row and column sum equal to zero. Our 3 × 3 example is reduced optimally, i.e. it is completely linearized to the assignment problem

$$\min_{X \in \Pi} \operatorname{tr} \begin{pmatrix} 84 & 34 & 84 \\ 84 & 34 & 84 \\ 109 & 40 & 109 \end{pmatrix} X^t.$$

The approximation $s(A) \leq m(A)$ is usually very good. For highly structured problems, however, both values might be completely different. The traveling salesman problem (TSP) illustrates such a case. Let X_c denote the cyclic permutation

$$X_c = \begin{pmatrix} 0 & 1 & 0 & \cdot & \cdot & \cdot & 0 \\ 0 & 0 & 1 & 0 & \cdot & \cdot & 0 \\ & & \ddots & & & & \\ 0 & \cdot & \cdot & \cdot & 0 & 1 & 0 \\ 0 & \cdot & \cdot & \cdot & \cdot & 0 & 1 \\ 1 & 0 & \cdot & \cdot & \cdot & \cdot & 0 \end{pmatrix}$$

The symmetric TSP based on the matrix $A = A^t$ is described by the QAP with matrices A and X_c. Consider the equivalent symmetric QAP with A and $B = (X_c + X_c^t)/2$. The spread of B is bounded by $s(B) \leq m(B) = \sqrt{n}$. The reduction entries are $r_k = -(n-1)^{-1}$ and $e_k = 2^{-1}(n-1)^{-1}$ which gives the insignificant reduction to $m(\bar{B}) = (n^2 - 3n)^{1/2}(n-1)^{-1/2}$. The eigenvalues of X_c are the roots of unity $\lambda_k = \exp(2\pi k \sqrt{-1}/n)$ for $k = 0, 1, \ldots, n-1$. One obtains for B the eigenvalues

$$\lambda_k(B) = (\lambda_k + \lambda_k^*)/2 = \cos(2\pi k/n)$$

where λ_k^* denotes the conjugate complex number. Consequently, we have $s(B) \leq 2$ for all n, which characterizes symmetric TSPs as an easy class of symmetric

QAPs. We were curious to find out the performance of the reduction (16) in connection with the Gilmore-Lawler bound (12). The results are given in Table 1 under GLB_4. It is interesting to note that this bound usually gives the overall best values for random QAPs, as demonstrated in Table 1 (ii) and in other similar test runs.

The eigenvalue bound EVB_1 is obtained as follows. Again, we reduce A, B, C to $\bar{A}$, $\bar{B}$, $\bar{C}$ by means of (16). Let $\bar{A}$ and $\bar{B}$ have the eigenvalues $\bar{\lambda}$ and $\bar{\mu}$. Then define

$$EVB_1 = \langle \bar{\lambda}, \bar{\mu} \rangle_- + \min_{X \in \Pi} \operatorname{tr} \bar{C}X^t. \tag{17}$$

As shown in Table 1, this bound does not provide an improvement for QAPs of small sizes. But for the larger QAPs of size $n = 20$ and, in particular, $n = 30$, the bound turned out to be significantly better.

6.3. Almost Linear QAPs

Consider the purely quadratic assignment problem with symmetric matrices A and B that have zero main diagonals, as is the case for applications. The eigenvalue form is simplified to

$$\operatorname{tr} AXBX^t = \operatorname{tr}(\bar{A}X\bar{B} + \bar{C})X^t = \bar{\lambda}^t S(X)\bar{\mu} + \langle c, Xd \rangle \tag{18}$$

where $c_i = 2e_i$ and $d_i = \Sigma_j b_{ij}$.

Consequently, we obtain instead of (17)

$$EVB_1 = \langle \bar{\lambda}, \bar{\mu} \rangle_- + \langle c, d \rangle_-. \tag{19}$$

Rendl [42] introduced the following measure for the *degree of linearity*

$$L := (\langle \bar{\lambda}, \bar{\mu} \rangle_+ - \langle \bar{\lambda}, \bar{\mu} \rangle_-)/(\langle c, d \rangle_+ - \langle c, d \rangle_-). \tag{20}$$

A «small» ratio indicates a small influence of the quadratic term compared to the magnitude of the linear term. If this is the case, it is suggested to rank the scalar products $\langle c, Xd \rangle$ in increasing order and scan the corresponding permutations. In this way, a QAP with very small L may be solved completely or at least the eigenvalue bound (19) may be improved.

The ranking could be done by ordering the solutions of an assignment problem with costs $(c_i d_j)$. Murty [39] proposes a procedure that has a time complexity $O(kn^3)$ to obtain the k-best solutions. Ranking directly the scalar products yields in Rendl [42] a complexity $0(n \log n + (n + \log k)k)$ which is a considerable improvement over the classical bound.

The special form (18) also shows that ranking the scalar products $\langle c, Xd \rangle$ is not advisable if the row sums of the matrices A and B take on only few different values. The classical test examples by Nugent et al. [40] are of this type. This may be one of the reasons why this class of problems cannot be reduced and linearized with great success.

A large value L, on the other hand, indicates a negligible linear term. An improvement of the eigenvalue bound $\langle \bar{\lambda}, \bar{\mu} \rangle_-$ for the quadratic term is obtained as follows. One has $\bar{\lambda}^t S(X) \bar{\mu} = \Sigma_i \Sigma_j \bar{\lambda}_i \bar{\mu}_j \langle x_i, X y_j \rangle^2$ where x_i and y_j represent the eigenvectors of the reduced matrices. The entries of $S(X)$ may be bounded by $\ell_{ij} \leq \langle x_i, X y_j \rangle^2 \leq u_{ij}$ with

$$u_{ij} = \max\{\langle x_i, y_j \rangle_-^2, \langle x_i, y_j \rangle_+^2\} \quad \text{and}$$

$$\ell_{ij} = \begin{cases} 0 & \text{if } \langle x_i, y_j \rangle_- \text{ and } \langle x_i, y_j \rangle_+ \text{ have different signs} \\ \min\{\langle x_i, y_j \rangle_-^2, \langle x_i, y_j \rangle_+^2\} & \text{otherwise} \end{cases} \quad (21)$$

Consider the capacitated transportation problem

$$\min \sum_i \sum_j (\bar{\lambda}_i \bar{\mu}_j) s_{ij}$$

$$\text{s.t.} \sum_j s_{ij} = 1 \quad (22)$$

$$\sum_i s_{ij} = 1$$

$$\ell_{ij} \leq s_{ij} \leq u_{ij}.$$

Replacing $\langle \bar{\lambda}, \bar{\mu} \rangle_-$ in (19) by the optimal value of (22) yields the improved eigenvalue bound EVB_2 in Table 1.

A complete new branch and bound algorithm to solve QAPs may be based on the capacitated transportation problem (22). For this problem, the costs $(\bar{\lambda}_i \bar{\mu}_j)$ are highly structured and remain unchanged throughout the enumeration. However, if more indices are assigned at the nodes of the enumeration tree, the length of the flow interval $[\ell_{ij}, u_{ij}]$ is more and more reduced. A computer code for this approach will be implemented in the near future.

7. Concluding Remark

Evaluating the quality of the various reduction and bounding methods seems difficult. There is no apparent common trend for the different types and sizes of problems. Perhaps this is characteristic for a combinatorial problem of such extreme difficulty. Apart from providing an elegant theoretical frame, the trace form also supplies eigenvalue related bounds which are competitive with the Gilmore-Lawler bounding technique. There is, in particular, some hope to obtain a better asymptotic behavior with increasing problem sizes.

Table 1. Lower Bounds for QAPs.

(i) Test Data by Nugent et al. (ii) Random Data.

	Size n	Best Known Value	Gilmore-Lawler Bounds				Eigenvalue Bounds	
			GLB_1	GLB_2	GLB_3	GLB_4	EVB_1	EVB_2
(i)	5	50	50	49	48	49	47	48
	6	86	82	82	77	76	70	73
	7	148	137	137	131	130	123	124
	8	214	186	186	175	174	160	164
	12	578	493	493	464	463	446	448
	15	1150	963	963	919	918	927	934
	20	2570*	2057	2057	1963	1960	2075	2085
	30	6124*	4539	4558	4324	4320	4982	5005
(ii)	8	7706	6479	5880	6359	6551	6495	6507
	12	18180	14698	14461	14499	14973	14423	14479
	15	37408*	31833	31127	31968	31969	31413	31477

* Values not known to be optimal.

Table II. Deterioration of Gilmore-Lawler Bounds.

Size n	5	6	7	8	12	15	20	30
Relative Error (%)	0.0	4.6	7.4	13.1	14.7	17.1	$\leqslant 20.0$	$\leqslant 25.6$

References

[1] E. Balas and J.B. Mazzola, «Quadratic 0 - 1 Programming by a New Linearization», *Presented at the TIMS/ORSA Meeting, Washington, D.C., 1980.*
[2] M.S. Bazaraa and M.D. Sherali, «Benders' Partitioning Scheme Applied to a New Formulation of the Quadratic Assignment Problem», *Naval Research Logistics Quarterly* 27, 29 - 41, 1980.
[3] E.S. Buffa, G.C. Armour and T.E. Vollmann, «Allocating Facilities with CRAFT», *Harvard Business Review* 42, 136 - 158, 1962.
[4] R.E. Burkard, «Die Störungsmethode zur Lösung quadratischer Zuordnungsprobleme», *Operations Research Verfahren* 16, 84 - 108, 1973.
[5] R.E. Burkard and T. Bönniger, «A Heuristic for Quadratic Boolean Programs With Applications to Quadratic Assignment Problems», *European Journal of Operational Research* 13, 374 - 386, 1983.
[6] R.E. Burkard and U. Derigs, *Assignment and Matching Problems: Solution Methods with FORTRAN--Programs*, Springer, Berlin, 1980.
[7] R.E. Burkard and J. Offermann, «Entwurf von Schreibmaschinentastaturen mittels quadratischer Zuordnungsprobleme», *Zeitschrift für Operations Research* 21, B121 - B132, 1977.
[8] R.E. Burkard and F. Rendl, «A Thermodynamically Motivated Simulation Procedure for Combinatorial Optimization Problems», *European Journal of Operational Research* 17, 169 - 174, 1984.
[9] R.E. Burkard and L.H. Stratmann, «Numerical Investigation on Quadratic Assignment Problems», *Naval Research Logistics Quarterly* 25, 129 - 148, 1978.
[10] N. Christofides, A. Mingozzi and P. Toth, «Contributions to the Quadratic Assignment Problem», *European Journal of Operational Reserach* 4, 243 - 247, 1980.
[11] K. Conrad, *Das quadratische Zuweisungsproblem und zwei seiner Spezialfälle*, Mohr-Siebeck, Tübingen, 1971.

[12] J.W. Dickey and J.W. Hopkins, «Campus Building Arrangement Using TOPAZ», *Transportation Research* 6, 59 - 68, 1972.
[13] C.S. Edwards, «The Derivation of a Greedy Approximator for the Koopmans-Beckmann Quadratic Assignment Problem», *Proc. CP77 Combinatorial Prog. Conf.*, Liverpool, 55 - 86, 1977.
[14] C.S. Edwards, «A Branch and Bound Algorithm for the Koopmans-Beckmann Quadratic Assignment Problem», *Mathematical Programming Study* 13, 35 - 52, 1980.
[15] A.N. Elshafei, «Hospital Lay-out as a Quadratic Assignment Problem», *Operational Research Quarterly* 28, 167 - 179, 1977.
[16] G. Finke, R.E. Burkard and F. Rendl, «Eigenvalue Approach to Quadratic Assignment Problems», *Presented at 5th Symposium on Operations Research, Osnabrück, West Germany*, 1984.
[17] R.L. Francis and J.A. White, *Facility Layout and Location*, Prentice-Hall, Englewood Cliffs, N.J., 1974.
[18] A.M. Frieze and J. Yadegar, «On the Quadratic Assignment Problem», Discrete Applied Mathematics 5, 89 - 98, 1983.
[19] N. Gaffke and O. Krafft, «Matrix Inequalities in the Löwner Ordering», in B. Korte (ed.), *Modern Applied Mathematics - Optimization and Operations Research*, North-Holland, 576 - 622, 1982.
[20] J.W. Gavett and N.V. Plyter, «The Optimal Assignment of Facilities to Locations by Branch and Bound», *Operations Research* 14, 210 - 232, 1966.
[21] A.M. Geoffrion and G.W. Graves, «Scheduling Parallel Production Lines With Changeover Costs: Practical Applications of a Quadratic Assignment/LP Approach», *Operations Research* 24, 595 - 610, 1976.
[22] P.C. Gilmore, «Optimal and Suboptimal Algorithms for the Quadratic Assignment Problem», *SIAM Journal on Applied Mathematics* 10, 305 - 313, 1962.
[23] M. Hanan and J.M. Kurtzberg, «A Review of the Placement and Quadratic Assignment Problems», *SIAM Review* 14, 324 - 342, 1972.
[24] D.R. Heffley, «Assigning Runners to a Relay Team», in S.P. Ladany and R.E. Machol, eds., *Optimal Strategies in Sports*, North Holland, Amsterdam, 169 - 171, 1977.
[25] C.H. Heider, «A Computationally Simplified Pair Exchange Algorithm for the Quadratic Assignment Problem», Paper No. 101, Center for Naval Analyses, Arlington Va., 1972.
[26] L. Kaufman and F. Broeckx, «An Algorithm for the Quadratic Assignment Problem Using Benders' Decomposition», *European Journal of Operational Research* 2, 204 - 211, 1978.
[27] S. Kirkpatrick, C.D. Gelatti, Jr. and M.P. Vecchi, «Optimization by Simulated Annealing», *Science* 220, 671 - 680, 1983.
[28] T.C. Koopmans and M.J. Beckmann, «Assignment Problems and the Location of Economic Activities», *Econometrica* 25, 53 - 76, 1957.
[29] J. Krarup and P.M. Pruzan, «Computer-aided Layout Design», *Mathematical Programming Study* 9, 75 - 94, 1978.
[30] A.M. Land, «A Problem of Assignment with Interrelated Costs», *Operational Research Quarterly* 14, 185 - 198, 1963.
[31] E.L. Lawler, «The Quadratic Assignment Problem», *Management Science* 9, 586 - 599, 1963.
[32] J.D.C. Little, K.G. Murty, D.W. Sweeney and C. Karel, «An Algorithm for the Travelling Salesman Problem», *Operations Research* 11, 972 - 989, 1963.
[33] M. Markus and H. Minc, *A Survey of Matrix Theory and Matrix Inequalities*, Allyn and Bacon, Boston, 1964.
[34] T. Mautor and D. Savoye, *Etudes d'heuristiques pour le problème d'affectation quadratique*, Mémoire, diplôme d' Ingénieur I.I.E., 1983.
[35] E.J. McCormick, *Human Factors Engineering*, McGraw-Hill, New York, 1970.
[36] P.B. Mirchandani and T. Obata, «Locational Decisions with Interactions Between Facilities: the Quadratic Assignment Problem - a Review», *Working Paper* PS-79-1, Rensselaer Polytechnic Institute, Troy, N.Y., 1979.
[37] L. Mirsky, «The Spread of a Matrix», *Mathematika* 3, 127 - 130, 1956.
[38] H. Müller-Merbach, *Optimale Reihenfolgen*, Springer, Berlin, 158 - 171, 1970.
[39] K.G. Murty, «An Algorithm for Ranking All the Assignments in Order of Increasing Cost», *Operations Research* 16, 682 - 687, 1968.
[40] C.E. Nugent, T.E. Vollmann and J. Ruml, «An Experimental Comparison of Techniques for the Assignment of Facilities to Locations», *Operations Research* 16, 150 - 173, 1968.
[41] J.F. Pierce and W.B. Crowston, «Tree Search Algorithms for Quadratic Assignment Problems», *Naval Research Logistics Quarterly* 18, 1 - 36, 1971.
[42] F. Rendl, «Ranking Scalar Products to Improve Bounds for the Quadratic Assignment Problem», *European Journal of Operational Research* 20, 363 - 372, 1985.

[43] C. Roucairol, «A Reduction Method for Quadratic Assignment Problems», *Operations Research Verfahren* 32, 183 - 187, 1979.
[44] C. Roucairol, «Un nouvel algorithme pour le problème d'affectation quadratique», *R.A.I.R.O.* 13, 275 - 301, 1979.
[45] C. Roucairol, «An Efficient Branching Scheme in Branch and Bound Procedures», *TIMS XXV*, Copenhagen, 1984.
[46] S. Sahni and T. Gonzalez, «P-complete Approximation Problems», *Journal of ACM* 23, 555 - 565, 1976.
[47] L. Steinberg, «The Backboard Wiring Problem: a Placement Algorithm», *SIAM Review* 3, 37 - 50, 1961.
[48] I. Ugi, J. Bauer, J. Brandt, J. Friedrich, J. Gasteiger, C. Jochum and W. Schubert, «Neue Anwendungsgebiete für Computer in der Chemie», *Angewandte Chemie* 91, 99 - 111, 1979.
[49] D.H. West, «Algorithm 608: Approximate Solution of the Quadratic Assignment Problem», *ACM Transactions on Mathematical Software* 9, 461 - 466, 1983.

Classification by Subjects

Problem formulations:
1, 2, 5, 10, 13, 14, 18, 26, 28, 31.

Applications:
7, 12, 15, 17, 21, 24, 28, 29, 35, 47, 48.

Complexity and exact algorithms:
2, 4, 20, 22, 23, 26, 30, 31, 36, 40, 41, 46.

Heuristics:
2, 3, 5, 8, 9, 22, 25, 27, 38.

Reductions:
4, 11, 13, 14, 18, 22, 31, 32, 43, 44.

Computer Codes:
5, 6, 34, 45, 49.

Eigenvalue approach:
16, 19, 33, 37, 39, 42.

Gerd Finke
Technical University of Nova Scotia
P.O. Box 1000
B3J 2X4 - Halifax, N.S.
Canada

Rainer E. Burkard
Franz Rendl
Institut für Mathematik
Technische Universität Graz
A - 8010 Graz
Austria

ORDER RELATIONS OF VARIABLES IN 0-1 PROGRAMMING

Peter L. HAMMER and Bruno SIMEONE

1. Introduction

In this paper we present old and new results concerning two important preorders (i.e. reflexive and transitive relations) in the set of variables of a 0-1 (linear or nonlinear) programming problem:

a) the ordinary order relation $x_i \leq x_j$, which means

"x_j is 1 when x_i is 1"

b) the preorder $x_i \preccurlyeq x_j$ defined by the property

"every feasible vector x such that $x_i = 0$ and $x_j = 1$, remains feasible when x_i is replaced by 1, and x_j is replaced by 0".

Accordingly, the paper is divided into two parts. Part I deals with the relation $\leq$. In Section 2, we recall some basic facts about 0-1 programs. In Section 3 we discuss the generation and the use of order constraints $x_i \leq x_j$ in linear and nonlinear 0-1 programming. In Section 4 we consider the problem of maximizing a linear function of n binary variables subject to order constraints (only). It is well known (Picard [20]) that this problem is equivalent to finding a maximum weight closure of a directed graph. We exhibit a direct reduction of the maximum weight closure problem to a minimum cut one (an indirect reduction was given in Picard [20]); moreover, we give a linear-time algorithm for the special case when the graph is a rooted tree. In Section 5 we determine the minimum number of order constraints which, due to the fact that $x_i \leq x_j$ implies $x_i x_j = x_i$, cause a quadratic function f of n binary variables to become linear. We prove that such number is always $n - 1$, independent of the number of terms appearing in f. We introduce a class of directed (but not necessarily rooted) spanning trees of a graph — the palm trees — which generalize depth-first-search trees, and show that there is a natural one-to-one correspondence between palm trees and minimum cardinality sets of order constraints linearizing f. We also give two different characterizations of palm trees. Finally, we extend these results to arbitrary pseudo-boolean functions.

Part II deals with the preorder $\preccurlyeq$ and with the related regularity property. In Section 6 the concepts of regular sets and regular boolean functions are recalled. In Section 7 we present the results of Chvátal and Hammer [3] showing the close connection between regular set packing problems and threshold graphs. On this basis, efficient graph-theoretic algorithms can be obtained for recognizing and for solving regular set packing probles.

The case of regular set covering problems is harder, but they can still be recognized and solved in polynomial time, as shown by Peled and Simeone [19]. Their results are presented in Section 8, where we also propose an improved algorithm for solving regular set covering problems.

The theoretical importance of regular set covering problems is enhanced by the result of Hammer, Johnson and Peled [11] according to which every (linearly or nonlinearly constrained) 0-1 program with linear objective function can be formulated as a regular set covering problem, under a mild assumption on the objective function. Their result is presented in Section 9.

PART I

2. Generalities on 0-1 programming

Throughout this paper, we assume that the reader has some familiarity with boolean functions (see e.g. Rudeanu [24], Hammer [8]) and with graphs (see e.g. Berge [2], Lovász [17]).

Let B denote the set $\{0, 1\}$, and let $B^n = \overbrace{B \times \ldots \times B}^{n \text{ times}}$ be the *binary n-cube*. A *pseudo-boolean function* is any mapping $h : B^n \to R$ (= the reals). Any pseudo-boolean function can be represented as a multilinear function of the n binary variables $x_1, x_2, \ldots, x_n$:

$$h(x_1, \ldots, x_n) \equiv h(x) = \sum_{T \in \mathcal{C}} a_T \prod_{j \in T} x_j,$$

where $\mathcal{C}$ is a collection of distinct subsets of $N = \{1, \ldots, n\}$ and $a_T \neq 0$ for all $T \in \mathcal{C}$. Moreover, the representation is unique.

The general *0-1 programming* problem is

$$\left. \begin{array}{ll} \text{maximize} & f(x) \\ \text{subject to} & g_i(x) \leq b_i \ (i = 1, \ldots, m) \\ & x \in B^n \end{array} \right\} \quad (1)$$

where $f, g_1, \ldots, g_m$ are pseudo-boolean functions of n variables and $b_1, \ldots, b_m$ are real numbers. In particular, when $f, g_1, \ldots, g_m$ are linear, one has a *linear 0-1 programming* problem, which in matrix form is usually written as

$$\left. \begin{array}{ll} \text{maximize} & cx \\ \text{subject to} & Ax \leq b \\ & x \in B^n \end{array} \right\} \quad (2)$$

where c is a row n-vector, A an $m \times n$ matrix and b a column m-vector.

Let $X \equiv \{x \in B^n : g_i(x) \leq b_i, i = 1, \ldots, m\}$ be the feasible set of (1), let φ_X be the characteristic function of X and let ρ_X be the complement of φ_X (i.e. $\rho_X(x) = 0$ iff $x \in X$). Then (1) can be formulated as

$$\left. \begin{array}{ll} \text{maximize} & f(x) \\ \text{subject to} & \rho_X(x) = 0 \\ & x \in B^n \end{array} \right\} \quad (3)$$

The function ρ_X is called the *resolvent* of X. In the linear case, i.e. when $X \equiv \{x \in B^n : Ax \leq b\}$, the resolvent has the following expression (see Granot and Hammer [7])

$$\rho_X(x) = \bigvee_{i=1}^{m} \bigvee_{C \in M_i} \prod_{j \in C} x_j^{\alpha_{ij}}, \tag{4}$$

where

(i) M_i is the set of all *minimal covers* of the inequality

$$\sum_{j=1}^{n} a_{ij} x_j \leq b_i \quad (i = 1, \ldots, m)$$

i.e. the collection of all the minimal $C \subseteq N \equiv \{1, \ldots, n\}$ with the property

$$\sum_{j \in C} |a_{ij}| > b_i - \sum_{j=1}^{n} \min\{0, a_{ij}\},$$

(ii) $\alpha_{ij} = 1$ if $a_{ij} \geq 0$ and $\alpha_{ij} = 0$ if $a_{ij} < 0$, and,

(iii) as usual, $\xi^\beta = \begin{cases} \xi & \text{if } \beta = 1 \\ \overline{\xi} = 1 - \xi & \text{if } \beta = 0 \end{cases}$

Surprisingly, the 0-1 program (1), under the only assumption that the objective function f is linear, can always be represented in the form (2), where in addition the elements of A belong to the set $\{0, -1, 1\}$. Such a representation can be easily obtained once the resolvent of the feasible set of (1) is available.

3. Generation and use of order constraints

In the present Section, for a given 0-1 programming problem, we consider the following questions:

1) How to detect pairs (x_i, x_j) of variables, such that the relation $x_i \leq x_j$ holds for all *feasible* points x? (Section 3.1)

2) How to detect pairs (x_i, x_j) of variables, such that the relation $x_i \leq x_j$ holds for all *optimal* points x? (Section 3.2)

Once any such relation $x_i \leq x_j$ has been discovered, it can be exploited in several ways:

(i) *fixing or eliminating variables*

If, for example, it is known by some other means that, beside the relation $x_i \leq x_j$, the inequality $x_i + x_j \leq 1$ must also hold, then the variable x_i is forced to have the value 0. Similarly, if one has $x_i \leq x_j$ and $x_i + x_j \geq 1$, then the variable

x_j is forced to 1; if $x_i \leq x_j$, and $x_i \geq x_j$ also holds, then one has $x_i = x_j$ and one of the variables x_i, x_j can be eliminated, etc.

(ii) *adding cutting planes*

The constraint $x_i \leq x_j$ can be added as a cutting plane in enumerative methods for solving the given 0-1 program. See, for example, Hammer and Nguyen [12] and Spielberg [25].

(iii) *lowering the degree of the polynomials appearing in the constraints and/or in the objective function*

Suppose that one such polynomial has a monomial containing the product $x_i x_j$ as a factor. If the constraint $x_i \leq x_j$ holds, then one can replace $x_i x_j$ by x_i (see also Section 5).

3.1. Generation of order constraints in APOSS

Given a linear 0-1 program (2), the APOSS method of Hammer and Nguyen [12] makes use of boolean manipulations to produce order constraints $x_i \leq x_j$ (and also $\bar{x}_i \leq x_j$ or $x_i \leq \bar{x}_j$). Before outlining that procedure, we briefly recall the classical *consensus* method of Quine [23]. Given a boolean function φ in disjunctive normal form

$$\varphi = T_1 \vee \ldots \vee T_m$$

where each *term* T_k is the product of a finite number of *literals* (i.e. uncomplemented or complemented variables), an *implicant* of φ is any monomial I such that $I \leq \varphi$; the implicant I is said to be *prime* if there is no proper factor J of I such that $I \leq J \leq \varphi$. One can obtain all the prime implicants of φ by the following *consensus method*:

Starting from the list of all terms T_k in (5), repeat the following two steps as many times as possible:

Consensus: if in the current list there exist two terms of the form xC and $\bar{x}D$, where x is a variable and there is no variable which is complemented in C and uncomplemented in D or viceversa, then add to the list the term CD (after eliminating repeated literals from CD);

Absorption: if in the current list there are two terms C and D such that C is a (proper or improper) factor of D, then delete D from the list.

The terms appearing in the final list are precisely the prime implicants of φ.

Let k be an integer ≥ 1. The *k-truncated consensus method* is similar to the consensus method described above, the only difference being that a consensus step is performed only when the number of literals in both xC and $\bar{x}D$ is $\leq k$. While the consensus method requires in general an exponential number of operations, the k-truncated consensus method has a $O(n(m + n^k)^2)$ complexity and hence is polynomial for any *fixed k*.

APOSS essentially applies a 3-truncated consensus method to the resolvent (4) of the feasible set of (2) in order to obtain a set of (not necessarily prime) implicants of length ≤ 3. If, in particular, one obtains in this way the implicant $x_i \bar{x}_j$, then the constraint $x_i \leq x_j$ must be satisfied by every feasible solution x, since $0 \leq x_i \bar{x}_j \leq \rho_X(x) \leq 0$. More generally, the method detects also constraints of the form $x_i + x_j \leq 1$ or $x_i + x_j \geq 1$ — although not necessarily all of them. The study of systems of inequalities of the form $x_i \leq x_j$, $x_i + x_j \leq 1$ and $x_i + x_j \geq 1$ is undertaken in Johnson and Padberg [14].

3.2. Generation of order constraints via second derivatives

Given the *unconstrained quadratic 0-1 optimization problem*

$$\max_{x \in B^n} x^T Q x \qquad (6)$$

where Q is a symmetric $n \times n$ matrix, we describe a simple procedure, due to Hammer and Hansen [9], for detecting order constraints $x_i \leq x_j$ which must necessarily be satisfied by every binary point maximizing the pseudo-boolean function $f(x) = x^T Q x$.

The *second (boolean) derivative* of f with respect to x_i and x_j ($i < j$) is defined as

$$\Delta_{ij}(x) = f(x_1, \ldots, \overset{i}{1}, \ldots, \overset{j}{0}, \ldots, x_n) - f(x_1, \ldots, \overset{i}{0}, \ldots, \overset{j}{1}, \ldots, x_n)$$

$$= q_{ii} - q_{jj} + 2 \sum_{k \neq i,j} (q_{ik} - q_{jk}) x_k.$$

For all $x \in B^n$, one has

$$\underline{\Delta}_{ij} \leq \Delta_{ij}(x) \leq \overline{\Delta}_{ij},$$

where

$$\underline{\Delta}_{ij} = q_{ii} - q_{jj} + 2 \sum_{k \neq i,j} \min\{0, q_{ik} - q_{jk}\}$$

and

$$\overline{\Delta}_{ij} = q_{ii} - q_{jj} + 2 \sum_{k \neq i,j} \max\{0, q_{ik} - q_{jk}\}.$$

Then, the following result holds:

Proposition 1. (Hammer and Hansen [9]):

a) if $\overline{\Delta}_{ij} < 0$, then one must have $x_i \leq x_j$ for all optimal solutions to (6);
b) if $\underline{\Delta}_{ij} > 0$, then one must have $x_i \geq x_j$ for all optimal solutions to (6).

4. Maximization of a linear function subject to order constraints

In the present Section we consider the problem of maximizing a linear function of n binary variables subject to order contraints on these variables:

$$\left. \begin{array}{ll} \text{maximize} & \sum_{i=1}^{n} w_i x_i \\ \text{subject to} & x_i \leq x_j, \quad (i,j) \in A \\ & x \in B^n, \end{array} \right\} \quad (7)$$

where $A \subseteq N \times N$, with $N = \{1, \ldots, n\}$. Let D be the digraph (N, A). A subset X of N is said to be a *closure* if $i \in X \Rightarrow j \in X$ for every successor j of i.

Let us assign a weight w_i to each vertex i of D. For all $Y \subseteq N$, let $W(Y) = \Sigma_{i \in Y} w_i$. As remarked in Picard [20], problem (7) is equivalent to finding a maximum weight closure of N. Picard also pointed out that problem (7) is reducible to the maximization of a quadratic supermodular function, which in its turn is reducible to a minimum cut problem (Balinski [1], Picard and Ratliff [21]) and hence can be solved by a max-flow algorithm.

A direct reduction of (7) to a minimum cut problem can be obtained as follows. Without loss of generality, we may assume that the property

(P) "$w_i > 0$ for every source i and $w_j < 0$ for every sink j"

holds. If one had $w_h \leq 0$ for some source h, then there would be an optimal solution x^* to (7) such that $x_h^* = 0$ and hence variable x_h could be eliminated from (7). The case of sinks is similar.

Now, let $N^+ \equiv \{i \in N : w_i \geq 0\}$, $N^- \equiv \{j \in N : w_j < 0\}$ and let us augment D by introducing two dummy vertices s and t and dummy edges (s, i) for all $i \in N^+$ and (j, t) for all $j \in N^-$. The augmented digraph obtained in this way will be denoted D^*. Notice that, because of (P), s is the only source and t the only sink of D^*. Next, we assign a capacity w_i to each edge (s, i) $(i \in N^+)$, a capacity $-w_j$ to each edge (j, t) $(j \in N^-)$, and a capacity $+\infty$ to all remaining edges of D^*, i.e. those in A.

A *cut* in D^* is any bipartition $\tau = \{S, \overline{S}\}$ of the vertex-set of D^* such that $S \ni s$, $\overline{S} \ni t$. The *capacity* $c(\tau)$ of τ is the sum of the capacities of all edges (i, j) such that $i \in S$ and $j \in \overline{S}$.

Theorem 2. *There is a one-to-one correspondence between the finite capacity cuts*

τ of D^* and the closures X of D. For any such cut τ and for the corresponding closure X, one has

$$c(\tau) + w(X) = w(N^+) = \text{constant}. \tag{8}$$

Proof. Clearly the mapping which associates with each cut $\tau = \{S, \overline{S}\}$ the set $X = S - \{s\}$ is a bijection. If the capacity of τ is finite one has

$$i \in X, \quad (i,j) \in A \Rightarrow j \in X,$$

otherwise the edge (i,j) would have capacity $+\infty$. Hence X is a closure. Conversely, if X is a closure there is no edge $(i,j) \in A$ such that $i \in S$ and $j \in \overline{S}$. Hence the cut τ corresponding to X has finite capacity. Moreover,

$$c(\tau) = \sum_{i \in (N-X) \cap N^+} w_i - \sum_{j \in X \cap N^-} w_j$$

$$w(X) = \sum_{i \in X \cap N} w_i + \sum_{j \in X \cap N^-} w_j.$$

Adding up the above two identities, one obtains (8). □

The above theorem immediately implies the reducibility of (7) to a minimum cut problem. It follows that problem (7) can be solved in $0(n^3)$ time via, say, the max-flow algorithm of Karzanov [16].

We shall now describe an $0(n)$ time algorithm for solving problem (7) in the special case when D is a rooted tree T. For convenience, in this case we denote problem (7) by $\mathcal{P}(T; w)$.

Lemma 3. *Let k be a leaf of T such that $w_k \geq 0$. Then there exists an optimal solution x^* to $\mathcal{P}(T; w)$ such that $x_k^* = 1$.*

Proof. Obvious.

Definition. A vertex h of T is called a *pedicle* if all its sons are leaves.

Lemma 4. *Let h be a pedicle and let L be the set of sons of h. If $w_j < 0$ for all $j \in L$, then for all optimal solutions x^* to $\mathcal{P}(T; w)$ one has $x_h^* = x_j^*$ for all $j \in L$.*

Proof. Let x^* be an arbitrary optimal solution to $\mathcal{P}(T; w)$. If $x_h^* = 1$ then also $x_j^* = 1$ for all $j \in L$ because of the order constraints in (7). If $x_h^* = 0$ and $x_j^* = 1$ for some $j \in L$, then $y^* = x^* - u_j$ (where u_j is j-th unit vector) is a feasible solution such that $\sum_{i=1}^{n} w_i y_i^* < \sum_{i=1}^{n} w_i x_i^*$, contradicting the optimality of x^*. □

Given the tree T with vertex-weights w_i, one of the following two cases must occur:
1) There exists a leaf k such that $w_k \geq 0$;
2) There exists a pedicle h such that $w_j < 0$ for all j in the set L of sons of h.

In case 1) $\mathscr{P}(T; w)$ is reducible, in view of Lemma 3, to $\mathscr{P}(T'; w')$, where $T' = T - k$ and w' is the restriction of w to T'. (REDUCTION 1)

In case 2) $\mathscr{P}(T; w)$ is reducible, in view of Lemma 4, to $\mathscr{P}(T''; w'')$, where

$$T'' = T - L \text{ and } w_i'' = \begin{cases} w_i & \text{if } i \neq h \\ w_h + \Sigma_{j \in L} w_j & \text{if } i = h \end{cases} \quad \text{(REDUCTION 2)}$$

In both cases $\mathscr{P}(T; w)$ can be reduced to a problem on a tree with less vertices. Thus, starting from T and applying each time either REDUCTION 1 or REDUCTION 2, one can recursively solve $\mathscr{P}(T; w)$ by reducing it to problems on smaller and smaller trees, until the trivial tree with one vertex is obtained.

The resulting algorithm can be implemented so as to run in $O(n)$ time. Here are the details (we follow the terminology of Tarjan [26]).

Definition. A *breadth-first order* in N is a linear order σ in N such that
  a) if the depth of i is smaller than the depth of j, then $i \sigma j$;
  b) if i and j are brothers and $i \sigma k \sigma j$, then k is a brother of i and j.

Algorithm

Let T be given by its successor list.

Step 1. For $k = 1, \ldots, n$, let $q(k)$ be the k-th vertex in a breadth-first order in N. Set $s(i) := w_i$, $x_i := 0$ for all $i \in N$;

Comment: $w(\cdot)$ is the vector of vertex weights in the current tree; x will eventually contain the optimal solution to $\mathscr{P}(T; w)$.

Step 2.
  For $k := n$ **down to** 1 **do**
    begin
      $i := q(k)$;
      Let S_i be the set of all sons j of i such that $s(j) < 0$;
      If $S_i \neq \phi$ **then** $s(i) := s(i) + \Sigma_{j \in S_i} s(j)$;
      If $s(i) \geq 0$ **then** $x_i := 1$;
    end

Step 3:
  For $k := 1$ **to** n **do**
    begin
      $i := q(k)$;
      If $x_i = 1$ **then** $x_j := 1$ for all successors j of i;
    end;

Comment: At the end of STEP 2, the vector x might not be feasible. The purpose of STEP 3 is to ensure that the order contraints in (7) are satisfied.

Example. Consider the vertex-weighted tree T of Fig. 1 (recall that an edge $(i) \rightarrow (j)$ corresponds to the order constraint $x_i \leq x_j$).

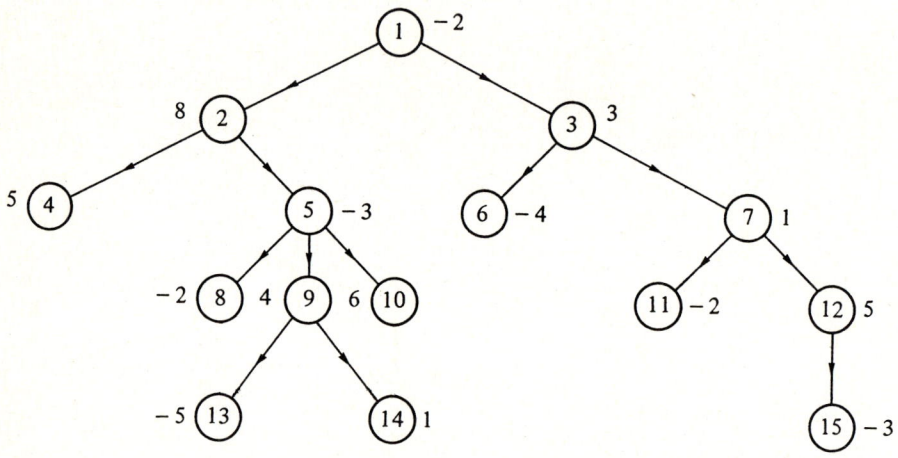

Fig. 1

Step 1. In this case the linear order defined by $q(k) = k$ for all k is already a breadth-first-order.

Step 2.

$k = q(k)$	$s(k)$	x_k
15	-3	0
14	1	1
13	-5	0
12	2	1
11	-2	0
10	6	1
9	-1	0
8	-2	0
7	-1	0
6	-4	0
5	-6	0
4	5	1
3	-2	0
2	2	1
1	-4	0

Step 3.
$x_2 = 1 \Rightarrow x_5 = 1$;
$x_5 = 1 \Rightarrow x_8 = x_9 = 1$;
$x_9 = 1 \Rightarrow x_{13} = 1$;
$x_{12} = 1 \Rightarrow x_{15} = 1$.

Thus, an optimal solution is (0, 1, 0, 1, 1, 0, 0, 1, 1, 1, 0, 1, 1, 1, 1).

Clearly each *Step* 1, 2, and 3 requires $O(n)$ time. Hence the overall complexity of the algorithm is also $O(n)$.

5. Linearization of a pseudo-boolean function via order constraints

As we have already mentioned in Section 3, if ξ and η are binary variables, one has

$$\xi \leq \eta \Leftrightarrow \xi\eta = \xi. \tag{9}$$

The equivalence (9) suggests the following question: given a quadratic function $f(x) = x^T Q x$ of n binary variables (where Q is a symmetric matrix), what is the smallest number of order constraints $\xi \leq \eta$ which make f linear?

If f has m terms, clearly m order constraints — one for each term — are always enough. However, by exploiting the transitivity of $\leq$, one can usually do better. For example, let

$$f(x) = 2x_1 x_2 + 3x_1 x_3 + x_2 x_4 + 7x_3 x_4 + x_3 x_5 + 6x_4 x_6 + 4x_5 x_6. \tag{10}$$

Every point x satisfying the 5 constraints

$$x_1 \leq x_3, \ x_3 \leq x_4, \ x_4 \leq x_2, \ x_4 \leq x_6, \ x_6 \leq x_5 \tag{11}$$

satisfies also, by transitivity, the constraints

$$x_1 \leq x_2, \ x_3 \leq x_5.$$

For any such x, one has

$$f(x) = 2x_1 + 3x_1 + x_4 + 7x_3 + x_3 + 6x_4 + 4x_6$$
$$\equiv 5x_1 + 8x_3 + 7x_4 + 4x_6.$$

Thus, 5 order constraints are enough in order to linearize f.

We shall require that the constraint $x_i \leq x_j$ can be selected only if $q_{ij} \neq 0$, i.e. if the term $q_{ij} x_i x_j$ is actually present in f. As we shall see, such requirement will cause no loss of optimality.

Given f, let us introduce the undirected graph $G = (N, E)$, where $N \equiv \{1,...,n\}$ and $E \equiv \{\langle i, j \rangle : q_{ij} \neq 0\}$. Without loss of generality, we may assume that G is connected. Our problem can then be formulated as follows.

Problem ($\mathscr{L}$). Find a minimum cardinality subset $F \subseteq E$ and an orientation $\vec{F}$

of all edges in F such that, whenever two vertices i and j are adjacent in G, in the digraph $D = (N, \vec{F})$ there is either a directed path from i to j or a directed path from j to i.

Then, clearly,

$$x_i \leq x_j \quad \text{for all} \quad (i,j) \in \vec{F}$$

is a minimum cardinality set of order constraints linarizing f. The digraph D will be called an *optimal solution* to problem ($\mathscr{L}$).

Lemma 5. $|F| \geq n - 1$.

The proof follows immediately from the fact that the partial graph (N, F) must be connected.

The above Lemma suggests that perhaps D is a directed spanning tree of G. In our example, the graph G corresponding to the function f given by (10) is shown in Fig. 2(a) and one can take as D the rooted tree of Fig. 2(b).

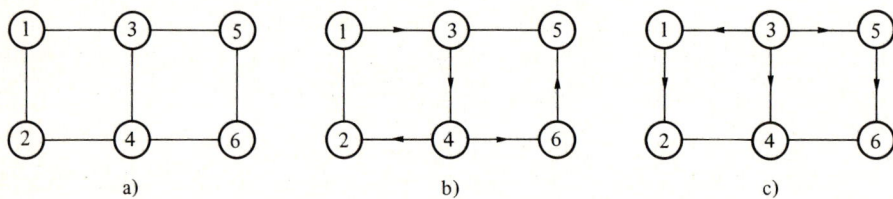

Fig. 2

In fact, the order constraints corresponding to D are precisely those given by (11), and we have already seen that they linearize f. Moreover D is optimal by Lemma 5.

However, not every directed spanning tree works, even if we require that it is a rooted tree. For example, the one shown in Fig. 2(c) does not (there is no directed path from 2 to 4 or from 4 to 2).

A class of optimal solutions to ($\mathscr{L}$) is given by the following theorem (for the terminology, the reader is referred to Tarjan [26]).

Theorem 6. *All depth-first-search trees of G are optimal solutions to ($\mathscr{L}$). Moreover, a spanning rooted tree of G is an optimal solution to ($\mathscr{L}$) only if it is a depth-first-search tree.*

Proof. Tarjan [26] proved that a spanning rooted tree T of G is a depth-first-search tree if and only if has the property that whenever two vertices i and j of G are adjacent then there is in T either a directed path from i to j or one from j to i. In view of the fact that, by Lemma 5, every optimal solution to ($\mathscr{L}$) must have at least $n-1$ edges, Tarjan's result immediately implies the theorem. □

Definition. A directed spanning tree T of G is a *palm tree* of G if, whenever two vertices i and j are adjacent in G, there is in T a directed path either from i to j or from j to i.

The term of «palm tree» was introduced by Tarjan, who in addition required T to be rooted. According to our definition, though, a palm tree does not have to be rooted. For example if G itself is a tree, every orientation of G is a palm tree. A less trivial example is shown in Fig. 3.

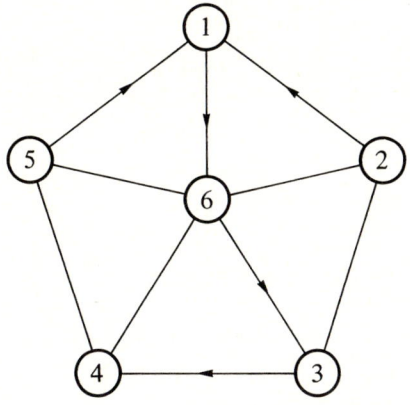

Fig. 3

An immediate consequence of Theorem 6 is that the equality $|F| = n-1$ must hold in Lemma 5. Hence,

Corollary 7. *The optimal solutions to* ($\mathscr{L}$) *are precisely the palm trees of G.*

A closer look at Fig. 3 reveals that, if we remove the source 5, then $T-\{5\}$ is a palm tree of $G-\{5\}$. Moreover, the only successor of vertex 5 in T, namely vertex 1, is an ancestor of 4 and 6, the other two vertices adjacent to 5 in G. More generally, palm trees admit the following recursive characterization.

Theorem 8. *A directed spanning tree T of G is a palm tree of G if and only if there exists in T a source s such that*
    (a) *if the outdegree of s in T is > 1, then s is an articulation point of G;*
    (b) *if the connected components of $T-s$ are $T_1, \ldots, T_r$, then T_h is a palm tree of the subgraph G_i of G induced by the vertex-set of T_h.*

(c) let v_h be the (only) vertex of T_h which is a successor of s in T. Then every vertex $v \neq v_h$ of T_h which is adjacent to s in G must be a descendant of v_h in T_h.

Proof. The «if» part is straightforward. Let us prove the «only if» part. Let s be any source of T (s exists because T has no circuits). If i is a vertex of G_h and j is a vertex of G_k ($h \neq k$), then i and j cannot be adjacent in G, for otherwise there would be in T a directed path either from i to j or from j to i. Such path would necessarily include s: but this is impossible, since s is a source. Hence (a) holds and $G_1, \ldots, G_r$ are precisely the connected components of $G - s$.

If p and q are adjacent in G_h then there is a directed path in T either from p to q or from q to p. Such path cannot include the source s: hence it must be within G_h and thus (b) holds.

Finally, if $v \neq v_h$ is a vertex of G_h adjacent to s, then there must be a directed path from s to v in T (no path from v to s can exist). Since s is a source and an articulation point, the vertex following s along the path is necessarily v_h, the only successor of s in T_h. Moreover the path must be within T_h, and thus (c) follows. □

Another characterization of palm trees in given below.

Theorem 9. *A directed spanning tree T of G is a palm tree of G if and only if, for every cycle C of G containing exactly one unoriented edge, all the oriented edges of C have the same direction (i.e. no two of them have the same head or the same tail).*

Proof. If). Let $\langle i, j \rangle$ be any unoriented edge of G. Since T is spanning, there is a path P in T connecting i and j. The path P together with the edge $\langle i, j \rangle$ forms a cycle whose only unoriented edge is $\langle i, j \rangle$. It follows that P is a directed path either from i to j or from j to i.

Only if). Suppose that T is a palm tree and let C be any cycle containing only one unoriented edge $\langle i, j \rangle$. In the tree T there must be a directed path Q, say, from i to j (the other case is similar). Let q_k ($k = 0, 1, \ldots, l$ where l is the length of Q) be the k-th vertex along Q; in particular $q_0 = i$ and $q_l = j$.

Let P be the path formed by the oriented edges of C. Then P must be simple, i.e. cannot have repeated vertices, otherwise T would have a cycle. Suppose that P were not directed. Then there would be some vertex of P which does not belong to Q.

Let

$$r = \max\{k : q_0, q_1, \ldots, q_k \text{ lie on } P\}, \quad (r < l)$$
$$s = \min\{h : r < h \leq l, \quad q_h \text{ lies on } P\}.$$

Then the two subpaths $Q(r, s)$ and $P(r, s)$ together would form a cycle of T

(see Fig. 4), a contradiction. Hence P is directed and T is a palm tree. □

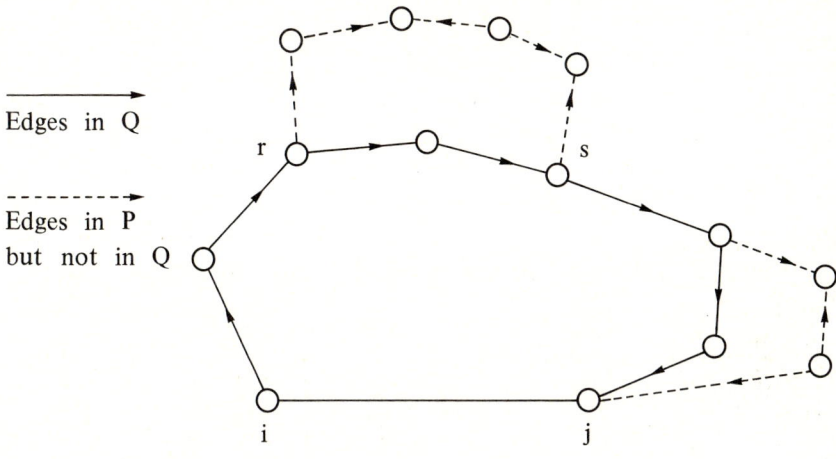

Fig. 4

Theorem 6 and Corollary 7 imply that the smallest number of order constraints which linearize a quadratic function f of n binary variables is $n-1$ and hence is independent of the number of terms in f.

So far, we have insisted in selecting only constraints $x_i \leq x_j$ such that $q_{ij} \neq 0$. If we allow for the presence of arbitrary order constraints $x_i \leq x_j$ (even when $q_{ij} = 0$) then the smallest number of order constraints linearizing f is still $n-1$, and a class of optimal solutions is given by the palm trees of the complete graph K_n. Notice that T is a palm tree of K_n iff T is a depth-first-search tree of K_n and also iff T is a directed hamiltonian path of K_n.

More generally, one sees that $n-1$ is the smallest number of order constraints linearizing an arbitrary pseudo-boolean function in n variables

$$f(x) = \sum_{T \in \mathscr{C}} a_T \prod_{j \in T} x_j,$$

(where $\mathscr{C}$ is a collection of subsets of $N = \{1, \ldots, n\}$ and $a_T \neq 0$ for all $T \in \mathscr{C}$), subject to the requirement that the order constraint $x_i \leq x_j$ may be selected only if $x_i x_j$ is a factor of some monomial $\Pi_{j \in T} x_j$, with $T \in \mathscr{C}$. In this case, the optimal solutions correspond to the palm trees of the graph

$$G \equiv \{\langle i, j \rangle : i, j \in N, \exists T \in \mathscr{C} \text{ such that } \{i, j\} \subseteq T\}.$$

PART II

6. The preorder x ≪ y and the concept of regularity

Let X be a given subset of B^n. One can define in the set of variables $\{x_1, \ldots, x_n\}$ a preorder (= reflexive and transitive relation) by «$x_i \stackrel{X}{\preccurlyeq} x_j$» if and only if either $i = j$ or $\forall x \in B^n$ such that $x_i = 0$ and $x_j = 1$ one has $x + u_i - u_j \in X$, where, as usual, u_k is the k-th unit vector $(0, 0, \ldots, \overset{k}{1}, \ldots, 0)$. When the set X is understood from the context, we simply write $x_i \preccurlyeq x_j$. We write $x_i \sim x_j$ when $x_i \stackrel{X}{\preccurlyeq} x_j$ and $x_j \stackrel{X}{\preccurlyeq} x_i$. If $\preccurlyeq$ happens to be a *total* (or linear) preorder then the set X is said to be *regular*. A prime example of regular set is the feasible set $X \equiv \{x \in B^n : a_1 x_1 + \ldots + a_n x_n \leqslant b\}$ of a knapsack problem: in this case, one has $x_i \preccurlyeq x_j$ iff $a_i \leqslant a_j$.

Similar concepts can be defined for boolean functions φ. One writes $x_i \stackrel{\varphi}{\preccurlyeq} x_j$ iff $\forall x \in B^n$ such that $x_i = 0, x_j = 1$ one has $\varphi(x) \leqslant \varphi(x + u_i - u_j)$. The boolean function φ is said to be *regular* if all pairs of variables are comparable in $\stackrel{\varphi}{\preccurlyeq}$ (¹). Clearly a set X is regular if and only if its characteristic function φ_X is regular.

7. Threshold functions, threshold graphs and regular set packing problems

The concept of regularity was first introduced by Winder [27] in connection with the study of threshold boolean functions. A boolean function $\varphi : B^n \to B$ is said to be *threshold* if there exist n «weights» $w_1, \ldots, w_n$ and a «threshold» t such that

$$\varphi(x) = 1 \quad \text{iff} \quad w_1 x_1 + \ldots + w_n x_n > t$$

(the weights and the threshold can be arbitrary real numbers). The $n + 1$-vector $(w_1, \ldots, w_n; t)$ is then called a *separator* of φ.

All threshold functions are regular, but the converse does not have to hold. However, Chvátal and Hammer [3] proved that quadratic positive boolean functions are threshold if and only if they are regular. Note that quadratic positive boolean functions are just the incidence functions $\varphi(x) = \bigvee_{(i,j) \in E} x_i x_j$ of undirected graphs $G = (V, E)$.

Those graphs whose incidence function is threshold were called *threshold graphs* by Chvátal and Hammer [3], who gave several characterizations of them.

(¹) In the usual definition of regularity, it is customary to require also the monotonicity of φ.

Given a graph $G = (V, E)$ and a vertex i of G, we denote by $N(i)$ the neighborhood $\{j \in V : \langle i, j \rangle \in E\}$ of i.

Theorem 10. (Chvátal and Hammer [3]). *For a graph $G = (V, E)$ the following statements are equivalent:*
1) *G is threshold.*
2) *There is a hyperplane separating the characteristic vectors of the stable sets of G from the remaining vectors of B^n.*
3) *For every pair (i, j) of distinct vertices of G, either*

$$N(i) - \{j\} \subseteq N(j) - \{i\} \quad \text{or} \quad N(j) - \{i\} \subseteq N(i) - \{j\}.$$

4) *There are no 4 vertices a, b, c, d such that*

$$\langle a, b \rangle \in E, \quad \langle c, d \rangle \in E, \quad \langle a, c \rangle \notin E, \quad \langle b, d \rangle \notin E.$$

5) *V can be partitioned into a clique K and a stable set $I \equiv \{i_1, \ldots, i_s\}$ where*

$$N(i_1) \supseteq N(i_2) \supseteq \ldots \supseteq N(i_s).$$

6) *There exists in G a vertex v which is either universal (i.e. its degree is $n - 1$) or isolated, and every induced subgraph of G has the same property.*

Further characterizations of threshold graphs in terms of their degree sequence were provided by Golumbic [6] and Hammer, Ibaraki and Simeone [10]. Many generalizations of threshold graphs have been investigated in the literature.

The interest of threshold graphs in 0-1 programming stems from their close connection with regular set-packing problems.

A set-packing problem

$$\left. \begin{array}{ll} \text{maximize} & cx \\ \text{subject to} & Ax \leq e \\ & x \in B^n \end{array} \right\} \tag{12}$$

(where A is a $m \times n$ 0-1 matrix, c is a positive row n-vector and $e = (1, \ldots, 1)$) is called *regular* if its feasible set $X \equiv \{x \in B^n : Ax \leq e\}$ is regular.

Theorem 11

(i) (Chvátal and Hammer [3]). *A set-packing problem is regular iff its feasible set C is the feasible set of some knapsack problem, i.e. $X \equiv \{x \in B^n : ax \leq b\}$ for some $a \in R^n_+$, $b \in R_+$.*

(ii) (Chvátal and Hammer [3], Orlin [18]). *There is a polynomial-time algorithm for finding the above knapsack problem or proving that none exists.*

(iii) *A regular set packing problem can be solved in polynomial time.*

Proof. Given the matrix A in (12), let G be the graph with vertex set $N = \{1, \ldots, n\}$

and edge set

$$E = \left\{ \langle i,j \rangle : i,j \in N, \sum_{k=1}^{m} a_{ki} a_{kj} \geq 1 \right\}.$$

It is well-known, and easy to see, that $x \in X$ if and only if x is the characteristic vector of some stable set of G. It follows that the resolvent of X is given by

$$\rho_X(x) = \bigvee_{\langle i,j \rangle \in E} x_i x_j.$$

Since the complement $\bar{\varphi} = 1 - \varphi$ of a boolean function φ is regular iff φ is regular, the above result of Chvátal and Hammer implies that the set packing problem (12) is regular if and only if G is threshold. Hence (i) follows.

In their paper, Chvátal and Hammer described a polynomial-time algorithm for bulding a separator $(w_1, \ldots, w_n; t)$ of the incidence function of G or proving that none exists. An efficient algorithm for computing an integral separator with smallest threshold t was subsequently given by Orlin [18].

From the above discussion it follows that (12) is equivalent to finding a maximum weight stable set of the threshold graph G. But G has very few maximal stable sets. Actually, by Theorem 10, the only maximal stable sets of G are I itself and all sets

$$\{k\} \cup (I - N(k)), \quad k \in K.$$

Therefore, G has at most n maximal stable sets and hence the maximum weight stable set problem in G can be solved in polynomial time. □

8. Regular set covering problems

We have seen in the previous section that there are efficient graph-theoretic algorithms for recognizing and for solving regular set packing problems. The key property which allows one to use graph-theoretic tools is the fact that the feasible set of such a problem has a *quadratic* resolvent.

Unfortunately, this property is lost for regular set covering problems (i.e. set covering problems whose feasible set is regular). Moreover, in the regular set covering case it is no longer true that the feasible set is always the feasible set $\{x \in B^n : a_1 x_1 + \ldots + a_n x_n \geq b\}$ of some knapsack problem. Thus, the set covering case is much more complex than the set packing case. Nevertheless, it turns out that there are polynomial-time algorithms
  a) for recognizing when a given set covering problem is regular;
  b) for solving regular set covering problems;
  c) for recognizing those set covering problems whose feasible set is also the

feasible set of some knapsack problem.

8.1. Recognition of regular set covering problems

Let

$$\left. \begin{array}{l} \text{minimize} \quad cy \\ \text{subject to} \quad Ay \geq e \\ \qquad\qquad y \in B^n \end{array} \right\} \tag{13}$$

be a regular set covering problem, where A is a $m \times n$ binary matrix, c is a positive row n-vector and e is the column m-vector whose components are all ones.

For convenience, we prefer to work on the equivalent problem

$$\left. \begin{array}{l} \text{maximize} \quad cx \\ \text{subject to} \quad Ax \leq b \\ \qquad\qquad x \in B^n \end{array} \right\} \tag{14}$$

where $b = Ae' - e$ and e' is the column n-vector $(1, \ldots, 1)$. As usual, we denote by X the feasible set of (14).

We shall say that x' *lies above* x if $x \leq x'$.

Let us denote by $A(i;j)$ the minor of A obtained by dropping all rows having a 1 in the j-th column and then suppressing both the i-th and the j-th column.

Theorem 12. *One has $x_i \precsim x_j$ if and only if every row of $A(i;j)$ lies above some row of $A(j;i)$.*

The proof is based on a well-known result of Quine (see Theorem 4 in Peled and Simeone [19]).

In order to find out whether X is regular one does not have to check the condition of Theorem 12 for all pairs (i, j). The following theorem, essentially due to Winder [27], gives a necessary condition for the relation $x_i \precsim x_j$ to hold.

Theorem 13. *Let C be the $n \times n$ matrix whose element c_{kj} is equal to the number of rows of A having exactly k 1's and having a 1 in column j.*

If $x_i \sim x_j$, then the i-th column and the j-th column of C are equal. If $x_i \precsim x_j$, but $x_i \not\sim x_j$, then the i-th column of C is lexicographically smaller than the j-th column.

To use this theorem, we sort the columns of C lexicographically in $O(n^2 \log n)$ time so that the $\pi(1)$-th column is largest, the $\pi(2)$-th is second largest, and so on. Then, using Theorem 12, we check the conditions $x_{\pi(i+1)} \precsim x_{\pi(i)}$ for all $i = 1, \ldots, n-1$; the feasible set X is regular if and only if all such conditions hold. It follows that there is a polynomial-time algorithm for recognizing regular set covering problems (notice that the feasible set of (13) is regular if and only if X is such).

8.2. Solution of regular set covering problems

We now turn our attention to the solution of regular set covering problems. We shall assume that the feasible set X of (14) is regular and that the column of A are renumbered so that $x_1 \succcurlyeq x_2 \succcurlyeq \ldots \succcurlyeq x_n$. Moreover, without loss of generality, we shall assume that no row of A lies above another row, and that A contains both 0's and 1's.

A *maximal feasible point* (MFP) of (14) is any $x \in X$ such that $x' \in X, x \leqslant x' \Rightarrow$ $\Rightarrow x = x'$. A *minimal infeasible point* (MIP) is any $x \notin X$ such that $x' \notin X, x' \leqslant$ $\leqslant x \Rightarrow x = x'$. It is well known, and easy to see, that the MIPs are precisely the rows of A.

Let us introduce a linear order in B^n as follows. For any $x \in B^n$, the *support* of x is the set $\mathrm{supp}(x) \equiv \{j : x_j = 1\}$. The *positional representation* of x is the n-vector whose components are the elements of $\mathrm{supp}(x)$ followed by 0's. Let $x, x' \in B^n$. We shall say that x' *follows* x ($x \prec x'$) if the positional representation of x' is lexicographically greater than the positional representation of x. We denote by $\mathrm{succ}(x)$ the immediate successor of x in the above defined linear order. We need some additional notations (as usual, u_k denotes the k-th unit vector):

$$b(x) = \begin{cases} 0 & \text{if } x = 0 \\ \max\{j : x_j = 1\} & \text{if } x \neq 0 \end{cases}$$

$$d(x) = \begin{cases} 0 & \text{if } x = 0 \text{ or } x_1 = \ldots = x_{b(x)} = 1 \\ \max\{j : x_{j-1} = 0, x_j = 1\}, & \text{else} \end{cases}$$

$\mathrm{fill}(x) = x + u_{b(x)+1}$ (undefined if $b(x) = n$)

$\mathrm{brs}(x) = x - u_{b(x)} + u_{b(x)+1}$ (undefined if $b(x)$ is 0 or n)

$\mathrm{trunc}(x) = x - \Sigma\{u_j : d(x) \leqslant j \leqslant b(x)\}$, where $u_0 = 0$.

For example, if $x = (1\ 0\ 1\ 0\ 0\ 0\ 1\ 1\ 0\ 0)$, one has $b(x) = 8, d(x) = 7$, $\mathrm{fill}(x) =$ $= (1\ 0\ 1\ 0\ 0\ 0\ 1\ 1\ 1\ 0)$, $\mathrm{brs}(x) = (1\ 0\ 1\ 0\ 0\ 0\ 1\ 0\ 1\ 0)$, $\mathrm{trunc}(x) = (1\ 0\ 1\ 0\ 0\ 0\ 0\ 0\ 0\ 0)$.

It is easy to see that

$$\mathrm{succ}(x) = \begin{cases} \mathrm{fill}(x), & \text{if } x_n = 0 \\ \mathrm{brs}(x - u_n), & \text{if } x_n = 1 \end{cases}$$

A *shelter s* is a MIP such that brs (*s*) is feasible or undefined.

We are now ready to describe an algorithm, due to Peled and Simeone [19], which generates in polynomial time al MFPs, starting from the list of all MIPs. The algorithm is a modification of an earlier algorithm of Hammer, Peled and Pollatschek [13], who did not analyze the running time, but empirically observed its linear dependence on m.

From the list of MIPs (i.e. the rows of A) one preliminarily produces the shelters. The list of shelters, sorted according to the linear order $\prec$ and followed by a dummy shelter, constitutes the input of Peled and Simeone's algorithm. The idea behind the algorithm is to scan all the points of B^n according to the linear order $\prec$, «jumping» over large intervals that cannot contain MFPs.

Hop-Skip-and-Jump Algorithm

$s :=$ first shelter on the list;
START : $x := 0$
while true **do**
  begin {outer while}
    while $x \neq s - u_{b(s)}$ **do** {inner while}
      if $x_n = 0$
        then FILL-UP : $x :=$ fill (x)
        else SKIP : **begin** {skip}
            output x;
            $y :=$ trunc (x);
            if $y = 0$ **then** stop **else** $x :=$ brs (y)
        end; {skip}
    {end inner while}
    LEAP : **begin** {leap}
      if $s_n = 0$ **then** HOP : $x :=$ brs (s)
        else JUMP:
            begin {jump}
            output x;
            if $s = u_n$ **then** stop **else** $x :=$ succ (s)
        end;{jump}
      $s :=$ next shelter on the list
    end {leap}
  end {outer while}

Remark. If s is the dummy shelter, then $x \neq s - u_{b(s)}$ is considered to be true.

Example. Let

$$A = \begin{bmatrix} 1 & 1 & 1 & 0 & 0 & 0 \\ 1 & 1 & 0 & 1 & 1 & 0 \\ 1 & 0 & 1 & 1 & 1 & 0 \\ 1 & 1 & 0 & 1 & 0 & 1 \\ 0 & 1 & 1 & 1 & 1 & 1 \end{bmatrix};$$

then the MFP and the shelters (the latter are starred), ordered according to $\prec$, are

* (1 1 1 0 0 0)

   (1 1 0 1 0 0)

* (1 1 0 1 0 1)

   (1 1 0 0 1 1)

* (1 0 1 1 1 0)

   (1 0 1 1 0 1)

   (1 0 1 0 1 1)

   (1 0 0 1 1 1)

   (0 1 1 1 1 0)

* (0 1 1 1 1 1)

   (0 1 1 1 0 1)

   (0 1 1 0 1 1)

   (0 1 0 1 1 1)

   (0 0 1 1 1 1)

Theorem 14. (Peled and Simeone [19]). *The HOP-SKIP-and-JUMP algorithm runs in* $O(n^3 m)$ *time and outputs precisely the MFPs of (14). The number of MFPs is at most* $mn + m + n = O(nm)$.

Given the matrix A, the sorted list of shelters can be generated, in $O(nm \log_2 m)$ time, as shown in Hammer, Peled and Pollatschek [13]. (Note that $\log_2 m = O(n)$).

An immediate consequence of Theorem 14 is that one can solve (14), and hence also the regular set covering problem (13), in $O(n^3 m)$ time simply by generating all the MFPs of (14) via the above algorithm and evaluating the objective function in each of them.

We shall now describe an improved algorithm for solving (14). The algorithm relies on a characterization of the MFPs between any two consecutive shelters (see Lemma 15 below) and on an efficient way to find the best among such MFPs (see Lemma 16).

Consider the points x and s obtained immediately after START (in this case $x = 0$ and s is the first shelter) or after each execution of LEAP in the HOP-SKIP-and-JUMP algorithm. As shown by Peled and Simeone [19], x is feasible and s is the first shelter following x in $\prec$.

Lemma 15 ([2]). *Define:*

$$p = \begin{cases} 0, & \text{if } s \text{ is the dummy shelter} \\ \min\{j : x_j \neq s_j\}, & \text{else;} \end{cases}$$

$$S \equiv \{j : p < j \text{ and } x_j = 1\};$$

(a) *If $S \neq \phi$ and $S \equiv \{j_1, \ldots, j_h\}$, where $j_1 > j_2 > \ldots > j_h$, then the points*

$$x^{(1)} = x + \Sigma\{u_j : j > j_1\}$$

and

$$x^{(k)} = x - u_{j_k} + \Sigma\{u_j : j > j_k \text{ and } x_j = 0\}, \quad (2 \leqslant k \leqslant h)$$

are MFPs;

(b) *If $s_n = 1$, then $z = s - u_n$ is a MFP (note: if s is the dummy shelter, then $s_n = 1$ is considered to be false);*

(c) *The points $x^{(1)}, \ldots, x^{(h)}$ in (a) (if any) and the point z in (b) (if any) are the only MFPs preceding s and following the previous shelter r (if any) in the linear order $\prec$.*

The proof follows from an analysis of the HOP-SKIP-and-JUMP algorithm.

Example. With reference to the previous example, let $s = (0\ 1\ 1\ 1\ 1\ 1)$. Then one has

r	1	0	1	1	1	0	shelter
$x = x^{(1)}$	1	0	1	1	0	1	
$x^{(2)}$	1	0	1	0	1	1	
$x^{(3)}$	1	0	0	1	1	1	MFPs
z	0	1	1	1	1	0	
s	0	1	1	1	1	1	shelter

([2]) Here and in the sequel we make the convention that the result of the summation over the empty set is zero and that the integer integral $[a, b]$ is empty when $a > b$.

$(p = 1; \; j_1 = 6, \; j_2 = 4, \; j_3 = 3)$.

Lemma 16. *Let S, j_k and $x^{(k)}$ be defined as in Lemma 15 and assume that $h = |S| \geq 2$. Let $\delta_k = \Sigma\{c_j : j_k < j < j_{k-1}, x_j = 0\}$, $2 \leq k \leq h$. Let $\Delta_1 = 0$, $\Delta_2 = \delta_2 - c_{j_2}$, and $\Delta_k = \Delta_{k-1} + \delta_k + c_{j_{k-1}} - c_{j_k}$, $2 < k \leq h$. Then*

$$cx^{(k)} - cx^{(1)} = \Delta_k.$$

In view of Lemmas 15 and 16, one can compute

$$\max\{c_1 y_1 + \ldots + c_n y_n : r \prec y \prec s \text{ and } y \text{ is a MFP}\}$$

in $O(n)$ time. It follows that one can solve (13) in $O(nm)$ time, provided that the sorted list of shelters is available.

In conclusion, the solution of the regular set covering problem (13) can be computed in $O(nm \log_2 m)$ time –– the time required in order to obtain the sorted list of shelters.

An elegant algorithm for generating all the MFPs of a regular set covering problem has been recently proposed by Crama [4].

8.3. Regular set covering problems and knapsack problems

Given a set covering problem (13), we consider the following question: are there a non-negative row n-vector a and a non-negative 'threshold' b such that, for every $y \in B^n$,

$$Ay \geq e \quad \Leftrightarrow \quad ay \geq b \quad ? \tag{15}$$

(Equivalently: is the feasible set Y of (13) also the feasible set of a knapsack problem?) On the basis of the results of last Section, one can answer this question is polynomial time, as pointed out by Peled and Simeone [19].

A necessary condition for the existence of such a and b is that (13) is regular. As shown in Sec. 8.1. one can check the regularity of (13) in polynomial time. Assume then, that (13) is regular. It is not difficult to see that there exist a and b satisfying (15) if and only if there exist $w \geq 0$, $t \geq 0$ such that

$$wy \geq 1 + t \quad \text{for all } y \in Y$$
$$wy \leq t \quad \text{for all } y \notin Y.$$

Such w and t exist if and only if the linear program in the $n + 1$ variables $w_1, \ldots, w_n, t$

$$\left. \begin{array}{ll} \text{minimize} & w_1 + \ldots + w_n \\ \text{subject to} & wy - t \geq 1 \quad \text{for all the minimal feasible points } y \\ & wy - t \leq 0 \quad \text{for all the maximal infeasible points } y \\ & w \geq 0, t \geq 0 \end{array} \right\} \tag{16}$$

has a feasible solution.

The key observation is that the maximal infeasible points of (13) are the complements of the rows of A, and hence their number is m; on the other hand, the minimal feasible points are the complements of the MFPs of (14), and hence they are at most $mn + m + n$ by Theorem 14. Thus, using the linear programming algorithm of Karmarkar [15], one can solve (16) in polynomial time. Taking into account the fact that all the coefficients of (16) are 0, 1 or -1 and thus the input size L of (16) is $0(m\,n^2)$ bits, it follows that one can recognize those set covering problems (13) such that (15) holds for some a, b in $0(m^2 n^{7.5})$ time.

9. Equivalence between regular set covering problems and 0-1 programs with linear objective function

From a theoretical point of view, one important reason for studying regular set covering problems ensues from the following result of Hammer, Johnson and Peled [11], which asserts the equivalence between arbitrary (linearly or nonlinearly constrained) 0-1 programs with linear objective function and regular set covering problems, provided that a mild condition on the objective function is met.

Consider the 0-1 program with linear objective function

$$\left.\begin{array}{ll} \text{minimize} & c_1 x_1 + \ldots + c_n x_n \\ \text{subject to} & g_i(x) \leq b_i, \quad i = 1, \ldots, m \\ & x \in B^n, \end{array}\right\} \quad (17)$$

where $g_1, \ldots, g_m$ are arbitrary pseudo-boolean functions. Upon possible complementation and re-indexing of some variables, one may always assume that $c_1 \geq c_2 \geq \ldots \geq c_n \geq 0$. Suppose however that the stronger condition

$$c_1 > c_2 > \ldots > c_n > 0 \qquad (18)$$

is met. Then, Hammer, Johnson and Peled [11] have shown that

Theorem 17. *There exists a regular set covering problem*

$$\left.\begin{array}{ll} \text{minimize} & c_1 x_1 + \ldots + c_n x_n \\ \text{subject to} & Ax \geq e \\ & x \in B^n \end{array}\right\} \quad (19)$$

with the same objective function as (17) and having the same optimal solutions as (17).

The fact that, according to Theorem 14, one can solve (19) in polynomial time is not in conflict with the well-known fact (see e.g. Garey and Johnson [5]) that (17) is NP-complete. In fact, there is in general a *size blow-up* effect, which may cause the number of rows in A to be exponential in the input size of (17).

The matrix A can be constructed by the following procedure.

Step 1. Determine the resolvent $\rho_X(x)$ of the feasible set X of (19).
If all g_i are linear, then ρ_X is given by the expression (4). In general ρ_X can be computed by the method described in Granot and Hammer [7].

Step 2. Let $\psi(y)$ be the boolean function defined by $\psi(y) = \overline{\rho}_X(y)$.
That is, $\psi(y) = 0$ iff $(e - y) \in X$.

Step 3. Find (e.g. by the consensus method) all prime implicants of ψ.

Step 4. Let ψ_{mon} be the union of all those prime implicants of ψ which do not involve complemented variables. It is easy to see that the boolean function ψ_{mon} is monotone nondecreasing, i.e. $y \leq y'$ implies $\psi_{\text{mon}}(y) \leq \psi_{\text{mon}}(y')$.

Step 5. Starting from ψ_{mon}, construct a new boolean function ψ_{reg} according to the following procedure

$\psi_{\text{reg}} := \psi_{\text{mon}}$
while there are two indexes i, j such that $i < j$ and $y_i \not\succ y_j$ with respect to ψ_{reg}
    begin
        write $\psi_{\text{reg}}(y)$ as $\alpha y_i y_j \vee \beta y_i \vee \gamma y_j \vee \delta$, where α, β, γ, and δ do not depend on y_i and y_j
        $\psi_{\text{reg}}(y) := (\alpha \vee \gamma) y_i y_j \vee \beta y_i \vee (\beta\gamma) y_j \vee \delta;$
    end

The function ψ_{reg} output by the above procedure can be shown to be regular (actually $i < j$ implies $y_i \succ y_j$ with respect to ψ_{reg}) and nondecreasing.

Step 6. Write ψ_{reg} in disjunctive normal form

$$\psi_{\text{reg}}(y) = \bigvee_{T \in \mathscr{C}} \prod_{j \in T} y_j.$$

Step 7. Let $\mathscr{C} = \{T_1, \ldots, T_p\}$. Define the $p \times n$ binary matrix $A = [a_{ij}]$ by

$$a_{ij} = \begin{cases} 1 & \text{if } j \in T_i \\ 0 & \text{if } j \notin T_i \end{cases}$$

Then A is the desired matrix.

The correctness of this procedure rests on the following key result, due to Hammer, Johnson and Peled [11].

Theorem 18. *Under the assumption that (18) holds,*
x^* *is an optimal solution to* min $\{cx : Ax \geq e, x \in B^n\}$
⇔ $y^* = e - x^*$ *is an optimal solution to* max $\{cy : \psi_{\text{reg}}(y) = 0\}$
⇔ y^* *is an optimal solution to* max $\{cy : \psi_{\text{mon}}(y) = 0\}$
⇔ y^* *is an optimal solution to* max $\{cy : \psi(y) = 0\}$
⇔ x^* *is an optimal solution to (17).*

Example. Consider the 0-1 linear programming problem

$$\text{minimize} \quad 20x_1 + 16x_2 + 10x_3 + 7x_4 + 5x_5 + 2x_6 \tag{20}$$

$$\text{subject to} \quad -7x_1 + 6x_2 - 9x_3 + 4x_4 + 3x_5 - x_6 \leq -3 \tag{21}$$

$$-3x_1 - 4x_2 - 6x_3 - x_4 - 5x_5 - 2x_6 \leq -11 \tag{22}$$

$$x \in B^6.$$

Step 1. The resolvent of (21) is

$$\rho_1 = \bar{x}_1\bar{x}_3 \vee x_2\bar{x}_3 \vee \bar{x}_3 x_4 x_5 \vee \bar{x}_1 x_2 x_4 \vee \bar{x}_1 x_2 x_5 \vee \bar{x}_1 x_4 x_5 \bar{x}_6.$$

The resolvent of (22) is

$$\rho_2 = \bar{x}_3\bar{x}_5 \vee \bar{x}_1\bar{x}_2\bar{x}_3 \vee \bar{x}_1\bar{x}_2\bar{x}_5 \vee \bar{x}_1\bar{x}_3\bar{x}_6 \vee \bar{x}_2\bar{x}_3\bar{x}_4 \vee \bar{x}_2\bar{x}_3\bar{x}_6 \vee \bar{x}_2\bar{x}_5\bar{x}_6 \vee \bar{x}_1\bar{x}_4\bar{x}_5\bar{x}_6.$$

The resolvent of the system (21), (22) is $\rho = \rho_1 \vee \rho_2$.

Step 2.

$$\psi(y) = y_1 y_3 \vee \bar{y}_2 y_3 \vee y_3 \bar{y}_4 \bar{y}_5 \vee y_1 \bar{y}_2 \bar{y}_4 \vee y_1 \bar{y}_2 \bar{y}_5 \vee y_1 \bar{y}_4 \bar{y}_5 y_6 \vee y_3 y_5 \vee y_1 y_2 y_3 \vee$$
$$\vee y_1 y_2 y_5 \vee y_1 y_3 y_6 \vee y_2 y_3 y_4 \vee y_2 y_3 y_6 \vee y_2 y_5 y_6 \vee y_1 y_4 y_5 y_6.$$

Step 3. After the execution of the consensus method we get

$$\psi(y) = y_3 \vee y_1 \bar{y}_2 \bar{y}_4 \vee y_1 y_2 y_5 \vee y_1 \bar{y}_2 \bar{y}_5 \vee y_1 \bar{y}_2 y_6 \vee y_1 \bar{y}_4 y_5 \vee y_1 \bar{y}_4 y_6 \vee$$
$$\vee y_1 y_5 y_6 \vee y_2 y_5 y_6, \tag{23}$$

where $y_3, \ldots, y_2 y_5 y_6$ are the prime implicants of ψ.

At this point, before going on with the procedure, we notice that some simplifications are possible, and we shall indeed implement them. As a matter of fact by Theorem 18 the given linear 0-1 program is equivalent to the maximization of $20y_1 + 16y_2 + 10y_3 + 7y_4 + 5y_5 + 2y_6$ subject to $\psi(y) = 0$, where ψ is given by (23). But one must have $y_3 = 0$ for all feasible solutions and $y_4 = 1$ for all optimal solutions to the latter problem. Hence we shall set $y_3 = 0$ and $y_4 = 1$ (implying $x_3 = 1$ and $x_4 = 0$) and ψ simplifies to

$$\psi(y) = y_1 y_2 y_5 \vee y_1 \bar{y}_2 \bar{y}_5 \vee y_1 \bar{y}_2 y_6 \vee y_1 y_5 y_6 \vee y_2 y_5 y_6.$$

Step 4. $\psi_{\text{mon}} = y_1 y_2 y_5 \vee y_1 y_5 y_6 \vee y_2 y_5 y_6$.

Step 5 and 6.
Since $y_2 \neq y_5$,
$$y_1 y_2 y_5 \vee y_1 y_5 y_6 \vee y_2 y_5 y_6 \rightarrow y_1 y_2 y_5 \vee y_2 y_5 y_6.$$
Since $y_1 \neq y_2$ in the new function,
$$y_1 y_2 y_5 \vee y_2 y_5 y_6 \rightarrow y_1 y_2 y_5.$$
The last function, being symmetric, is regular.
Hence $\psi_{\text{reg}} = y_1 y_2 y_5$. From the absence of y_6 we conclude that $y_6 = 1$ in all optimal solutions.

Step 7. The required set covering problem is

$$\left.\begin{aligned}\text{minimize} \quad & 20x_1 + 16x_2 + 5x_5 \\ \text{subject to} \quad & x_1 + x_2 + x_5 \geq 1 \\ & x_1, x_2, x_5 = 0 \text{ or } 1\end{aligned}\right\} \qquad (24)$$

The only optimal solution to (24) is given by $x_1 = x_2 = 0$, $x_5 = 1$. It follows that the only optimal solution to the initial problem is (0, 0, 1, 0, 1, 0).

Acknowledgements

Our research was supported by the Air Force Office of Scientific Research Grant AFOSR 0271 to Rutgers University and by the NSF Grant ECS 85 03212. We are indebted to Prof. Nicola Santoro for having pointed out to us the connection between depth-first-search and the linearization of pseudoboolean functions by order constraints, and to Dr. Susan Bubeck for the term «pedicle».

References

[1] M.L. Balinski, «On a Selection Problem», *Management Science* 17, 230 - 231, 1970.
[2] C. Berge, *Graphs and Hypergraphs*, North-Holland, Amsterdam, 1973.
[3] V. Chvátal and P.L. Hammer, «Aggregation of Inequalities in Integer Programming», *Annals of Discrete Mathematics* 1, 145 - 162, 1977.
[4] Y. Crama, «Dualization of a Regular Boolean Function», *Technical Report, Rutgers University*, May 1986.
[5] M.R. Garey and D.S. Johnson, *Computers and Intractability: a Guide to the Theory of NP-Completeness*, Freeman, S. Francisco, 1979.
[6] H.C. Golumbic, «Threshold Graphs and Synchronizing Parallel Processes», *Colloquia Mathematica Societatis Janos Bolyai (Combinatorics)*, 18, 331 - 352, 1978.
[7] F. Granot and P.L. Hammer, «On the Role of Generalized Covering Problems», *Cahiers du Centre d'Etudes de Recerche Opérationnelle* 16, 277 - 289, 1974.
[8] P.L. Hammer, «Boolean Elements in Combinatorial Optimization», *Annals of Discrete Mathematics* 4, 51 - 71, 1979.
[9] P.L. Hammer and P. Hansen, «Logical Relations in Quadratic 0-1 Programming», *Revue Roumaine de Mathématiques Pures et Appliquées* 26, 421 - 429, 1981.
[10] P.L. Hammer, T. Ibaraki and B. Simeone, «Threshold Sequences», *SIAM Journal on Algebraic and Discrete Methods* 2, 39 - 49, 1981.
[11] P.L. Hammer, E.L. Johnson and U.N. Peled, «Regular 0-1 Programs», *Cahiers du Centre d'Etudes de Recherche Opérationnelle* 16, 267 - 276, 1974.
[12] P.L. Hammer and S. Nguyen, «APOSS-A Partial Order in the Solution Space of Bivalent Programs», in N. Christofides, ed., *Combinatorial Optimization*, Wiley, New York, 1979.
[13] P.L. Hammer, U.N. Peled and M.A. Pollatschek, «An Algorithm to Dualize a Regular Switching Function», *IEEE Transactions on Computers* C - 28, 238 - 243, 1979.
[14] E.L. Johnson and M. Padberg, «Degree-Two Inequalities and Biperfect Graphs», *Technical Report, Universität Bonn, Institut für Okonometrie und Operations Research*, 1981.
[15] N. Karmarkar, «A New Polynomial Algorithm for Linear Programming», *Combinatorica* 4, 373 - 396, 1984.
[16] A.V. Karzanov, «Determining the Maximal Flow in a Network by the Method of Preflows», *Soviet Mathematics Doklady* 15, 434 - 437, 1974.
[17] L. Lovász, *Combinatorial Problems and Exercises*, North-Holland, Amsterdam, 1979.
[18] J. Orlin, «The Minimal Integral Separator of a Threshold Graph», *Annals of Discrete Mathematics* 1, 415 - 419, 1977.
[19] U.N. Peled and B. Simeone, «Polynomial-Time Algorithms for Regular Set-Covering and Threshold Synthesis», *Discrete Applied Mathematics* 12, 57 - 69, 1985.
[20] J.C. Picard, «Maximal Closure of Graph and Applications to Combinatorial Problems», *Management Science* 22, 1268 - 1270, 1976.
[21] J.C. Picard and H. Ratliff, «Minimum Cuts and Related Probles», *Networks* 5, 357 - 370, 1975.
[22] W.V. Quine, «Two Theorems about Truth Functions», *Boletín de la Sociedad Matematica Mexicana* 1, 64 - 70, 1953.
[23] W.V. Quine, «A Way of Simplifying Truth Functions», *American Mathematical Monthly* 52, 627 - 631, 1955.
[24] S. Rudeanu, *Boolean Functions and Equations*, North-Holland, Amsterdam, 1974.
[25] K. Spielberg, «Minimal Preferred Variable Reduction Methods in Zero-One Programming», *IBM Philadelphia Scientific Center Report n. 320 - 3013*, July 1972.
[26] R.E. Tarjan, «Depth-First Search and Linear Graph Algorithms», *SIAM Journal on Computing* 1, 146 - 160, 1972.
[27] R.O. Winder, «Threshold Logic», *Ph.D. Dissertation, Department of Mathematics, Princeton University*, 1962.

Peter L. Hammer
RUTCOR
Rutgers University
New Brunswick, NJ 08903
USA

Bruno Simeone
RUTCOR
Rutgers University
New Brunswick, NJ 08903
USA
on leave from
Department of Statistics
University of Rome «La Sapienza»
Rome
Italy

SINGLE FACILITY LOCATION ON NETWORKS*

Pierre HANSEN, Martine LABBÉ, Dominique PEETERS
and Jacques-François THISSE

1. Introduction

We consider the problem of selecting one point of a network in order to optimize one or several functions which are distance-dependent with respect to given points of the network. The problem is motived by a number of potential applications. For example: a plant is set up at some point of a transportation system to minimize production and shipment costs; an emergency service unit is located in a rural area to minimize the maximal intervention time to population centers; a switching center is located in a communication network to minimize transmission costs from and toward peripheral units.

Network location theory can be traced back to Jordan [62] who obtained a characterization of the median set of a tree (see 3.2 below). However, this was only a side-result of a study devoted to automorphisms of quadratic forms. Much more recently, Ore [83] provided some properties of the median of a graph in his book *Theory of Graphs*. But it is the seminal paper by Hakimi [35] on «Optimal location of switching centers and the absolute centers and medians of a graph» which gave its real start to network location theory. [Notice that results similar to those of Hakimi have been obtained independently by Guelicher [34]]. Since then the theory has undergone a phasis of rapid development. A book by Handler and Mirchandani [50], called *Location on Networks*, presents the state of the art in the late 70's. In a clear and well-documented survey published in 1983, Tansel, Francis and Lowe [94, 95] cite 117 references. Many more entries can be found in the recent bibliography of Domschke and Drexl [21].

This survey has two purposes. First, we present concisely the main models, theorems and algorithms for the location of a *single* facility on a network. Thus our paper is narrower is scope than Tansel, Francis and Lowe who are also concerned with the multi-facility case, but attempts to be more technical in that

---

* The research of the first author was supported by the Air Force Office of Scientific Research Grant No. AFOSRO271 to Rutgers University, that of the second author by the Action de Recherche Concertée of the Belgian Government under contract 84/8965 and that of the fourth author by the National Science Foundation grant SES 85-02886 to the University of Pennsylvania. The authors thank J. Krarup for his comments on a first draft of the paper.

many theorems and algorithms are explicitely stated. We also provide an extensive bibliography on single facility location on networks. Second, a substantial part of the survey is devoted to results recently found and approaches that have emerged during the last five years. This includes various extensions of the median and center problems (see 3.4 and 4.4 below), as well as the location of a facility by voting and competitive processes (see 5.3 and 5.4 below).

Proofs are omitted, due to the abundance of material. The interested reader is referred to Handler and Mirchandani [50] and Labbé [65] where a large number of them can be found. Again for brevity, algorithms are described in a somewhat informal way. Their computational complexity (all those considered are polynomial) as well as the data structures necessary for their implementation are mentioned. [Background material on these topics is given in Aho, Hopcroft and Ullman [1], and Garey and Johnson [28]].

The remainder of the paper is organized as follows. In the next section, we give the definition of a network and of some related concepts; we also state basic properties of the distance function associated with a network, that are useful for the subsequent analysis. The two main classes of location problem, i.e., median (or minisum) and center (or minimax) problems, are studied in sections 3 and 4 respectively. In the former case, the objective in locating the facility is to minimize the (weighted) sum of distances between the facility and a given set of clients located along the network. In the latter one, the objective is to minimize the largest of these distances. Both cases subdivide into several subcases according to the fact that locations of the facility and/or the clients are restricted to vertices or may be anywhere on the network (i.e., also at inner points of arcs). Various extensions of the median and center problems are considered in 3.4 and 4.4. respectively. In the foregoing two sections, the location decision is assumed to be made by a *single* agent (the decision-maker) having *one* well-defined objective. In section 5, we consider different extensions of this problem in which either the decision-maker is still single but has several *conflicting* criteria or the location decision is the outcome of a collective action in which *different* agents pursue their own interest. Bicriterion and multicriteria problems are taken up in 5.1 and 5.2 respectively. The concepts of cent-dian and of k-centrum, which correspond to different combinations of the median and center, play a prominent role. Sub--sections 5.3 and 5.4 are devoted to problems studied very recently: (i) the voting location problem, in which it is asked to find a location such that no other one is preferred to it by a strict majority of clients (5.3); (ii) the competitive location problem, in which it is required to find a locational configuration for a given set of facilities such that no facility would attract more clients at an alternate location (5.4). Miscellaneous further problems are mentioned and some conclusions are drawn in section 6.

2. The model

2.1. Description of the model

The model can be viewed as consisting of two «sides», a transportation side and a demand side. We first consider the transportation side. The following definitions are useful (Berge [8]): a topological arc is the image of [0, 1] by a continuous mapping f from [0, 1] to R^n such that $f(\theta) \neq f(\theta')$ for any $\theta \neq \theta'$ in [0, 1]; a rectifiable arc is a topological arc of a well-defined length. A *network* is then defined as a subset N of R^n which satisfies the following conditions: (i) N is the union of a finite number of rectifiable arcs; (ii) any two arcs intersect at most at their extemities; (iii) N is connected. The set of *vertices* of the network is made of the extremities of the arcs defining N; it is denoted by $V = \{v_1, \ldots v_n\}$. Points v_i correspond to transportation nodes (crossroads, railway junctions, ...) and tips (demand points uniquely connected to the network) of the real space we intend to model. The set of arcs defining the network is denoted by A; an arc $[v_i, v_j] \in A$ iff in the real space there is a transportation line (road, railway, ...) linking the sites corresponding to v_i and v_j and passing through no other sites corresponding to points of N. The length of the arc $[v_i, v_j] \in A$ is given and denoted by l_{ij} (see Berge [8] for the definition of the length of an arc). Each point $x \in N$ belongs to some arc of A but x may not be a vertex. For any two points $x_1, x_2 \in [v_i, v_j]$ the subset of points of N between and including x_1 and x_2 is a *subarc* $[x_1, x_2]$. Let f_{ij} be the mapping defining $[v_i, v_j]$ and θ_{ij} the inverse of f_{ij}; to each point $x \in [v_i, v_j]$ corresponds one and only one value $\theta_{ij}(x)$ in [0, 1]. Then the length of $[x_1, x_2]$ is $|\theta_{ij}(x_1) - \theta_{ij}(x_2)| l_{ij}$. A *path* $P(x_1, x_2)$ joining $x_1 \in N$ and $x_2 \in N$ is a minimal connected subset of N containing x_1 and x_2. The length of a path is equal to the sum of the lengths of all its constituent arcs and subarcs. The *distance* $d(x_1, x_2)$ between $x_1 \in N$ and $x_2 \in N$ is equal to the length of a shortest path joining x_1 and x_2. Clearly d is a metric on N. We say that $x \in N$ is *between* $x_1 \in N$ and $x_2 \in N$ iff $d(x_1, x) + d(x, x_2) = d(x_1, x_2)$; let $B(x_1, x_2)$ be the set of points in N between x_1 and x_2. Without loss of generality, we assume that there is no redundant arc, that is, $B(v_i, v_j) = [v_i, v_j]$ for all pairs of vertices such that $[v_i, v_j] \in A$.

The demand side of the model is as follows. There is a set U of clients and *client* $u \in U$ is described by his/her location in N and his/her demand per unit of time which is a given positive number. Two cases are considered. In the first one U is a *finite* set; for any subset $\bar{U}$ of U, $|\bar{U}|$ stands for the number of elements of $\bar{U}$. Without loss of generality, we assume that client $u \in U$ is located at $v(u) \in V$ and has a unit demand. It is possible that $v(u) = v(\bar{u})$ for $u \neq \bar{u}$ is which case clients u and $\bar{u}$ are at the same vertex. To simplify notation, we then set

$$w_i = |\{u \in U; v(u) = v_i\}|, \quad \text{for} \quad i = 1, \ldots, n;$$

of course, $\sum_{i=1}^{n} w_i = |U|$. In the second case, U is a *continuum*. More specifically, we assume that clients are uniformly distributed along the network N and have a unit demand density.

2.2. Basic properties of the distance on N

The distance between a given point in N and a variable point along an arc of A obeys the following properties:

Theorem 2.1. *Let $\bar{x}$ be any given point of N and $x = f_{ij}(\theta) \in [v_i, v_j] \in A$. Then $d[\bar{x}, f_{ij}(\theta)]$ is a function of θ*
  (i) *continuous and concave on $[0, 1]$;*
  (ii) *linearly increasing with slope l_{ij} on $[0, \bar{\theta}_{ij}]$ and linearly decreasing with slope $-l_{ij}$ on $[\bar{\theta}_{ij}, 1]$, where*

$$\bar{\theta}_{ij} = \frac{l_{ij} + d(\bar{x}, v_j) - d(\bar{x}, v_i)}{2 l_{ij}}.$$

More specific properties of the distance are obtained when the network is a *tree* T, i.e., a network such that for any two distinct points in N there is a single path joining them.

Theorem 2.2. (Dearing, Francis and Lowe [20]). *Let $\bar{x}$, x_1 and x_2 be any three points of a tree T. If $x_3 \in P(x_1, x_2)$, then, along $P(x_1, x_2)$, $d(\bar{x}, x_3)$ is*
  (i) *piecewise linear in $d(x_1, x_3)$ with at most two pieces, the slopes of which are $+1$ and -1;*
  (ii) *convex in x_3, i.e., $d(\bar{x}, x_3) \leq \alpha d(\bar{x}, x_1) + (1 - \alpha) d(\bar{x}, x_2)$ where $d(x_3, x_2) = \alpha d(x_1, x_2)$.*

Given that the objective functions considered in location theory are often convex in distances, Theorem 2.2 allows one to obtain stronger properties and algorithms with lower complexity for trees than for general networks.

Let v_k be any vertex of N. The point

$$x_{ij}(v_k) = f_{ij}\left[\frac{l_{ij} + d(v_k, v_j) - d(v_k, v_i)}{2 l_{ij}}\right]$$

of $[v_i, v_j] \in A$ (when it exists) is called a *bottleneck point*. The distance between v_k and $x_{ij}(v_k)$ via v_i is equal to that via v_j: there are two shortest paths linking v_k and $x_{ij}(v_k)$. Let B denote the set of bottleneck points associated with N; since any vertex generates at most one bottleneck point on each arc, $|B| = O(|V||A|)$. Clearly, if N is a tree then $|B| = 0$. A subarc $[x_1, x_2]$ delimitated by two successive vertices or bottleneck points of an arc $[v_i, v_j]$ is called a *segment*. Along a segment the distance from any vertex v_k is either linearly increasing or linearly decreasing. A direct consequence of this observation is as follows:

Theorem 2.3. *Let x' and x'' be any two points in the interior of a segment. Then the trees of shortest paths from x' and x'' to the vertices of N are identical except for the roots x' and x''.*

Let Q be the set of *points equidistant* from any two vertices v_k, v_l via v_i and v_j in V (or v_j and v_i) respectively: $Q = \{x \in N; [v_i, v_j] \in A \text{ with } x \in [v_i, v_j] \text{ and } v_k, v_l \in V \text{ exist such that either } d(v_k, v_i) + d(v_i, x) = d(v_l, v_j) + d(v_j, x) \text{ or } d(v_l, v_i) + d(v_i, x) = d(v_k, v_j) + d(v_j, x)\}$. Requiring $v_k = v_l$ yields the set of bottleneck points which is therefore a subset of Q. In a general network $|Q| = O(|V|^2|A|)$, while in a tree $|Q| = O(|V|^2)$. A subarc $[x_1, x_2]$ delimited by two successive vertices or equidistant points of an arc $[v_i, v_j]$ is called a *subsegment*.

Theorem 2.4. *Let x' and x'' be any two points in the interior of a subsegment. Then the rankings of vertices by order of increasing distances from x' and x'' are identical.*

We now consider the distance between a given arc of A and a variable point along an arc of A. The *remoteness* of $x \in [v_i, v_j] \in A$ relative to $[v_k, v_l] \in A$ is defined by $\bar{d}(x, [v_k, v_l]) = \underset{y \in [v_k, v_l]}{\text{Max}} d(x, y)$. Obviously, $\bar{d}(x, [v_k, v_l]) = [l_{kl} + d(x, v_l) + d(x, v_k)]/2$ if $x \notin [v_k, v_l]$ and $\bar{d}(x, [v_k, v_l]) = \text{Max}\{d(x, v_k), d(x, v_l)\}$ if $x \in [v_k, v_l]$. Theorem 2.1 is to be replaced by the following one:

Theorem 2.5. (Frank [26]). *Let $[v_k, v_l]$ be a given arc of A and $x = f_{ij}(\theta) \in [v_i, v_j] \in A$. If $[v_k, v_l] \neq [v_i, v_j]$ then $\bar{d}(f_{ij}(\theta), [v_k, v_l])$ is:*
  (i) *continuous and concave on $[0, 1]$;*
  (ii) *piecewise linear with at most three pieces having slopes l_{ij}, 0 and $-l_{ij}$.*
If $[v_k, v_l] = [v_i, v_j]$ then $d(f_{ij}(\theta), [v_k, v_l])$ is
  (i) *continuous on $[0, 1]$;*
  (ii) *concave on $[0, 1/2]$ and $[1/2, 1]$.*

Typical examples of remoteness functions are given in Figures 2.1 and 2.2 for $[v_k, v_l] \neq [v_i, v_j]$ and $[v_k, v_l] = [v_i, v_j]$ respectively. In the former case, the func-

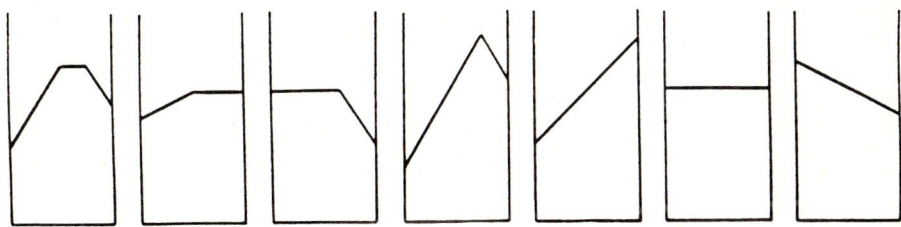

Fig. 2.1

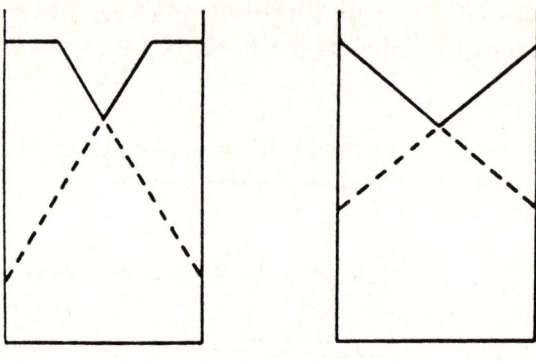

Fig. 2.2

tion is called a *hat function*; in the latter case, the function can be viewed as the upper envelope of two degenerate hat functions.

3. Median problems

3.1. Notation

Handler and Mirchandani [50] have proposed a four-symbol notation for the designation of problems, specifying: (i) the subset of N in which facilities may be located (N of V); (ii) the subset of N containing clients' locations (N or V); (iii) the number of facilities to be located (1 or m); (iv) the type of network considered (N for a general network or T for a tree). As we consider only single facility location, the third element will always be 1. We add to this notation a fifth symbol to denote the objective function: Σ for the sum of distances to the clients and μ for the maximum of these distances. For example, when the facility can be located everywhere in the network and when clients are located at the vertices of the network, the corresponding median problem is denoted by $(N/V/1/N/\Sigma)$; when clients are distributed along the network, it becomes $(N/N/1/N/\Sigma)$.

3.2. The median and absolute median problems: $(V/V/1/N/\Sigma)$ and $(N/V/1/N/\Sigma)$

A point $m \in N$ is an *absolute median* iff

$$F(m) = \sum_{i=1}^{n} w_i d(v_i, m) \leq F(x) = \sum_{i=1}^{n} w_i d(v_i, x), \quad \forall x \in N.$$

An absolute median is also called a Weber point, by reference to Weber [98]'s use of the weighted sum of distances in continuous location theory, or a centroid. The set of absolute medians, called the *absolute median set*, is denoted by M. The set of vertices of N belonging to M, called the *vertex absolute median set*, is denoted $M_v = M \cap V$. Restricting m and x to belong to V, we similarly define a *median* and the *median set*.

A direct consequence of the concavity of $d(v_i, f_{kl}(\theta))$ for each arc $[v_k, v_l] \in A$ [Theorem 2.1] and of the theorem of minimization of a concave function [Berge [8]] is as follows:

Theorem 3.1. (Hakimi [35]). *The set V of vertices contains an absolute median. Furthermore, when $|U|$ is odd, any absolute median is a vertex.*

The vertex absolute median set M_v is thus equal to the median set. Finding M_v, and hence solving the median problem, is done in $O(|V|^2)$ operations by the algorithm described below.

Algorithm 3.1.
  (i) Set $F(m) = \sum_{i=1}^{n} w_i d(v_i, v_1)$ and $M_v = \{v_1\}$.
  (ii) Compute $F(v_k) = \sum_{i=1}^{n} w_i d(v_i, v_k)$ for $k = 2, \ldots, n$. If $F(v_k) > F(m)$, leave $F(m)$ and M_v unchanged; if $F(v_k) = F(m)$, set $M_v = M_v \cup \{v_k\}$; if $F(v_k) < F(m)$, set $F(m) = F(v_k)$ and $M_v = \{v_k\}$.

It follows from the concavity of $d[v_i, f_{kl}(\theta)]$ that if m is an absolute median interior to an arc $[v_k, v_l]$ then $F(x)$ is constant on $[v_k, v_l]$; hence any point of this arc is an absolute median. This allows us to determine M in $O(|V||A|)$ operations when $|U|$ is even (as stated above $M = M_v$ when $|U|$ is odd):

Algorithm 3.2.
  (i) Compute M_v by Algorithm 3.1 and set $M = M_v$.
  (ii) Consider in turn each arc $[v_k, v_l] \in A$ such that both v_k and v_l belong to M_v; compute $F(x_{kl})$ where x_{kl} is the middle point of $[v_k, v_l]$; if $F(x_{kl}) = F(m)$, add the arc $[v_k, v_l]$ to M.

The median set has been studied by Slater [90] and Barthélemy [6] in the special case of graphs whose arcs have a unit length and vertices a unit weight. The main result obtained is as follows:

Theorem 3.2. (Slater [90]). *For any graph G there exists a graph H such that the subgraph of H induced by its median set is isomorphic to G.*

Though this result deals with graphs only, it is important for our purpose because it implies that nothing can be said about the structure of the set M_v in general networks.

Theorem 3.1 has been extended in a variety of directions. Levy [68] considers the case of concave functions of distance, which allows for scale economies in transportation. Capacity constraints on arcs, and thus congestion in transportation, are taken up by Wendell and Hurter [99]. Mirchandani [80] deals with directed arcs and, therefore, asymmetries in transportation. Louveaux, Thisse and Beguin [70] work with a multi-modal transportation system, and study the impact of fixed transportation costs on the absolute median. Finally, different formulations of the multi-facility location problem for which the optimality property of the vertices holds, have been examined by Hakimi [36], Levy [68], Goldman [29], Hakimi and Maheshwari [38], and Wendell and Hurter [99].

Properties more specific than Theorem 3.1 can be obtained for particular weight distributions and/or particular network configurations.

Theorem 3.3. (Witzgall [101]). *If* $w_k \geq |U|/2$, *then* $v_k \in M_v$. *Furthermore, when* $w_k > |U|/2$, $M = M_v = \{v_k\}$.

Thus v_k is an absolute median when a majority of clients are established at v_k.

Conditions for a connected subset of N to contain an absolute median have been investigated by Goldman and Witzgall [33] and Goldman [30]. The best existing result is as follows:

Theorem 3.4. (Goldman and Witzgall [33]). *Let* $S \subset N$. *If*
  (i) *S is gated, i.e., there exists a function g from N-S to S such that for each $x \in N$-S and each $s \in S$ we have* $d(x, s) = d(x, g(x)) + d(g(x), s)$; *and*
  (ii) $\Sigma_{\{k; v_k \in S\}} w_k \geq |U|/2$,
then S contains an absolute median. Furthermore, when the inequality is strict in condition (ii), any absolute median belongs to S.

Condition (i) means that a shortest path linking $x \in N$-S and $s \in S$ passes through a «gate» $g(x) \in A$ which depends on x but not on s; condition (ii) states that a majority of clients are located in S. Condition (i) is satisfied when there exists an isthmus connecting two disjoint subsets of the network. Goldman [31] has found an upper bound on $\underset{s \in S}{\text{Min}} \ F(s)$ when conditions (i) and/or (ii) are relaxed within some given tolerances. Furthermore, as condition (i) is always satisfied when S is a singleton, Theorem 3.3 turns out to be a special case of Theorem 3.4.

A kind of converse proposition is now given.

Theorem 3.5. (Hansen, Thisse and Wendell [57]). *Let σ be a permutation defined on the set U of clients. If $x \in N$ is between $v(u)$ and $v(\sigma(u))$ for all $u \in U$, then $x \in M$.*

The above theorem is applicable when x is such that $N - \{x\}$ is disconnected and none of the connected components of $N - \{x\}$ contains a majority of clients. Notice that this result also follows from Theorem 3.4 since x is the gate of the connected components of $N - \{x\}$. Finally, Theorem 3.3 can be obtained from Theorem 3.5 for a permutation σ such that $v(\sigma(u)) = v_k$ whenever $v(u) \neq v_k$.

The special case of a tree has attracted the attention of several authors, including Jordan [62], Witzgall [101], Zelinka [102], Slater [87] and Mitchell [82]. Assuming that there is no vertex with at most two adjacent arcs and a zero weight, we can summarize their results as follows:

Theorem 3.6. *Let T be a tree. If $|U|$ is odd, then $M = M_v = \{v_k\}$ for some vertex v_k of T; if $|U|$ is even, then either $M = M_v = \{v_k\}$ for some vertex v_k of T or $M = [v_k, v_l]$ and $M_v = \{v_k, v_l\}$ for some pair v_k, v_l of adjacent vertices of T.*

Conversely, Barthélemy [7] has shown that if for all vectors of nonnegative weights of the vertices of a graph G, the subgraph of G induced by its median set is a tree, then G is a tree. Goldman [30] has shown how to solve $(V/V/1/T/\Sigma)$ and $(T/V/1/T/\Sigma)$ in $O(|V|)$ operations. The algorithm is based on the following observation which can be derived from Theorem 3.4: given any partitioning of T into two connected subsets T_1 and T_2, T_i contains an absolute median iff $\Sigma_{\{k; v_k \in T_i\}} w_k \geqslant |U|/2$.

Algorithm 3.3. (Goldman [30]).
  (i) If $T = \{v_l\}$, then $M = M_v = \{v_l\}$ and stop.
  (ii) Select a pending vertex v_k of T, i.e., a vertex with a single adjacent vertex v_l.
  (iii) If $w_k > |U|/2$, then $M = M_v = \{v_k\}$ and stop; if $w_k = |U|/2$, then $M = [v_k, v_l]$, $M_v = \{v_k, v_l\}$ and stop; otherwise delete v_k, set $w_l = w_l + w_k$ and return to (i).

Observe that the arc lengths play no role in the above algorithm.

3.3. The continuous median problem: $(N/N/1/N/\Sigma)$

A point $m_c \in N$ is a *continuous median* iff

$$\bar{F}(m_c) = \sum_{[v_i, v_j] \in A} \bar{d}(m_c, [v_i, v_j]) \leqslant \bar{F}(x) =$$

$$= \sum_{[v_i, v_j] \in A} \bar{d}(x, [v_i, v_j]), \quad \forall x \in N.$$

The set of continuous medians, called the *continuous median set*, is denoted by M_c. Let Q_e denote the set of *middle points* of the arcs defining N: $Q_e = \{x_{ij} \in [v_i, v_j] \in A; d(v_i, x_{ij}) = d(x_{ij}, v_j)\}$; of course $Q_e \subset Q$.

The counterpart of Theorem 3.1 for the continuous median problem, obtained from Theorem 2.5, is as follow:

Theorem 3.7. (Hansen and Labbé [53]). *The set $V \cup Q_e$ contains a continuous* median.

Minieka [74] has provided a necessary condition for the middle point x_{ij} of $[v_i, v_j]$ to be a continuous median: $|\overline{F}(v_i) - \overline{F}(v_j)| \leq d(v_i, v_j)$.

The following algorithm, based on the above results, allows us to determine $M_c \cap (V \cup Q_e)$ in $O(|A|^2)$ operations.

Algorithm 3.4.
(i) Compute $\overline{F}(v_i)$ for $i = 1, \ldots, n$.
(ii) For each arc $[v_i, v_j] \in A$, check if $|\overline{F}(v_i) - \overline{F}(v_j)| \leq d(v_i, v_j)$. If so, compute $\overline{F}(x_{ij})$; otherwise, proceed to the next arc.
(iii) Let $\overline{F}_{\min}$ be the smallest of the values $\overline{F}(v_i)$ and $\overline{F}(x_{ij})$ computed in steps (i) and (ii). Set $M_c \cap (V \cup Q_e) = \{v_i \in V \text{ and } x_{ij} \in Q_e; \overline{F}(v_i) = \overline{F}_{\min} \text{ and } \overline{F}(x_{ij}) = \overline{F}_{\min}\}$.

As in 3.2, the case of a tree is easier to deal with.

Theorem 3.8. (Hansen and Labbé [53]). *The continuous median set of a tree consists in either*
(i) *the middle point of a single arc, or*
(ii) *a vertex and, possibly, one or two subarcs delimitate by this vertex and the middle point of one or two adjacent arcs.*

The four possible types of solutions are depicted in Figure 3.1.

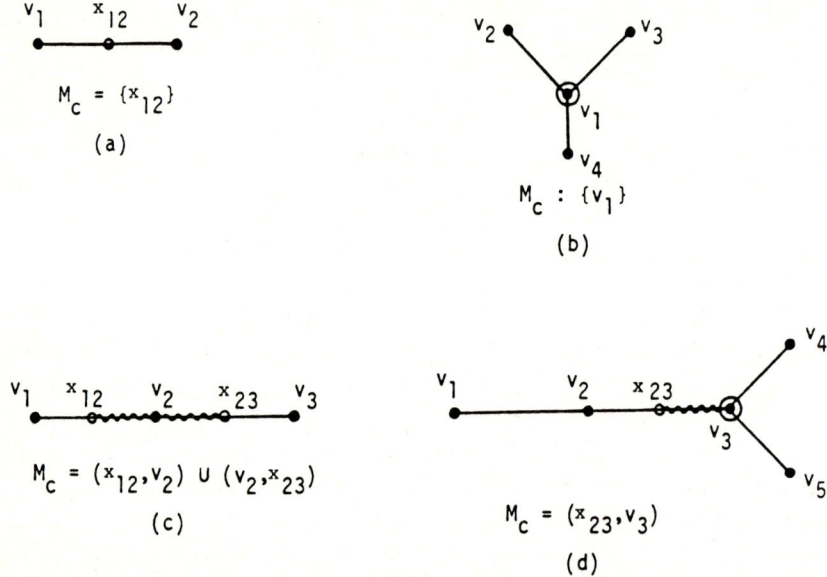

Fig. 3.1

The continuous median set of a tree can be determined in $O(|V|)$ operations by the following algorithm.

Algorithm 3.5.
(i) Set $M_c = \phi$ and $w_i = 1$ for $i = 1, \ldots, n$.
(ii) If $T = \{v_k\}$, then add v_k to M_c and stop.
(iii) Select a pending vertex v_k and let v_l be its adjacent vertex.
(iv) If $w_k > |A|/2 + 1$, add v_k to M_c and stop; if $w_k = |A|/2 + 1$, add the half-arc $[v_k, x_{kl}]$ and stop; if $w_k = (|A| + 1)/2$, add the middle point x_{kl} of $[v_k, v_l]$ and stop; if $w_k = |A|/2$, add the half-arc $[x_{kl}, v_l]$.
(v) Delete v_k and $[v_k, v_l]$, set $w_l = w_l + w_k$ and return to (ii).

3.4. Extensions of median problems

(i) In the real world transportation costs are often nonlinear functions of the distance covered, whence the need to investigate the median problem with such functions. As mentioned above, the optimality property of the vertices remains valid for concave functions. However, this is no longer true for convex functions. Shier and Dearing [85] have studied the special case of distances raised to the power p with $1 \leq p < \infty$, i.e., $F_p(x) = \Sigma_{i=1}^n w_i [d(v_i, x)]^p$. Using the directional derivatives of $F_p(x)$ along arcs, these authors have been able to get the following necessary and sufficient condition for local optima.

Theorem 3.9. (Shier and Dearing [85]). *A point $x^* \in N$ is a local minimizer of $F_p(x)$ iff for every vertex v_k adjacent to x we have*

$$\sum_{i \in I_k} w_i [d(v_i, x^*)]^{p-1} \leq \sum_{i \in \bar{I}_k} w_i [d(v_i, x^*)]^{p-1}$$

where $I_k = \{i = 1, \ldots, n; v_k$ is between v_i and $x^*\}$.

Goldman (personal communication) has obtained an $O(|V|)$ algorithm to find the point minimizing the sum of squared distances to the vertices of a tree. No algorithm seems to have been proposed as yet for general networks. The more general case of increasing functions of distance (neither concave nor convex) could be solved by adapting the Big-Square-Small-Square algorithm of Hansen, Peeters, Richard and Thisse [55] designed for the solution of a similar problem in continuous location theory.

(ii) The conditional median problem arises when, given that some facilities are already established along the network, an additional facility is to be located with the aim of minimizing the weighted sum of distances in the facility system. Let $L \subset N$ be the set of location of the incumbant facilities. A point $x^* \in N$ is a *conditional absolute median* iff

$$F(L \cup \{x^*\}) = \sum_{i=1}^{n} w_i \underset{y \in L \cup \{x^*\}}{\text{Min}} d(v_i, y) \leqslant$$

$$\leqslant F(L \cup \{x\}) = \sum_{i=1}^{n} w_i \underset{y \in L \cup \{x\}}{\text{Min}} d(v_i, y), \quad \forall x \in N.$$

This problem reduces to the absolute median problem after an adequate transformation of the vertex-to-vertex distances has been done. Hence

Theorem 3.10. (Minieka [75]). *The set V of vertices contains a conditional absolute median.*

Hansen and Labbé [53] have similarly extended Theorem 3.7 to the case of the conditional continuous median.

(iii) Another variant of the median problem obtains when the summation in the objective function is taken on the nearest points of the elements of a family of subsets of N. Let $\mathcal{S} = \{S_1, \ldots, S_m\}$ be a family of subsets of N and let $V_k = V \cap S_k$. A point $x^* \in N$ is an *absolute $\mathcal{S}$-median* iff

$$F_{\mathcal{S}}(x^*) = \sum_{k=1}^{m} \underset{v_i \in V_k}{\text{Min}} d(v_i, x^*) \leqslant F_{\mathcal{S}}(x) = \sum_{k=1}^{m} \underset{v_i \in V_k}{\text{Min}} d(v_i, x), \quad \forall x \in N.$$

Theorem 3.11. (Slater [91]). *The set V of vertices contains an absolute $\mathcal{S}$-median.*

Additional results are given in Slater [92] for the case of trees.

(iv) When clients situated along the path to more distant clients can be served without extra cost, the problem becomes one of determining a point of N minimizing the sum of the distances to the pendant vertices of the shortest path spanning tree in N, i.e., a *pendant-median* (Minieka [77]). If follows from Theorem 2.3 that the set of vertices and of bottleneck points contains a pendant-median. In the case of a tree, this problem is equivalent to the absolute median problem in which the pendant vertices of T have weights equal to one and all other vertices have weights equal to zero. Solving the pendant-median problem for a general network is still an open problem.

(v) Stochastic median problems have been studied by several authors. Frank [25] has considered the case where the weights w_i are replaced by nonnegative random variables W_i. A first solution-concept is a point $x^* \in N$, called an *absolute expected median*, such that

$$E[F(x^*)] = \sum_{i=1}^{n} E(W_i) d(v_i, x^*) \leqslant E[F(x)] = \sum_{i=1}^{n} E(W_i) d(v_i, x), \quad \forall x \in N.$$

Obviously, V contains an absolute expected median. Another solution concept proposed by Frank is a point of the network which, given some positive number q, minimizes the probability that the weighted sum of the distances is larger than or equal to q. The optimality property of the vertices does not hold for such a point and a good algorithm to determine it is still to be found. The special case of correlated normal demands is studied in Frank [26]. Mirchandani and Odoni [81] have taken up a more general problem in which both the weights and the network are random. More specifically, the weight distribution $(W_1, \ldots, W_n)$ and the network undergo probabilistic transitions among a finite number H of states. Each state $h = 1, \ldots, H$ corresponds to a particular realization of the vertex weights and arc lengths, and has a probability p_h which is independent of the previous state. Denoting by $d^h(x, y)$ the distance between x and y and by w_i^h the weight of vertex v_i in state h, a point x^* of N is an *absolute expected median* iff

$$E[F(x^*)] = \sum_{h=1}^{H} p_h \sum_{i=1}^{n} w_i^h \, d^h(v_i, x^*) \leq$$

$$\leq E[F(x)] = \sum_{h=1}^{H} p_h \sum_{i=1}^{n} w_i^h \, d^h(v_i, x), \quad \forall x \in N.$$

The above authors then show the set of vertices contains an absolute expected median.

Berman and Odoni [12] consider an even more general case: they allow the facility to be reposited on the network in order to respond to major changes in the network and/or the weights. A relocation has an associated cost which is assumed to be an increasing and concave function of the distance separating the initial and the new locations. Furthermore, Berman and Odoni suppose that the transition probabilities are described by a Markovian process. Again, the set of vertices is shown to contain a sequence of locations minimizing total costs.

Berman and Larson [10] attack the problem from a different perspective: the availability of the facility to the clients is a random variable. A further extension is considered by Berman, Larson and Chiu [11] who deal with queueing aspects when the service units are busy at the moment a demand occurs. A detailed discussion of these questions is contained in Berman, Chiu, Larson and Odoni [9].

(vi) When clients' demand depends on the distance to the facility, the problem is to locate the facility in order to maximize the total demand. Let $D_i[d(v_i, s)]$ be the demand issued at vertex v_i. A point $x^* \in N$ is a *maximum demand point* iff

$$D(x^*) = \sum_{i=1}^{n} D_i[d(v_i, x^*)] \geq D(x) = \sum_{i=1}^{n} D_i[d(v_i, x)], \quad \forall x \in N.$$

When functions D_i are decreasing and convex in distance, the set V of vertices contains a maximum demand point. On the other hand, when the D_i are not convex, no maximum demand point may belong to V. However, in the special case where the demand from v_i is constant within some specified range R_i and zero beyond R_i, a maximum demand point can be found in the finite set defined by the vertices of N and the points x belonging to the arcs of A such that $d(v_i, x) = R_i$ (Church and Meadows [16]). This result has been generalized by Minieka [78] to the case of a continuum of clients, when the facility is to be located with the aim of maximizing the total length of the arcs and subarcs convered within some given range R.

In the same vein, notice that various economic models dealing with the location and price policy of a profit-maximizing firm are discussed in Hanjoul and Thisse [51], Hurter and Martinich [61], and Louveaux and Thisse [71].

(vii) Finally, the problem of locating an «obnoxious» facility with the aim of maximizing the weighted sum of distances — the *maxisum problem* — has been considered by Zelinka [102], Church and Garfinkel [15], Minieka [79] and Ting [96]. In the case of a tree, Zelinka [102] shows that the set of pendant vertices contains an optimal location while Ting [96] provides a linear algorithm to find it. For general networks, Church and Garfinkel [15] show that the set $B \cup V$ of bottleneck points and vertices contains an optimal solution; Minieka [79] gives an upper bound on $F(x)$ when x is an interior point of an arc.

4. Center problems

4.2. The center and absolute center problems: $(V/V/1/N/\mu)$ *and* $(N/V/1/N/\mu)$

A point $c \in N$ is an *absolute center* iff

$$G(c) = \underset{i=1,\dots,n}{\text{Max}} \; d(v_i, c) \leqslant G(x) = \underset{i=1,\dots,n}{\text{Max}} \; d(v_i, x), \quad \forall x \in N.$$

The set of absolute centers, called the *absolute center set*, is denoted by C. The value of the objective function G at any point of C is called the *absolute radius* r_a of the network N. Restricting c and x to belong to V, we similarly define a *center*, the *center set* and the *radius* r of the network.

The absolute centers are contained in a finite subset of N.

Theorem 4.1. (Minieka [73]). *Any absolute center belongs to the union of the set V of vertices and of the set $Q - B$ of equidistant points which are not bottleneck points.*

This property follows directly from the fact that along each arc $[v_i, v_j] \in A$ the objective function $G(x)$ is the upper envelope of a finite family of piecewise linear and continuous functions of $\theta_{ij}(x)$ (Theorem 2.1) and hence is itself piecewise linear and continuous. The break points of $G(x)$ interior to $[v_i, v_j]$ that are local maxima correspond to bottleneck points, and those that are local minima correspond to equidistant points which are not bottleneck points. A typical plot of $G(x)$ along an arc is shown in Figure 4.1.

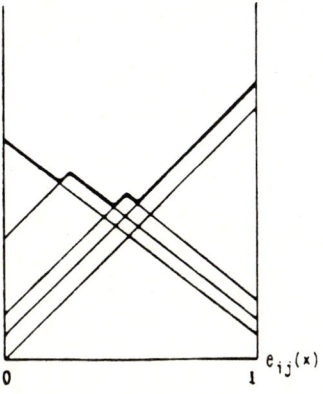

Fig. 4.1

The next result gives an upper bound on the number of local minima which is useful for establishing complexity properties.

Theorem 4.2. (Kariv and Hakimi [63]). *The function $G(x)$ has at most $|V| + 2$ local minima on each arc of A.*

This suggests the following which is at the basis of most existing ones.

Algorithm 4.1. (Hakimi [35]).
 (i) For each arc of A, plot $G(x)$ and determine the local minima of G.
 (ii) Select the smallest of the so-obtained values and let C be equal to the set of corresponding points.

Hakimi, Schmeichel and Pierce [39] show that this algorithm requires $O(|A| |V| \log|V|)$ operations when the data are stored in a stack during the search for the local minima of $G(x)$ on an arc. Using a 2 - 3 tree for the same purpose, Kariv and Hakimi [63] have obtained a complexity of $O(|A| |V| + |V|^2 \log|V|)$. Other procedures of complexity $O(|V|^3)$ are due to Minieka [76] and Cuninghame-Green [17].

The derivation of tight lower bounds of $G(x)$ is of practical importance for speeding up the algorithms. An example is provided by Halpern [43]. Let us consider the arc $[v_i, v_j]$ and denote by v_k (resp. v_l) the vertex such that $G(v_i) = d(v_i, v_k)$ (resp. $G(v_j) = d(v_j, v_l)$). In case of ties, v_k (resp. v_l) is the farthest vertex from v_j (resp. v_i). Halpern then shows that $m_{ij} = [l_{ij} + d(v_i, v_l) + d(v_j, v_k)]/2$ is a lower bound of $G(x)$ on $[v_i, v_j]$. Consequently, if $\bar{r}_a$ is the best known upper bound on r_a and if $m_{ij} > \bar{r}_a$ then the arc $[v_i, v_j]$ contains no absolute centers.

The center set is readily obtained from the matrix of distances between vertices: the radius r is the minimum of the maxima in all columns and the center set is composed of those vertices associated with no entry larger than r. Thus $O(|V|^2)$ operations are required. Notice that, unlike the median problem, the solutions of the center problem may not be solutions to the absolute center problem.

Specialized algorithms have been proposed for $(N/V/1/T/\mu)$ by Goldman [32], Dearing and Francis [19], Halfin [40], Handler [48] and Hedetniemi, Cockayne and Hedetniemi [60]. Handler's is as follows:

Algorithm 4.2. (Handler [48]).
(i) Select a vertex v_i of T and determine the vertex v_k farthest from v_i.
(ii) Determine the vertex v_l farthest from v_k and note $P(v_k, v_l)$ the path joining v_k and v_l.
(iii) The middle point of $P(v_k, v_l)$ is an absolute center of T.

As only two traversals of T are required, the complexity of this algorithm is $O(|V|)$. Because of the convexity of the distance on a tree (Theorem 2.2), the center of a tree is an endpoint of the arc containing an absolute center. Thus, Algorithm 4.2 also solves $(V/V/1/T/\mu)$.

4.2. *The continuous and general center problems:* $(N/N/1/N/\mu)$ *and* $(V/N/1/N/\mu)$

A point $c_c \in N$ is a *continuous center* iff

$$\bar{G}(c_c) = \underset{[v_i, v_j] \in A}{\text{Max}} \bar{d}(c_c, [v_i, v_j]) \leq \bar{G}(x) = \underset{[v_i, v_j] \in A}{\text{Max}} \bar{d}(x, [v_i, v_j]), \quad \forall x \in N.$$

The set of continuous centers, called the *continuous center set*, is represented by C_c. The value of $\bar{G}$ at any point of C_c is called the *continuous radius* r_c of the network. When c_c and x are restricted to be in V, one defines a *general center*, the *general center set* and the *general radius* r_g of the network.

Frank [27] and Minieka [74] have sketched algorithms for the continuous center. The principle is the same as for Algorithm 4.1: it consists of, first determining the upper envelope of the family of hat functions $\bar{d}(x, [v_k, v_l])$ for each arc $[v_i, v_j] \in A$ relative to all other $[v_k, v_l] \in A$, second finding the local minima of these envelope functions for each arc $[v_i, v_j]$ and, third selecting the best among the local minima. It immediately follows from Theorem 2.5 that $\bar{d}(x, [v_k, v_l])$ is continuous along $[v_i, v_j]$ with at most $2|A| + 2$ break points. Furthermore, the set of local minima may contain subarcs. Hansen, Labbé and Nicolas [54] have proposed an algorithm similar to Kariv and Hakimi's one for the weighted version of the absolute center problem (see 4.4.1). The principle is as follows. For each arc, one ranks the hat functions in order of increasing values of their highest points and builds their upper envelope by adding one hat function at a time: at each stage at most two new break points are created which can be located with respect to existing ones in $O(\log|A|)$ time when a 2-3 tree is used. The overall complexity of the algorithm is $O(|A|^2 \log|A|)$. Finally, different rules allowing for the elimination of arcs are presented in Labbé [66].

Solving $(V/N/1/N/\mu)$ is straightforward. First, one computes the remoteness of each vertex relative to all arcs, which can be easily done from the matrix of distances between vertices. Then, for each vertex we retain the largest value and the vertices for which the latter value is smallest. This can be achieved in $O(|A||V|)$ time, as noted by Minieka [76].

Finally, let us notice that because of the absence of cycles in a tree, the problems $(N/N/1/T/\mu)$ and $(V/N/1/T/\mu)$ have the same solutions as $(N/V/1/T/\mu)$ and $(V/V/1/T/\mu)$ respectively.

4.3. Comparison of radii and diameters

Let us now define various maximum distances between pairs of points in N: the *diameter* of the network

$$\delta = \underset{v_i, v_j \in V}{\text{Max}} d(v_i, v_j);$$

the *absolute diameter*

$$\delta_a = \underset{v_i \in V, y \in N}{\text{Max}} d(v_i, y);$$

and the *continuous diameter*

$$\delta_c = \underset{x, y \in N}{\text{Max}} d(x, y).$$

To determine δ_a, all pairs made of one vertex $v_i \in V$ and one arc $[v_k, v_l] \in A$ are considered in turn and the remoteness $\overline{d}(x, [v_k, v_l])$ is computed. The maximum of these values is equal to δ_a and is obtained in $O(|A||V|)$ operations. Chen and Garfinkel [13] describe an algorithm for computing δ_c in $O(|A|^2)$ time. It relies on the property that there always exists a pair of points among the vertices and the bottleneck points which determine δ_c. Finally, δ can be obtained by comparing distances between vertices in $O(|V|^2)$ operations.

The ratios of the radii and diameters of a network can be used in various tests for accelerating the algorithms presented above. Upper bounds are given in Table 4.1 (Hansen, Labbé and Nicolas [54]) and the examples of Figure 4.2 show that they are best possible.

Table 4.1. Bounds on the ratios of radii and diameters.

	r	r_a	r_g	r_c	δ	δ_a	δ_c
r	1	2	1	2	1	1	1
r_a	1	1	1	1	1	1	1
r_g	∞	∞	1	2	∞	1	1
r_c	∞	∞	1	1	∞	1	1
δ	2	2	2	2	1	1	1
δ_a	∞	∞	2	2	∞	2	1
δ_c	∞	∞	2	2	∞	2	1

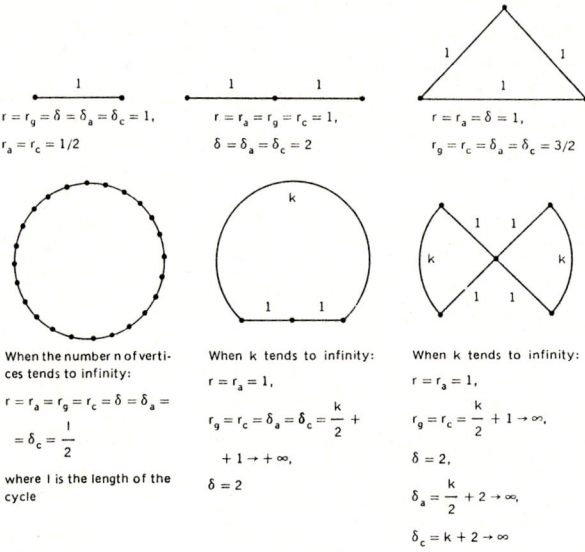

Fig. 4.2. Networks for which the bounds in Table 4.1 are the best possible.

4.4. Extensions

(i) When distances are weighted by positive constants w_i, the problem is to minimize the maximum weighted distance to the vertices of the network. A point $x^* \in N$ is an *absolute weighted center* iff

$$G_w(x^*) = \underset{i=1,\ldots,n}{\text{Max}} \, w_i d(v_i, x^*) \leq G_w(x) = \underset{i=1,\ldots,n}{\text{Max}} \, w_i d(v_i, x), \qquad \forall x \in N.$$

The minimal value of $G_w(x)$ is called the *absolute weighted radius* and is denoted by r_w.

Algorithm 4.1 can be applied for solving the above problem since $G_w(x)$ is still continuous and piecewise linear along each arc $[v_i, v_j] \in A$. However, the local minima of $G_w(x)$ now correspond to the points of $[v_i, v_j]$ which are equidistant from two vertices for the weighted distance. Furthermore, the slopes of $G_w(x)$ are no longer given by l_{ij} and $-l_{ij}$ but depend on the value of the weights. This complicates the search for the local minima and increases the complexity of the procedure. Hakimi, Schmeichel and Pierce [39] have implemented Algorithm 4.1 in $O(|A||V|^2 \log|V|)$ time with stacks as data structure, while Kariv and Hakimi [63] have obtained a complexity of $O(|A||V| \log|V|)$ using a 2-3 tree.

An entirely different approach has been proposed by Christofides [14]. Let $N(v_i, R)$ be the set of points of N whose weighted distance to v_i does not exceed some given number R. The weighted absolute radius is the smallest R such that $\bigcap_{i=1}^{n} N(v_i, R)$ is not empty. Chirstofides then suggests an iterative method which starts with R equal to a lower bound on r_w, computes the intersection $\bigcap_{i=1}^{n} N(v_i, R)$, increases R by a small amount if the intersection is empty and stops otherwise. A lower bound on r_w has been given by Dearing and Francis [19]:

$$\underset{i,j=1,\ldots,n}{\text{Max}} \; [w_i w_j / (w_i + w_j)] \, d(v_i, v_j).$$

Specific results for trees are due to Goldman [32], Dearing and Francis [19], Hakimi, Schmeichel and Pierce [39], and Kariv and Hakimi [63].

A second modification of the absolute center problem deals with the addition of positive constants a_i (addends) to the distance to $v_i \in V$. A point $x^* \in N$ is then said to be an *absolute center with addends* iff

$$G_a(x^*) = \underset{i=1,\ldots,n}{\text{Max}} \; [a_i + d(v_i, x^*)] \leq G_a(x) =$$

$$= \underset{i=1,\ldots,n}{\text{Max}} \; [a_i + d(v_i, x)], \qquad \forall x \in N.$$

This new problem can be solved by straightforward modifications of the procedures described in 4.1. In particular, the absolute centers with addends of the network N are equal to the absolute centers of the expanded network N' obtained from N by adding $|V|$ vertices v_i' and $|V|$ arcs $[v_i, v_i']$ whose length is equal to a_i.

In the case of a tree, Halfin [40] and Lin [69] have adapted Goldman [32] and Handler [48] procedures to solve the absolute center problem with addends in $O(|V|)$ operations. Mitchell [82] and Hedetniemi, Cockayne and Hedetniemi [60] have devised specific linear algorithms.

Finally, the case of nonlinear functions of distance has been studied by Dearing [18] and Francis [24]. Given n strictly increasing and continuous functions $f_1 \ldots f_n$, a point $x^* \in N$ is called a *nonlinear absolute center* iff

$$G_n(x^*) = \underset{i=1,\ldots,n}{\text{Max}} \; f_i[d(v_i, x^*)] \leq G_n(x) =$$

$$= \underset{i=1,\ldots,n}{\text{Max}} \; f_i[d(v_i, x)], \qquad \forall x \in N.$$

Christofides' iterative method can be easily adapted to find a good approximate solution (see also Hansen, Peeters, Richard and Thisse [55] for a similar approach in continuous location theory).

(ii) The counterpart of the conditional median is the conditional center. Let $L \subset N$ be the locations of the existing facilities. A point $x^* \in N$ is a *conditional absolute center* iff

$$G(L \cup \{x^*\}) = \underset{i=1,\ldots,n}{\text{Max}} \underset{y \in L \cup \{x^*\}}{\text{Min}} d(v_i, y)$$

$$\leqslant G(L \cup \{x\}) = \underset{i=1,\ldots,n}{\text{Max}} \underset{y \in L \cup \{x\}}{\text{Min}} d(v_i, y), \quad \forall x \in N.$$

Again, as $G(L \cup \{x\})$ is continuous and piecewise linear on each arc, the procedures described in 4.1 are readily adapted for finding the conditional absolute center (Minieka [75]). The same observation applies to the conditional continuous center.

(iii) A partial counterpart of the absolute $\mathscr{S}$-median is the $\mathscr{S}$-center. Let $\mathscr{S} = \{S_1, \ldots, S_m\}$ be a family of subsets of N and let $V_k = V \cap S_k$. A point $x^* \in V$ is called a $\mathscr{S}$-*center* iff

$$G(x^*) = \underset{k=1,\ldots,m}{\text{Max}} \underset{v_i \in V_k}{\text{Min}} d(v_i, x^*) \leqslant G(x) =$$

$$= \underset{k=1,\ldots,m}{\text{Max}} \underset{v_i \in V_k}{\text{Min}} d(v_i, x), \qquad \forall x \in V.$$

This problem is easily solved by inspection of the matrix of distances between vertices. Furthermore, the following characterization of the $\mathscr{S}$-center holds for a tree.

Theorem 4.3. (Slater [91]). *Let T be a tree and $\mathscr{S} = \{S_1, \ldots, S_m\}$ a family of subtrees of T.*

(1) If $\bigcap_{k=1}^{m} V_k \neq \phi$, then $\bigcap_{k=1}^{m} V_k$ is the set of $\mathscr{S}$-centers and is the vertex set of a subtree of T.

(2) If $\bigcap_{k=1}^{m} V_k = \phi$, then the set of $\mathscr{S}$-centers of T is equal to the center set of a subtree of T and, hence, consists of either one or two adjacent vertices.

(iv) We now consider two stochastic extensions of the absolute center problem. Let W_i be a nonnegative random variable associated with $v_i \in V$. Frank [25] has proposed the solution-concept of *absolute expected center*, defined as a point $x^* \in N$ such that

$$\underset{i=1,\ldots,n}{\text{Max}} E(W_i) d(v_i, x^*) \leqslant \underset{i=1,\ldots,n}{\text{Max}} E(W_i) d(v_i, x), \quad \forall x \in N.$$

Obviously, any method that solves the weighted absolute center, solves the absolute expected center as well. A second solution-concept, also considered by Frank, is a point of the network which, given some positive number q, minimizes the probability that the maximum weighted distance is larger than or equal to q. Finding such a point is usually hard, but when the variables W_i are discrete and independent, there exists an interval of local solutions on each arc and the problem can be solved by an Hakimi-like method. Frank [26] also discusses the case of joint multivariate normal distributions.

(v) Lastly, the case of an obnoxious facility has been addressed by Minieka [79]. The objective is to maximize the distance to the nearest vertex — the *maximin problem*. It is then easy to see that an optimal solution must be an equidistant point of the network.

5. Multi-criteria and multi-agents problems

5.1. Median-center problems

Comparing median and center problems, we can see that the former take into account the whole set of clients served by the facility, while the later deal only with the worst-off client. For this reason, median problems are generally seen as maximizing efficiency, and center problems as maximizing a criterion of distributional justice in the sense of Rawls [84]. Medians and centers do not generally coincide, hence efficiency and justice are often antagonistic goals, and it may be desirable to choose a solution which is a compromise between them. There exist a number of ways to express this compromise, reviewed in the rest of the subsection.

As noticed in Section 2, weights can be given to the vertices in both the median and center problems. It is easier to consider here the set U of clients instead. Let $F_U(x) = \Sigma_{u \in U} d(v(u), x)$ and $G_U(x) = \underset{u \in U}{\text{Max}}\, d(v(u), x)$ be the corresponding median and center functions; m and c denote respectively an absolute median and absolute center.

(i) Handler [49] has considered a set of points including the efficient points for the bicriterion problem defined by the simultaneous minimization of F_U and G_U. An α-*constrained medi-center* of N, denoted $m^*(\alpha)$, is defined as any solution of the problem

$$F_\alpha^* = \text{Min}\{F_U(x); G_U(x) \leq \alpha \text{ and } x \in N\}.$$

Obviously, the problem is feasible and non-trivial iff $\underline{\alpha} \leq \alpha \leq \overline{\alpha}$, where $\underline{\alpha} = G_U(c)$ and $\overline{\alpha} = G_U(m)$. We may formulate and associated problem, whose solution is denoted by $c^*(\beta)$, by reversing the role of the median and center objectives, i.e.,

$$G_\beta^* = \text{Min}\{G_U(x); F_U(x) \leq \beta \text{ and } x \in N\}.$$

The relevant range for β is the interval $[\underline{\beta}, \overline{\beta}]$, where $\underline{\beta} = F_U(m)$ and $\overline{\beta} = F_U(c)$.

A duality relationship between the two problems is as follows

Theorem 5.1. (Halpern [44]).

(a) *If* $\alpha \in [\underline{\alpha}, \overline{\alpha}]$ *and* $F_\alpha^* = \beta$, *then* $F_{G_\beta^*}^* = \beta$; *if* $\beta \in [\underline{\beta}, \overline{\beta}]$ *and* $G_\beta^* = \alpha$, *then* $G_{F_\alpha^*}^* = \alpha$.

(b) *When N is a tree, F_α^* is a one-one mapping from $[\underline{\alpha}, \overline{\alpha}]$ into $[\underline{\beta}, \overline{\beta}]$, G_β^* is a one-one mapping from $[\underline{\beta}, \overline{\beta}]$ into $[\underline{\alpha}, \overline{\alpha}]$, and $F_\alpha^* = (G_\beta^*)^{-1}$.*

Notice that the second part of the theorem does not hold for a general network and that the set of α-constrained medi-centers is equal to the set of efficient points for Min$\{F_U(x), G_U(x)\}$ when the network is a tree.

(ii) Halpern [41, 42, 44] has studied extensively the convex combination of the median and center functions. A point $x \in N$ is a λ-*cent-dian* ($0 \le \lambda \le 1$) iff

$$H_\lambda(x^*) = \lambda\, G_U(x^*) + (1-\lambda)\, \frac{F_U(x^*)}{|U|} \le H_\lambda(x) =$$

$$= \lambda\, G_U(x) + (1-\lambda)\, \frac{F_U(x)}{|U|}, \qquad \forall x \in N.$$

Hence $H_\lambda(x)$ is a weighted mean between the maximum distance and the average distance to the clients. The set of λ-cent-dians is a subset of the set of efficient points for Min$\{F_U(x), G_U(x)\}$.

The main properties of $H_\lambda(x)$ are as follows.

Theorem 5.2. (Halpern [42]). *For $x = f_{ij}(\theta) \in [v_i, v_j] \in A$ and a given value of $\lambda \in [0, 1]$, $H_\lambda(f_{ij}(\theta))$ is continuous and piecewise linear with a finite number of break points corresponding to the bottleneck points and the local minima of $G_U(x)$ on $[v_i, v_j]$.*

This result suggests an algorithm to determine all λ-cent-dians, which is simpler than that proposed by Halpern [42], although both could be implemented with the same complexity.

Algorithm 5.1.

(i) For each arc $[v_i, v_j] \in A$, determine the image I_{ij} of $[v_i, v_j]$ in the plane with coordinates $F_U(x)/|U|$ and $G_U(x)$. To this effect, (ia) find the set of bottleneck points on $[v_i, v_j]$; (ib) determine with Kariv and Hakimi's algorithm the set of local minima of $G_U(x)$; (ic) sort together the two lists so-obtained by increasing values of θ; (id) compute the values of $F_U(x)/|U|$ and $G_U(x)$ at all points in the list of (ic) and joint successive points by straight line segments.

(ii) Find the convex hull $\mathcal{H}$ of $\cup I_{ij}$, e.g. with Eddy [22]'s algorithm. The set of all λ-cent-dians is given by the extreme points and sides of the south--west boundary of $\mathcal{H}$ which coincide with extreme points and sides of some I_{ij}.

This algorithm has a complexity of $O(|A||V|\log|A||V|)$. It can be modified to determine all efficient points for Min $(F_U(x), G_U(x))$. Step (i) is unchanged;

step (ii) is replaced by a pairwise comparison of the linear segments of all I_{ij} to eliminate dominated points. This requires $O(|A|^2|V|^2)$ operations.

In the case of a tree the λ-cent-dian problem is easier to solve.

Theorem 5.3. (Halpern [41]). *The λ-cent-dians of a tree are all points on the path joining the absolute center and the closest median.*

Thus the cent-dian set can be found in $O(|V|)$ time. In addition, it is equal to the efficient point set for the bicriterion problem Min $(F_U(x), G_U(x))$.

(iii) Slater [88] introduces another concept that generalizes the median and center criteria. The *k-centrum* of a network, denoted by $C(N; k)$, is the subset of vertices such that the sum of the distances to the k farthest vertices is minimum; i.e., the set of solutions to the problem:

$$\min_{x \in V} \left\{ \sum_{v_i \in S} d(v_i, x); S \subset V, |S| = k \text{ and } d(v_i, x) \geq d(v_j, y), \forall x \in S \text{ and } \forall y \in V - S \right\}.$$

Of course, $C(N; |V|)$ is the median set, and $C(N; 1)$ is the center set. Slater [88, 89] investigates the structure of the k-centrum of a tree with unit arc length; in particular, he proves that $C(T; k)$ consists of either one or two adjacent vertices and that $C(T; k) \cap C(T, k+1) \neq \phi$. Andreatta and Mason [2] consider a general tree and show that $C(T; k)$ is a path, with $|C(T; k)| \leq 2$ if k is odd. They obtain some relationship between k-centrum and $(k + 1)$-centrum:

(i) if $C(T; k) \cap C(T; k+1) = \phi$, then $C(T; k)$ and $C(T; k+1)$ contain adjacent vertices;

(ii) $C(T; k) \cap C(T; k+1)$ induces an elementary path in T.

The same authors address the *absolute k-centrum* problem on a tree, and prove that it is connected and contained in a path (Andreatta and Mason [3]).

(iv) For completeness, let us also mention that another way of generalizing the concepts of center and median has been proposed by Slater [91]: the *k-nucleus* K_k is the set of vertices for which the sum of the distances to the balls of radius k around each other vertex is minimum; K_0 is the median set and K_r the center set.

5.2. Multi-criteria problems

A few papers only deal with location problems involving more than two criteria.

(i) The problem of finding the set of efficient points on a tree with regard to several objectives has been tackled by Lowe [72]. Let $f_1, \ldots, f_n$ be n convex functions defined on T and let $Q_i \subset T$ be the feasible domain for f_i. It is assumed that $Q = \cap_{i=1}^n Q_i$ is nonempty, compact and connected. The sets of optimal solutions to the problems $\underset{x \in T}{\text{Min}} \, f_i(x)$ and $\underset{x \in Q}{\text{Min}} \, f_i(x)$ are denoted by R_i^* and S_i^* respectively. The convexity of f_i implies that either $S_i^* = Q \cap R_i^*$ or S_i^* is the unique closest point in Q to R_i^* when $Q \cap R_i^* = \phi$. Finally, let $T^* = \cap_{i=1}^n S_i^*$ when this intersection is nonempty, and otherwise let T^* be the unique subset S of Q such that (a) $S \cap S_i^* \neq \phi$ for all i and (b) each extreme point s of S satisfies $\{s\} = S \cap S_i^*$ for at least one i.

Theorem 5.4. (Lowe [72]). *The set of efficient points for the multiobjective problem* Min $\{f_1(x), \ldots, f_n(x); x \in Q\}$ *is equal to* T^*.

Notice that since $F_U(x)$ and $G_U(x)$ are convex, Theorem 5.3 is a special case of Theorem 5.4. Lowe also provides an algorithm to find T^*.

(ii) Hansen, Thisse and Wendell [58] determine the efficient points for the simultaneous minimization of the distances to the clients. For a general network, the algorithm involves a comparison of the distance functions along pairs of segments to eliminate dominated points. Its complexity is $O(|U|^2 |A|^2 \max(|U|, \log|U||V|))$.

(iii) Halpern and Maimon [46] compare four objectives on a tree: the median function, the center function, the variance measure and the Lorenz measure. The *variance measure* at $x \in T$ is

$$V(x) = \sum_{u \in U} \left[d(v(u), x) - \frac{F_U(x)}{|U|} \right]^2.$$

The lowest possible value of $V(x)$ is zero and is obtained when all clients are equally distant to x. Any point where $V(x)$ is minimized is called a *variance point*. Halpern and Maimon [47] provide a linear algorithm to determine such a point on a tree. Let now $f(p|x)$ be the fraction of the total distance of x corresponding to the p nearest clients. The *Lorenz measure* is defined by

$$L(x) = \frac{2}{|U|} \int_0^{|U|} f(p|x) \, dp.$$

Clearly, $L(x)$ belongs to $[0, 1]$, the upper bound being reached when all clients are equidistant from x. The point l which maximizes $L(x)$ is called a *Lorenz point*. Halpern and Maimon [45] give a polynomial algorithm to find l on a tree. Both the variance and Lorenz points aim at minimizing dispersion in distances between clients and the facility.

Halpern and Maimon [46] define a certain number of divergence measures and perform a series of simulations for trees. Their major findings are as follows: (a) l is generally quite separated from m, c and s; (b) selecting l for locating the facility entails relatively high losses in all other three criteria, and selecting m, c or s

causes high losses in terms of the Lorenz measure; (c) relatively low losses in the median function occurs when c or s is chosen; (d) but, when m is chosen, moderate losses in the center function and variance measure are observed.

(iv) Finally, a general discussion of the multi-criteria approach in network location theory is given by Vincke [97].

5.3. Voting location problems

The set of efficient points for the simultaneous minimization of the distances to the clients contains, in general, more than one point. Hence there is no way to meet the wishes of all clients simultaneously by choosing a single location point. Consequently, the facility location must result from a compromise among clients. One possible institutional mechanism is to choose it by means of a *voting rule*. This motivates a new class of location problems, i.e., voting location problems. A natural solution-concept to such problems is a point $x^* \in N$, called a *Condorcet point*, for which no other point of the network is closer to a strict majority of clients:

$$|\{u \in U; d(v(u), x) < d(v(u), x^*)\}| \leq \frac{|U|}{2}, \quad \forall x \in N.$$

The following questions then suggest themselves: (i) does a Condorcet point exist and belong to the set of vertices? (ii) when do a planning solution (e.g., an absolute median) and a Condorcet point coincide? (iii) how much does the planning criterion (e.g., minimizing the weighted sum of distances) deteriorate when a Condorcet point is adopted and how many clients are dissatisfied by a location optimizing the planning criterion?

First, a Condorcet point may not exist. To see this, consider the network depicted in Figure 5.1. It is easy to check that any interior point of an arc is defeated by its nearest vertex and any vertex is beaten by an interior point of its opposite arc. Different conditions on clients' locations to guarantee the existence of a Condorcet point are discussed in Wendell and McKelvey [100], and Hansen, Thisse and Wendell [57, 59]. They include the assumption of Theorems 3.3 and 3.5. However, the question of what such conditions are

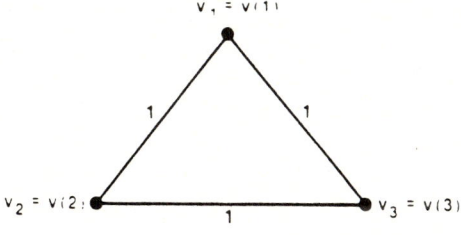

Fig. 5.1.

necessary for a Condorcet point to exist remains open. On the other hand, the class of networks on which there exists a Condorcet point for *any* distribution of clients has been characterized by Bandelt [4].

Furthermore, even if it exists, a Condorcet point does not necessarily belong to V as shown by the example of Figure 5.2 due to Wendell and McKelvey [100], and Hakimi [37]: Condorcet points exist and correspond to the subarc $[x_1, x_2]$ of $[v_2, v_5]$. However, when $|U|$ is odd, all the Condorcet points (if any) belong to V (Hansen and Thisse [56]).

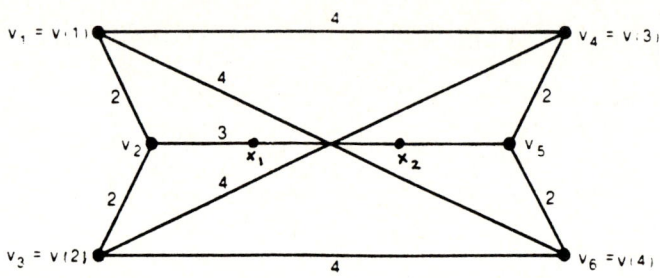

Fig. 5.2.

Second, Hansen and Thisse [56] have shown that the sets of Condorcet points and of absolute medians are equivalent on trees. A characterization of the networks for which this property holds has been obtained recently by Bandelt:

Theorem 5.5. (Bandelt [4]). *The sets of Condorcet points and of absolute medians on a network coincide iff*
 (i) *for any three vertices v_i, v_j and v_k there exists a unique $v_l \in V$ (with possibly $v_l \in \{v_i, v_j, v_k\}$) which is simultaneously between v_i and v_j, v_j and v_k, v_k and v_i;*
 (ii) *for any vertex v and any set $\overline{V}$ of vertices, $\cap\{B(v, \overline{v}); \overline{v} \in \overline{V}\} = \{v\}$ implies that $B(v, \overline{v}_1) \cap B(v, \overline{v}_2) = \{v\}$ for some $\overline{v}_1 \in \overline{V}$ and $\overline{v}_2 \in \overline{V}$.*

Third, and last, if the facility is established at a Condorcet point x^* the ratio $F(x^*)/F(m)$ cannot exceed 3. This result of Hansen and Thisse [56] has been refined by Labbé ($\lceil|U|/2\rceil$ denotes the smallest integer not less that $|U|/2$).

Theorem 5.6. (Labbé [66]). *Let x^* be a Condorcet point and m an absolute median. Then*

$$F(x^*)/F(m) \leq \frac{2|U| - \left(\left\lceil\dfrac{|U|}{2}\right\rceil + 1\right)}{\left\lceil\dfrac{|U|}{2}\right\rceil + 1}$$

Furthermore, the upper bound is the best possible.

Thus, locating the facility at a Condorcet point may lead to an important increase in total distance. Conversely, choosing an absolute median may dissatisfy almost all the clients. To see it, consider the example in Figure 5.3 due to Bandelt and Labbé [5]. Assuming that one client is located at v_1 and $m \geq 2$ clients at v_2 and v_3 respectively, we easily see that v_1 is the unique absolute median when $\epsilon < 1/m$. But then, $2m = |U| - 1$ clients are closer to x_{23}, the middle point of $[v_2, v_3]$, than to v_1.

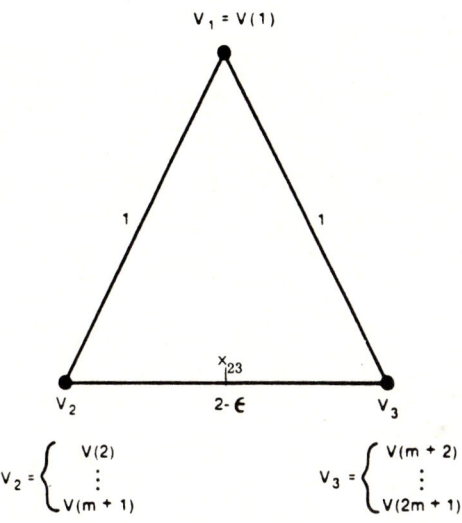

Fig. 5.3.

Additional results on the comparison between alternative voting and planning solutions (including the absolute center) can be found in Bandelt and Labbé [5], Hansen and Thisse [56], and Hansen, Thisse and Wendell [59].

Polynomial algorithms to find a Condorcet point, when $|U|$ is odd or even, are proposed in Hansen and Labbé [52]. The former case is easier to solve as all Condorcet points (if any) are at vertices. The algorithm involves, first, pairwise comparisons of vertices to eliminate those for which a majority of clients prefers another of them and, then, comparisons of vertices and points on arcs; the complexity is $O(|A||V||U|\log|U|)$. The latter case requires, in addition, pairwise comparisons of points on arcs and the complexity is $O(|A|^2|U|^3 \log|A||U|)$.

Finally, related solution-concepts applicable to the location of an obnoxious facility are studied by Labbé [64]. In particular, it is shown that, for a tree, a Condorcet point exists and belongs to the set of pendant vertices.

5.4. Competitive location problems

A competitive location problem arises when the number of clients of a facility depends not only upon its own location decision but also upon the locations chosen by *competing* facilities. This corresponds to a family of problems which are formulated within the framework of noncooperative game theory. Assuming that clients patronize the nearest facility, a natural solution to these problems is a locational pattern for the facilities such that given the locations of the others, no facility can be closer to more clients at an alternate location. To illustrate, consider the case of two facilities. A pair of locations $x_1^*, x_2^* \in N$ is called a *Nash equilibrium* iff, for $i, j = 1, 2$ and $i \neq j$,

$$|\{u \in U; d(v(u), x_i^*) < d(v(u), x_j^*)\}| +$$

$$+ \frac{1}{2} |\{u \in U; d(v(u), x_i^*) = d(v(u), x_j^*)\}| \geq$$

$$\geq |\{u \in U; d(v(u), x_i) < d(v(u), x_j^*)\}| +$$

$$+ \frac{1}{2} |\{u \in U; d(v(u), x_i) = d(v(u), x_j^*)\}|, \qquad \forall x_i \in N.$$

It is then easy to see that there exists a Nash equilibrium for the 2-competing facility location problem iff there exists a Condorcet point for the corresponding voting location problem. Furthermore, when a Nash equilibrium exists, there is such an equilibrium where the two facilities are located at the same point and this point is also a Condorcet point (Hansen, Thisse, and Wendell [59]). Thus, the results discussed in 5.2 still hold for the Nash equilibrium with two competing facilities.

Introducing the arrow of time, we now assume that facilities are located *sequentially* (and no longer simultaneously). In the 2-facility case, the problem for the *first* facility is to find a location which retains the maximum number of clients after the second facility has been established with the aim of maximizing its own clientele (Slater [86] and Hakimi [37]). Such a point is called a *Simpson point* or a $(1|1)$-centroid. It can be shown to exist for all networks and to be a Condorcet point when such a point exists (Hansen, Thisse and Wendell [59]). Consequently, the set of vertices may not contain a Simpson point. The set of Simpson points can be obtained by embedding the algorithm for Condorcet points with $|U|$ even in a Fibonacci search on the number of clients closer to another point than to a Simpson point; the complexity of the corresponding algorithm is $O(|U|^4 |A|^2 \log |U| |A| \log |U|)$ (Hansen and Labbé [52]).

Finally, notice that the problem of the *second* facility is merely one of finding a point of maximum demand, as discussed in 3.4, in which the range R_i for clients at v_i is equal to the distance to the first facility.

6. Miscellaneous problems and conclusions

Three new families of network location problems have recently attracted the attention of researchers. They have not been considered in the main body of the survey because they are only partially connected with mainstream location theory and they rely upon other chapters of Operations Research. They are:

(i) *Location-routing problems*, in which locating one or several facilities and routing of vehicles from these facilities to the clients are to be simultaneously decided on. The interested reader is referred to the survey of Laporte, Nobert and Arpin [67];

(ii) *Balanced-flow location problems*, in which a facility must be set up in order to minimize the maximum flow in any arc of the network when satisfying clients' demand (Eiselt and Pederzoli [23]);

(iii) *Path-location problems*, in which the purpose is to determine a simple path such as to minimize the sum of distances (or the maximum) distance from the clients to this path (Slater [93]).

Let us conclude with some general remarks. First, in most problems, one can identify a finite subset of N containing a solution. This subset is often the set of vertices for median-type problems; the set of vertices and equidistant points for center-type problems. Second, for all problems, polynomial algorithms can be derived to find a solution or the set of all solutions. When the finiteness property holds for the problems considered, complexity is usually low. For trees linear algorithms are frequent. Third, and last, whereas the basic median and center problems are well-studied, there remains much work to be done to unify and streamline the theory, and to obtain lower complexity algorithms than those known as yet.

References

[1] A. Aho, A. Hopcroft and J. Ullman, *The Design and Analysis of Computer Algorithms*, Addison-Wesley, Reading, Mass., 1974
[2] G. Andreatta and F. Mason, «k-Eccentricity and Absolute k-Centrum for a Probabilistic Tree», *European Journal of Operational Research* 19, 114 - 117, 1985.
[3] G. Andreatta and F. Mason, «Properties of the k-Centra in a Tree Network», *Networks* 15, 21 - 29, 1985.
[4] H.J. Bandelt, «Networks with Condorcet Solutions», *European Journal of Operational Research* 20, 314 - 326, 1985.
[5] H.J. Bandelt and M. Labbé, «How Bad Can a Voting Location Be?», *Social Choice and Welfare* (to appear).
[6] J.P. Barthélemy, «Médiane dans les graphes et localisation», *Cahiers du Centre d'Etudes de Recherche Opérationnelle* 25, 163 - 182, 1983.
[7] J.P. Barthélemy, «Caractérisation médiane des arbres», *Annals of Discrete Mathematics* 17, 39 - 46, 1983.
[8] C. Berge, *Espaces topologiques. Fonctions multivoques*, Dunod, Paris, 1966.
[9] O. Berman, S. Chiu, R.C. Larson and A.R. Odoni, «Location on Congested Networks», in R.L. Francis and P.B. Mirchandani, eds., *Discrete Location Theory*, Wiley, New York (to appear).

[10] O. Berman and R.C. Larson, «The Median Problem with Congestion», *Computers and Operations Research* 9, 119 - 126, 1982.
[11] O. Berman, R.C. Larson and S. Chiu, «Optimal Server Location on a Network Operating as an $M/G/1$ Queue», *Operations Research* 33, 746 - 771, 1985.
[12] O. Berman and A.R. Odoni, «Locating Mobile Servers on a Network with Markovian Properties», *Networks* 12, 73 - 86, 1982.
[13] C.E. Chen and R.S. Garfinkel, «The Generalized Diameter of a Graph», *Networks* 12, 335 - 340, 1982.
[14] N. Christofides, *Graph Theory: An Algorithmic Approach*, Academic Press, New York, 1975.
[15] R.L. Church and R.S. Garfinkel, «Locating an Obnoxious Facility on a Network», *Transportation Science* 12, 107 - 118, 1978.
[16] R.L. Church and M.E. Meadows, «Location Modelling Utilizing Maximum Service Distance Criteria», *Geographical Analysis* 11, 358 - 373, 1979.
[17] R.A. Cunninghame-Green, «The Absolute Center of a Graph», *Discrete Applied Mathematics* 7, 275 - 283, 1984.
[18] P.M. Dearing, «Minimax Location Problems with Nonlinear Costs», *Journal of Research of the National Bureau of Standards* B 82, 65 - 72, 1977.
[19] P.M. Dearing and R.L. Francis, «A Minimax Location Problem on a Network», *Transportation Science* 8, 333 - 343, 1974.
[20] P.M. Dearing, R.L. Francis and T.J. Lowe, «Convex Location Problems on Tree Networks», *Operations Research* 24, 628 - 642, 1976.
[21] W. Domschke and A. Drexl, *Location and Layout Planning: An International Bibliography*, Springer-Verlag, Berlin, 1985.
[22] W.F. Eddy, «A New Convex Hull Algorithm for Planar Sets», *ACM Transactions on Mathematical Sofware* 3, 398 - 403, 1977.
[23] H.A. Eiselt and G. Pederzoli, «A Location Problem in Graphs», *New Zealand Journal of Operational Research* 12, 49 - 53, 1984.
[24] R.L. Francis, «A Note on a Nonlinear Minimax Location Problem in Tree Networks», *Journal of Research of the National Bureau of Standards* B 82, 73 - 80, 1977.
[25] H. Frank, «Optimum Locations on a Graph with Probabilistic Demands», *Operations Research* 14, 409 - 421, 1966.
[26] H. Frank, «Optimum Locations on a Graph with Correlated Normal Demands», *Operations Reseach* 15, 552 - 557, 1967.
[27] H. Frank, «A Note on a Graph Theoretic Game of Hakimi's», *Operations Research* 15, 567 - 570, 1967.
[28] M. Garey and D. Johnson, *Computers and Intractability: A Guide to the Theory of NP-Completeness*, Freeman, San Francisco, 1979.
[29] A.J. Goldman, «Optimum Locations for Centers in a Network», *Transportation Science* 3, 352 - 360, 1969.
[30] A.J. Goldman, «Optimal Center Location in Simple Networks», *Transportation Science* 5, 212 - 221, 1971.
[31] A.J. Goldman, «Approximate Localization Theorems for Optimal Facility Placement», *Transportation Science* 6, 195 - 201, 1972.
[32] A.J. Goldman, «Minimax Location of a Facility in a Network», *Transportation Science* 6, 407 - 418, 1972.
[33] A.J. Goldman and C.J. Witzgall, «A Localization Theorem for Optimal Facility Placement», *Transportation Science* 4, 406 - 409, 1970.
[34] H. Guelicher, «Einige Eigenschaften optimaler Standorte in Verkehrsnetzen», *Schriften des Vereins für Sozialpolitik, Neue Folge* 42, 111 - 137, 1965.
[35] S.L. Hakimi, «Optimum Locations of Switching Centers and the Absolute Centers and Medians of a Graph», *Operations Research* 12, 450 - 459, 1964.
[36] S.L. Hakimi, «Optimum Distribution of Switching Centers in a Communication Network and Some Related Graph Theoretic Problems», *Operations Research* 13, 462 - 475, 1965.
[37] S.L. Hakimi, «On Locating New Facilities in a Competitive Environment», *European Journal of Operational Research* 12, 29 - 35, 1983.
[38] S.L. Hakimi and S.N. Maheshwari, «Optimum Locations of Centers in Networks», *Operations Research* 20, 967 - 973, 1972.
[39] S.L. Hakimi, E.F. Schmeichel and J.G. Pierce, «On p-Centers in Networks», *Transportation Science* 12, 1 - 15, 1978.
[40] S. Halfin, «On Finding the Absolute and Vertex Center of a Tree with Distances», *Transportation*

Science 8, 75 - 77, 1974.
[41] J. Halpern, «The Location of a Cent-Dian Convex Combination on an Undirected Tree», *Journal of Regional Science* 16, 237 - 245, 1976.
[42] J. Halpern, «Finding Minimal Center-Median Convex Combinations (Cent-Dian) of a Graph», *Management Science* 24, 534 - 544, 1978.
[43] J. Halpern, «A Simple Edge Elimination Criterion in a Search for the Center of a Graph», *Management Science* 25, 105 - 107, 1979.
[44] J. Halpern, «Duality in the Cent-Dian of a Graph», *Operations Research* 28, 722 - 735, 1980.
[45] J. Halpern and O. Maimon, «Equity Measures in Locational Decisions on Trees», *Mimeograph Series in Operations Research, Statistics and Economics* 254, Technion, Haifa, 1980.
[46] J. Halpern and O. Maimon, «Accord and Conflict Among Several Objectives in Locational Decisions on Tree Networks», in J.-F. Thisse and H.G. Zoller, eds., *Locational Analysis of Public Facilities*, North-Holland, Amsterdam, 301 - 314, 1983.
[47] J. Halpern and O. Maimon, «Equity Measures in Locational Decision Problems on Trees», *Operations Research*, (to appear).
[48] G.Y. Handler, «Minimax Location of a Facility in an Undirected Tree Graph», *Transportation Science* 7, 287 - 293, 1973.
[49] G.Y. Handler, «Medi-Centers of a Tree», *Working Paper No. 278/76, Faculty of Management, Tel-Aviv University*, 1976.
[50] G.Y. Handler and P. Mirchandani, *Location on Networks: Theory and Applications*, MIT Press, Campbridge, Mass., 1979.
[51] P. Hanjoul and J.-F. Thisse, «The Location of a Firm on a Network», in A.J. Hughes Hallet, ed., *Applied Decision Analysis and Economic Behavior*, Martinus Nijhoff, Dordrecht, 289 - 326, 1984.
[52] P. Hansen and M. Labbé, «Algorithms for Condorcet and Simpson Points on a Network», submitted for publication, 1984.
[53] P. Hansen and M. Labbé, «The Continuous p-Median of a Network», submitted for publication, 1985.
[54] P. Hansen, M. Labbé and B. Nicolas, «The Continuous Center Set of a Network», submitted for publication, 1985.
[55] P. Hansen, D. Peeters, D. Richard and J.-F. Thisse, «The Minisum and Minimax Location Problems Revisited», *Operations Research* 33, 1251 - 1265, 1985.
[56] P. Hansen and J.-F. Thisse, «Outcomes of Voting and Planning: Condorcet, Weber and Rawls Locations», *Journal of Public Economics* 16, 1 - 15, 1981.
[57] P. Hansen, J.-F. Thisse and R.E. Wendell, «Equivalence of Solutions to Network Location Problems», *Mathematics of Operations Research* (to appear).
[58] P. Hansen, J.-F. Thisse and R.E. Wendell, «Efficient Points on a Network», *Networks*, (to appear).
[59] P. Hansen, J.-F. Thisse and R.E. Wendell, «Location by Competitive and Voting Processes», in R.L. Francis and P.B. Mirchandani, eds., *Discrete Location Theory*, Wiley, New York, (to appear).
[60] S.M. Hedetniemi, E.J. Cockayne and S.T. Hedetniemi, «Linear Algorithms for Finding the Jordan Center and Path Center of a Tree», *Transportation Science* 15, 98 - 114, 1981.
[61] A.P. Hurter and J.S. Martinich, «Networks Production-Location Problems under Price Uncertainty», *European Journal of Operational Research* 16, 183 - 197, 1984.
[62] C. Jordan, «Sur les assemblages de lignes», *Zeitschrift fur die Reine und Angewandbte Mathematik* 70, 185 - 190, 1869. Reprinted in English translation in N. Biggs, E.K. Lloyd and R. Wilson, *Graph theory 1736 - 1936*, Oxford University Press, Oxford, 1976.
[63] O. Kariv and S.L. Hakimi, «An Algorithmic Approach to Network Location Problems I: The p-Centers», *SIAM Journal on Applied Mathematics* 37, 513 - 538, 1979.
[64] M. Labbé, «Equilibre de Condorcet pour le problème de localisation d'une installation polluante», *Cahiers du Centre d'Etudes de Recherche Opérationnelle* 24, 305 - 312, 1982.
[65] M. Labbé, *Essays in Network Location Theory*, Doctoral Dissertation, Universite Libre de Bruxelles. Published in *Cahiers du Centre d'Etudes de Recherche Opérationnelle* 27, 5 - 130, 1985.
[66] M. Labbé, «Outcomes of Voting and Planning in Single Facility Location Problems», *European Journal of Operational Research* 20, 299 - 313, 1985.
[67] G. Laporte, Y. Nobert and D. Arpin, «An Exact Algorithm for Solving a Capacitated Location--Routing Problem», in *Location Theory: Methodology and Applications*, J.C. Baltzer AG, Scientific Publishing Co., Basel, Switzerland, 1986 (to appear).
[68] J. Levy, «An Extended Theorem for Location on a Network», *Operational Research Quarterly* 18, 433 - 442, 1967.
[69] C.C. Lin, «On Vertex Addends in Minimax Location Problems», *Transportation Science* 9, 165 - 168, 1975.

[70] F. Louveaux, J.-F. Thisse and H. Beguin, «Location Theory and Transportation Costs», *Regional Science and Urban Economics* 12, 529 - 545, 1982.
[71] F. Louveaux and J.-F. Thisse, «Production and Location on a Network under Demand Uncertainty», *Operations Research Letters* 4, 145 - 151, 1985.
[72] T.J. Lowe, «Efficient Solutions in Multiobjective Tree Network Location Problems», *Transportation Science* 12, 298 - 316, 1978.
[73] E. Minieka, «The m-Center Problem», *SIAM Review* 12, 138 - 139, 1970.
[74] E. Minieka, «The Centers and Medians of a Graph», *Operations Research* 25, 641 - 650, 1977.
[75] E. Minieka, «Conditional Centers and Medians of a Graph», *Networks* 10, 265 - 272, 1980.
[76] E. Minieka, «A Polynomial Time Algorithm for Finding the Absolute Center of a Network», *Networks* 11, 351 - 355, 1981.
[77] E. Minieka, «Pendant-Medians», *Operations Research Letters* 2, 104 - 106, 1983.
[78] E. Minieka, «Radial Location Theory», *Networks* 13, 233 - 239, 1983.
[79] E. Minieka, «Anti-Centers and Anti-Medians of a Network», *Networks* 13, 359 - 365, 1983.
[80] P.B. Mirchandani, «Locational Decisions on Stochastic Networks», *Geographical Analysis* 12, 172 - 183, 1980.
[81] P.B. Mirchandani and A.R. Odoni, «Locations of Medians on Stochastic Networks», *Transportation Science* 13, 85 - 95, 1979.
[82] S.L. Mitchell, «Another Characterization of the Centroid of a Tree», *Discrete Mathematics* 24, 277 - 280, 1978.
[83] O. Ore, *Theory of graphs*, American Mathematical Society Colloquium Publications, Vol. XXXVIII, Providence, 1962.
[84] J. Rawls, *A Theory of Justice*, Harvard University Press, Cambridge, Mass., 1971.
[85] D.R. Shier and P.D. Dearing, «Optimal Locations for a Class of Nonlinear Single Facility Location Problems on a Network», *Operations Research* 31, 292 - 303, 1983.
[86] P.J. Slater, «Maximim Facility Location», *Journal of Research of the National Bureau of Standards* B 79, 107 - 115, 1975.
[87] P.J. Slater, «Central Vertices in a Graph», in *Proceedings of the 7th Southeastern Conference on Combinatorics, Graph Theory and Computing*, 487 - 489, 1976.
[88] P.J. Slater, «Centers to Centroids in Graphs», *Journal of Graph Theory* 2, 209 - 222, 1978.
[89] P.J. Slater, «Structure of the k-Centra in a Tree», in *Proceedings of the 9th Southeastern Conference on Combinatorics, Graph Theory and Computing*, 1978.
[90] P.J. Slater, «Medians of Arbitrary Graphs», *Journal of Graph Theory* 4, 389 - 392, 1980.
[91] P.J. Slater, «On Locating a Facility to Service Areas Within a Network», *Operations Research* 29, 523 - 531, 1981.
[92] P.J. Slater, «The k-Nucleus of a Graph», *Networks* 11, 233 - 242, 1981.
[93] P.J. Slater, «Locating Central Paths in a Graph», *Transportation Science* 16, 1 - 18, 1982.
[94] B.C. Tansel, R.L. Francis and T.J. Lowe, «Location on Networks: A Survey-Part I: The p-Center and p-Median Problems», *Management Science* 29, 482 - 497, 1983.
[95] B.C. Tansel, R.L. Francis and T.J. Lowe, «Location on Networks: A Survey-Part II: Exploiting the Tree Network Sturcture», *Management Science* 29, 498 - 511, 1983.
[96] S.S. Ting, «A Linear-Time Algorithm for Maxisum Facility Location on Tree Networks», *Transportation Science* 18, 76 - 84, 1984.
[97] P. Vincke, «Problèmes de localisation multicritères», *Cahiers du Centre d'Etudes de Recherche Opérationnelle* 24, 333 - 341, 1983.
[98] A. Weber, *Ueber den Standort der Industrien*, J.C.B. Mohr, Tübingen, 1909.
[99] R.E. Wendell and A.P. Hurter, «Optimal Locations on a Network», *Transportation Science* 7, 18 - 33, 1973.
[100] R.E. Wendell and R.D. McKelvey, «New Perspectives in Competitive Location Theory», *European Journal of Operational Research* 6, 174 - 182, 1981.
[101] C. Witzgall, «Optimal Location of a Central Facility: Mathematical Models and Concepts», *Report 8388, National Bureau of Standards*, 1964.
[102] B. Zelinka, «Medians and Peripherians of Trees», *Archivum Mathematicum* 4, 87 - 95, 1968.

Pierre Hansen
Ructor
Rutgers Center for Operations Research
Hill Center for Mathematical Sciences
Busch Campus
Rutgers University
New Brunswick, New Jersey 08903
U.S.A.

Martine Labbé
C.E.M.E.
University of Brussels
50 Av. F.D. Roosevelt CP139
1050 Brussels
Belgium

Dominique Peeters
Department of Geography
Catholic University of Louvain
1348 Louvain-la-Neuve
Belgium

Jacques-François Thisse
C.O.R.E.
Catholic University of Louvain
34 Voie du Roman Pays
1348 Louvain-la-Neuve
Belgium

EXACT ALGORITHMS FOR THE VEHICLE ROUTING PROBLEM*

Gilbert LAPORTE and Yves NOBERT

1. Introduction

In the broadest sense, the vehicle routing problem (VRP) can be described as the problem of designing optimal delivery routes from one or more depots to a set of geographically scatterred points, cities or customers. As noted elsewhere [11], the term delivery can sometimes be replaced by collection and there exist situations not involving the actual transportation of goods. Practical applications of the VRP are numerous and encompass many spheres of activity. Their economic importance cannot be overstated. Bodin et al. [5] quote a study which approximates annual distribution costs at $ 400 billion in the United States and £ 15 billion in the United Kingdom. Thus, any savings generated by improvements in route design methodology may be significant.

The VRP has received the attention of many Operations researchers over the last three decades. This interest is due in part to the practical importance of the problem, but also to its intrinsic difficulty. Only recently, the largest problems of any complexity which had been solved by exact algorithms contained approximately 30 points [5]. This figure has since been improved, but only in particular instances of the problem [9, 14, 21, 25, 31, 54-56, 58, 62]. The VRP constitutes a generalization of the travelling salesman problem (TSP) which consists of determining the shortest circuit or cycle passing through each of n points once and only once [18]. The TSP and the VRP are both NP-hard [65]. However, from a practical point of view, VRPs are in general much more difficult to solve than TSPs of the same size. This renders their study all the more challenging. Also, while much effort has been devoted to the solution of the TSP by exact methods, most algorithms for the VRP are heuristics, reflecting the relative difficulty of the problem. Promising avenues of research have been opened but many are still unexplored [67].

---

*The authors are grateful to the Canadian Natural Sciences and Engineering Research Council (grants A4747 and A5486) and to the Quebec Government (FCAC grant 80EQ04228) for their financial support.

The object of this paper is to present a survey of *exact* algorithms for the VRP, while emphasizing recent results and avoiding as far as possible, duplication of previous work. Several surveys and bibliographies have indeed been published on the VRP (see for example [4, 5, 8, 11, 28, 42, 59, 65, 67, 75, 78, 83, 86]), most of which have tended to concentrate on heuristic methods. The state-of--the-art paper of Bodin et al. [5] is fairly exhaustive in this respect. Also of major interest are the typology proposed by Bodin and Golden [4] and the analysis of research perspectives and prospects presented by Magnanti [67]. We adopt, whenever possible, the definitions and classifications of these authors.

1.1. Notation and definitions

Some notation and a more formal definition of the VRP are required at this point. Let $G = (N, E, C)$ be a graph where $N = \{1, \ldots, n\}$ is a set of nodes representing cities or customers, E is a set of arcs or edges (undirected arcs) and $C = (c_{ij})$ is a distance matrix associated with E. C is symmetrical if and only if $c_{ij} = c_{ji}$ for all $i, j \in N$. It satisfies the triangle inequality if and only if $c_{ij} + c_{jk} \geqslant c_{ik}$ for all $i, j, k \in N$. Problems fall into two broad categories: symmetrical problems (those for which C is symmetrical) and asymmetrical problems. Algorithms which exploit this distinction generally require that E be defined as a set of edges in the symmetrical case and as a set of arcs in the asymmetrical case. Also, depending on the type of algorithm, c_{ii} may be equal to 0, to infinity or simply left undefined. Let $R = \{1, \ldots, r\}$ $(r < n)$ be a set of depots. It is assumed that a fleet of m vehicles are based at the depots. The value of m is sometimes fixed a priori (it is then set equal to $\bar{m}$) or may constitute a decision variable. In general, one can impose bounds on m:

$$\underline{m} \leqslant m \leqslant \bar{m} \tag{1}$$

Similarly, there may be bounds on the number of vehicles assigned to each depot. Fixed costs may be imposed on the use of a depot or of a vehicle.

The VRP consists of establishing minimum cost vehicle routes in such a way that

(i) each city in N-R is visited exactly once by exactly one vehicle;
(ii) each vehicle starts and ends its journey at the *same* depot;
(iii) some side constraints on the number of routes and on their configurations are satisfied.

Before describing these constraints, it seems appropriate to bring some precision on the first condition. This condition is irrelevant when C satisfies the triangle inequality: in this case, it is never advantageous for a vehicle to deviate from the shortest route between any two cities and thus to visit any city more than once. However, most road networks are incomplete and multiple passages through the same city are more often than not necessary. This difficulty can be circumvented as follows. First replace each c_{ij} by the length of the shortest path from i to j and then establish the following distinction: each city must be *serviced* exactly once but may be *crossed* as many times as necessary if it lies on many shortest paths. However, it is sometimes inefficient to proceed this way: in some algorithms (see for example [31, 32, 73]), it is preferable to take advantage of the sparsity of the graph and to use the original c_{ij}'s while requiring that each city be visited *at least* once. Some examples of this approach will be presented.

The most common side conditions include:

(i) Capacity restrictions: a non-negative weight d_i is attached to each city i of N-R. (For $i \in R$, assume that $d_i = 0$). The sum of weights of any vehicle route may not exceed the vehicle capacity (see for example [1, 10, 13, 31, 32, 37, 46, 55, 58, 62]). (Capacitated VRPs will be referred to as CVRPs).

(ii) The number of cities on a route may be bounded above (this constitutes a particular case of (i)).

(iii) Total time restrictions: the length of the route of any vehicle may not exceed a prescribed bound L; this length is made up of intercity travel times and of stopping times (δ_i) at each city i included on the route [56, 62]. (Time or distance constrained VRPs will be referred to as DVRPs).

(iv) Time windows: city i may have to be visited within the time interval $[a_i, b_i]$ (This includes the two cases $a_i = -\infty$ and $b_i = \infty$) [20-23, 49, 77, 79, 80].

(v) Precedence relations between pair of cities: city i may have to be visited before city j [21, 79-81].

This list is by no means exhaustive and it is beyond the scope and purpose of this paper to describe every possible ramification of the VRP. This survey concentrates on formulations and methods for the more basic models, mentioning, where appropriate, important extensions for which exact algorithms have been developed. The following restrictions will be made.

(i) Fixed costs on depots and vehicles are not generally considered: their eventual inclusion in the models is in most cases straigthforward.

(ii) Unless otherwise specified (see section 4), there is only one depot located at city 1.

(iii) In most cases, all vehicles are identical and have the same capacity D.

(iv) All vehicles are assumed to drive at the same constant speed; in such a context, the c_{ij}'s may be interpreted as travel times whenever appropriate. The c_{ij}'s sometimes represent travel costs.

1.2. Classification of methods

It appears that all known exact algorithms for the VRP can be classified into one of the following categories:

(i) direct tree search methods,
(ii) dynamic programming (DP),
(iii) integer linear programming (ILP).

The latter category is very broad and accounts for most of the research effort of recent years. It will be subdivided into three sections, according to the classification suggested by Magnanti [67]:

iiia) set partitioning formulations,
iiib) vehicle flow formulations (by far the most widely used),
iiic) commodity flow formulations.

Several interesting relationships between these three LP formulations have been outlined by Magnanti. Readers are referred to [67] for full details.

2. Direct tree search algorithms

2.1. Branching on arcs

Direct tree search methods consist of sequentially building vehicle routes by means of a branch and bound tree. The paper by Christofides and Eilon [10] published in 1969 contains one of the earliest descriptions of such a method. It is based on the Little et al. algorithm [66] for the TSP and applies to CVRPs and DVRPs.

It will be useful to first present an ILP formulation of the TSP. Let x_{ij} $(i \neq j)$ be a 0-1 variable indicating whether arc (i,j) appears $(x_{ij} = 1)$ or not $(x_{ij} = 0)$ in the optimal solution and let $\overline{S} = N - S$. Then the TSP can be formulated as follows:

$$(\text{TSP}) \quad \text{minimize} \sum_{i,j \in N} c_{ij} x_{ij}$$

subject to

$$\sum_{i \in N} x_{ij} = 1 \quad (j \in N) \tag{2}$$

$$\sum_{j \in N} x_{ij} = 1 \quad (i \in N) \tag{3}$$

$$\sum_{i \in S, j \in \overline{S}} x_{ij} \geq 1 \quad (2 \leq |S| \leq n-2, S \subset N) \tag{4}$$

$$x_{ij} = 0, 1 \quad (i \neq j, i, j \in N). \tag{5}$$

This formulation was first proposed by Dantzig et al. [18]. Constraints (2), (3) and (5) describe an assignment problem while constraints (4) ensure that the graph associated with the optimal solution is connected. When constraints (2) and (3) are satisfied, constraints (4) are equivalent to

$$\sum_{i,j \in S} x_{ij} \leq |S| - 1 \qquad (2 \leq |S| \leq n-2, S \subset N) \tag{6}$$

This equivalence can be extended to more general situations (see [76]).
Two well known relaxations of the TSP are
 (i) the assignment problem (AP) characterized by constraints (2), (3), (5);
 (ii) the shortest spanning tree problem (SSTP) characterized by constraints (4) and (5); a valid lower bound for the TSP is provided by the value of the shortest spanning 1-tree, i.e. the shortest spanning tree to which is added the shortest remaining arc.

Efficient algorithms exist for solving the AP [69] and the SSTP [26, 50].

Thus, a valid bound on the TSP can be derived by solving either of these two problems or by computing a lower bound on their solution. Comparisons of these two bounds are given in [28].

In order to apply a TSP algorithm to the VRP, it is convenient to first transform the distance matrix C. This can be done by adjoining $\bar{m} - 1$ artificial depots to the set N of cities, where $\bar{m}$ is an upper bound on m satisfying

$$\bar{m} \geq \left\lceil \frac{\sum_{i \in N} d_i}{D} \right\rceil \tag{7}$$

Here and elsewhere, $\lceil t \rceil$ denotes the smallest integer greater than or equal to t if $t > 0$; (it is equal to 1 if $t \leq 0$).

Let N^* be the set including the original depot and its copies:

$$N^* = \{1, n+1, \ldots, n+m-1\} \tag{8}$$

and let $N' = N \cup N^*$.

The extended cost matrix $C' = (c'_{ij})$ is then defined by

$$c'_{ij} = \begin{cases} c_{ij} & (i, j \in N) \\ c_{i1} & (i \in N, j \in N^*) \\ c_{1j} & (i \in N^*, j \in N) \\ \lambda & (i, j \in N' - N). \end{cases} \tag{9}$$

In [10], λ is set equal to infinity in order to prohibit interdepot arcs from entering the solution. Lenstra and Rinnooy Kan [64] have later observed that $-\lambda$ can be interpreted as the cost of using a vehicle. Therefore,

$\lambda = \infty$ yields the minimum distance for $\bar{m}$ vehicles;

$\lambda = 0$ yields the minimum distance for at most $\bar{m}$ vehicles;

$\lambda = -\infty$ yields the minimum distance for the minimum number of vehicles.

Any *sequential* algoritms for the TSP, i.e. an algorithm which successively fixes arcs (variables) at 0 or 1 in a branch and bound tree (cf. Little et al. [66]) can be used for the CVRP or for the DVPR with the following modifications. Christofides and Eilon [10] fathom a branch in one of these cases:

(i) The total load for a vehicle exceeds its capacity D.

(ii) The total distance accumulated for a vehicle exceeds L.

(iii) The total weight of cities not already included in the tour exceeds the total capacity of the yet unused vehicles.

Instead of computing the assignment bound at every node of the search tree (as in [66]), Christofides and Eilon suggest using the shortest spanning 1-tree bound.

The algorithm was tested on two VRPs involving capacity restrictions only: a 6-city problem taken from [46] and the Dantzig and Ramser 13-city example [19] were solved in 1.5 and 5 minutes respectively on an IBM 7090.

2.2. Branching on routes

In the previous method, branches of the tree were created by including an arc in the solution or by excluding it from consideration. In [7], Christofides describes a depth first branch and bound algorithm based on a different philosophy. The class of VRPs to which the algorithm applies is very wide: problems can include capacity or distance restrictions, time windows, precedence conditions, stopping times, etc. In fact, the method performs better on tight problems as less branches need be explored.

Each level of the search tree corresponds to the definition of a new vehicle route. Thus the tree will have at most $\bar{m}$ levels. To each node of the tree, corresponds a set of vehicle routes. Tree construction can be summarized as follows. Let g represent a node of the search tree and define

$F(g)$: the set of unrouted (free) cities beyond node g;

$\bar{F}(g) = N - F(g)$: the set of routed cities at node g.

Forward branching from g is done by considering a city i of $F(g)$ and by generating the list of all feasible routes including i. Each of these routes corresponds to an immediate descendant node g'. It is apparent that it is advantageous to choose i in such a way that as few routes as possible are generated. The following criteria are used to eliminate from consideration potential routes passing through i:

(i) The route containing i violates at least one side constraint.

(ii) The unused vehicles cannot feasibly supply cities in $F(g')$ (the total weight of these cities exceeds the total unused vehicle capacity).

(iii) Define z^* : the cost of the best known feasible solution;

$c(S)$: the optimal cost of supplying cities of S ($S \subseteq N$);

$\underline{c}(S)$: a lower bound on $c(S)$;
$\overline{c}(S)$: an upper bound on $c(S)$.

Then node g' need not be created if

$$c(\overline{F}(g)) + c(\overline{F}(g') - \overline{F}(g)) + \underline{c}(F(g')) \geq z^*. \tag{10}$$

(iv) g' is dominated by another immediate descendant g'' to node g, i.e.

$$c(\overline{F}(g'') - \overline{F}(g)) + \overline{c}(\overline{F}(g'')) \leq c(\overline{F}(g') - \overline{F}(g)) + \underline{c}(\overline{F}(g')). \tag{11}$$

Note that c can be replaced by $\underline{c}$ in the left-hand side of (10) and in the right-hand side of (11), and by $\overline{c}$ in the left-hand side of (11).

If the VRP involves capacity or total distance constraints only, $c(S)$ may be computed by solving a TSP on S. Similarly, $\overline{c}(S)$ can be determined by using or adapting a TSP heuristic [41, 43] and the value of $\underline{c}(S)$ can be obtained from one of the known TSP bounds [28]. However, it is harder to determine these values in problems with time windows.

The largest problem reported solved to optimality with the algorithm contains 31 cities. No details are provided on the type of constraints or computing times.

2.3. k-degree centre trees and q-routes

The efficiency of any branch and bound algorithm (for a minimization problem) rests on the computation, at every node of the search tree, of sharp lower bounds on the value of the optimum. In [12], Christofides et al. derived such bounds for the VRP. The problem they consider is a VRP with exactly $\overline{m}$ vehicles and with capacity and time restrictions. In addition to travel times c_{ij} between cities, the authors also consider an unloading time δ_i at each city i. The problem is formulated as a «tree-index vehicle flow program» (see section 5) but no direct use is made of this model either in the computation of the bounds or in the algorithm which is a direct tree search procedure.

The first of these bounds is based on the fact that the value of the optimal VRP solution is bounded below by the total length of a *k-degree centre tree* (*k*-DCT), i.e. a tree where the degree of node 1 is k. Consider E, the set of all arcs. At the optimum, the set of arcs belonging to the solution can be partitioned into 3 sets:

(i) E_1 : arcs forming a k-DCT (where $k = 2\overline{m} - y$ and $y \leq \overline{m}$);
(ii) E_2 : y arcs incident to node 1,
(iii) E_3 : $\overline{m} - y$ arcs not incident to node 1.

Let x_l^t ($t = 1, 2, 3; l \in E$) be 0-1 variables taking the value 1 if and only if arc l belongs to E_t in the optimal solution, let c_l be the cost of arc l, E^i the set of all arcs incident to node i and $(S, \overline{S})$, the set of all arcs with one vertex in S and the other vertex in its complement $\overline{S}$. The authors then consider the multiple TSP [64] relaxation of the CVRP:

(MTSP)   minimize $\sum_{l \in E} c_l (x_l^1 + x_l^2 + x_l^3)$

subject to

$$\sum_{l \in (S,\bar{S})} x_l^1 \geq 1 \qquad (S \subset N, S \neq \phi) \qquad (12)$$

$$\sum_{l \in E^1} x_l^1 = 2\bar{m} - y \qquad (13)$$

$$\sum_{l \in E} x_l^1 = n - 1 \qquad (14)$$

$$\sum_{l \in E^1} x_l^2 = y \qquad (15)$$

$$\sum_{l \in E - E^1} x_l^3 = \bar{m} - y \qquad (16)$$

$$\sum_{l \in E^i} (x_l^1 + x_l^2 + x_l^3) = 2 \qquad (i = 2, \ldots, n) \qquad (17)$$

$$x_l^t = 0, 1 \qquad (t = 1, 2, 3; l \in E) \qquad (18)$$

$$y \geq 0 \text{ and integer.} \qquad (19)$$

Three problems are extracted from (MTSP) by only considering variables x_l^t ($t = 1, 2, 3$). In each case, constraints (17) are incorporated into the objective function in a Lagrangean fashion to yield

$$V^t(\lambda, y) = \sum_{l \in E} (c_l + \lambda_{\alpha(l)} + \lambda_{\beta(l)}) - 2 \sum_{i=1}^{n} \lambda_i \qquad (20)$$

where $\alpha(l)$, $\beta(l)$ are the two terminal vertices of arc l, $\lambda = (\lambda_i)$ is the vector of non-negative penalties associated with constraints (17) and $\lambda_1 = 0$. A lower bound for the solution of the VRP is therefore

$$LB1 = \max_{m_1 \leq y \leq \bar{m}} \left\{ \max_{\lambda \geq 0} \sum_{t=1}^{3} V^t(\lambda, y) \right\} \qquad (21)$$

where m_1 represents a lower bound on the number of single customer tours in the optimal VRP solution. Taking into account capacity and maximum time constraints, m_1 can be chosen so as to satisfy the following conditions:

(i) Suppose the cities are ordered in decreasing order of the d_i's. Then we we must have

$$(\overline{m} - m_1) D \geq \sum_{i=m_1+1}^{n} d_i \qquad (22)$$

(ii) Similarly, every city i contributes an amount of at least $u_i = \delta_i + 1/2 (c_{ii_1} + c_{ii_2})$ to the length of a route, where i_1 and i_2 are the two cities nearest to i. Then, if the cities are ordered in decreasing order of the u_i's, m_1 must satisfy

$$(\overline{m} - m_1) L \geq \sum_{i=m_1+1}^{n} u_i. \qquad (23)$$

Christofides et al. describe an efficient procedure to compute the $V^t(\lambda, y)$'s.

The second bound developed in [12] is based on q-routes, used by Houck et al. [48] in the case of the TSP. The bound, which applies to CVRPs. is derived as follows. Let W be the set of all possible weights that could exist on any vehicle route. Let the elements of W be ordered in ascending order and let $q(l)$ be the value of the lth element of W. Define $\psi_l(i)$ as the value of the least cost route

(i) passing through i;
(ii) starting and ending at the depot;
(iii) having no loop of the form (i_1, i_2, i_1);
(iv) having a total weight $q(l)$.

Such a route is called a q-route. Then, Christofides et al. prove that LB2 is a valid lower bound for the CVPR:

$$\text{LB2} = \sum_{i=1}^{n} \min_{l=1,\ldots,|W|} \left(\frac{\psi_l(i) d_i}{q(l)} \right). \qquad (24)$$

The authors show that $\psi_l(i)$ can be computed through a relatively simple recursion procedure. It is obvious that the effort involved in the computation of LB2 is directly related to $|W|$.

These two bounds (based on k-DCTs and on q-routes) were embedded in branch and bound algorithms and applied to CVRPs ranging from 10 to 25 cities. In the case of the first bound, two branching strategies were used: branching on arcs and branching on routes; with the second bound, only the first strategy was used. The results indicate that LB2 is generally superior to LB1: it certainly produces much sharper bounds at the root of the search tree. However, the best algorithm overall used LB1 in conjunction with a «branching on routes» strategy. All test problems were solved within 250 seconds on a CDC 7600 computer, using an FTN compiler.

Kolen et al. [49] later used the concept of q-routes to derive an exact branch and bound algorithm for the CVRP with time windows. Nine test problems involving from 6 to 15 cities were solved to optimality.

3. Dynamic programming

Dynamic programming (DP) has been applied to several types of VRPs. First, consider the general formulation provided by Eilon et al. [28].

3.1. General dynamic programming formulation for the VRP

Consider a VRP with a fixed number $\bar{m}$ of vehicles. Let $c(S)$ denote the cost of the optimal single route through the depot (city 1) and all the customers of a subset S of $N-\{1\}$. We wish to minimize

$$z = \sum_{j=1}^{\bar{m}} c(S_j) \tag{25}$$

over all feasible partitions $\{S_1, \ldots, S_{\bar{m}}\}$ of $N-\{1\}$. Let $f_k(U)$ be the minimum cost achievable using k vehicles and delivering to a subset U of $N-\{1\}$. Then the minimum can be determined through the following recursion formula:

$$f_k(U) = \begin{cases} c(U) & (k=1) \\ \min_{U^* \subseteq U \subseteq N-\{1\}} [f_{k-1}(U-U^*) + c(U^*)] & (k>1) \end{cases} \tag{26}$$

The solution cost is equal to $f_{\bar{m}}(N-\{1\})$ and the optimal partition will correspond to the optimizing subsets U^* in (26).

It is apparent that if $f_k(U)$ has to be computed for all values of k and for all subsets U of $N-\{1\}$, the number of computations required by the method is likely to be formidable in all but very small size problems. Efficient methods based on dynamic programming require a substantial reduction of the number of states by means of a mapping function or by using feasibility or dominance criteria. Consider for example the CVRP. In this problem, U and U^* must satisfy

$$\sum_{i \in N-\{1\}} d_i - (\bar{m}-k)D \leqslant \sum_{i \in U} d_i \leqslant kD \quad (k=1,\ldots,\bar{m}) \tag{27}$$

and

$$\sum_{i \in U} d_i - (k-1)D \leqslant \sum_{i \in U^*} d_i \leqslant D \quad (k=1,\ldots,\bar{m}) \tag{28}$$

3.2. State-space relaxation

State-space relaxation provides an efficient means of reducing the number of states. The method was introduced by Christofides et al. [13]. It provides a lower bound z^* on the cost z^* of the optimal solution. The optimum can then be reached by embedding the method in a branch and bound procedure.

Consider the general DP recursion

$$f_{0,i}(0,j) = \min_{k \in \Delta^{-1}(j)} [f_{0,i-1}(0,k) + c_i(k,j)] \tag{29}$$

where

$f_{0,i}(0,j)$ is the least cost of changing the system from state 0 at stage 0 to state j at stage i

$\Delta^{-1}(j)$ is the set of all possible states from which state j can be reached directly

$c_i(k,j)$ is the cost of changing the system from state k at stage $i-1$ to state j at stage i.

Let $g(\cdot)$ be a mapping from the state space S associated with (29) to a state space G with smaller cardinality and let $F^{-1}(g(j))$ be a set satisfying

$$k \in \Delta^{-1}(j) \Rightarrow g(k) \in F^{-1}(g(j)) \tag{30}$$

Recursion (29) then becomes

$$f_{0,i}(g(0), g(j)) = \min_{t \in F^{-1}(g(j))} [f_{0,i-1}(g(0), t) + \overline{c}_i(t, g(j))] \tag{31}$$

where

$$\overline{c}_i(t, g(j)) = \min [c_i(k,l) : g(k) = t, g(j) = g(l)]. \tag{32}$$

It results that

$$f_{0,i}(g(0), g(i)) \leqslant f_{0,i}(0, i). \tag{33}$$

This relaxation is useful only if

(i) $F^{-1}(\cdot)$ can be easily determined: this will be so if $g(\cdot)$ is *separable* so that given $g(U)$ and r, $g(U - \{r\})$ can be computed;

(ii) $g(\cdot)$ must be such that the optimization of (32) is over a small domain or that a good lower bound on $\overline{c}_i(t, g(j))$ can be computed.

Using various formulations and several choices of $g(\cdot)$, Chistofides at al. have applied state-space relaxation to the TSP and to a variety of VRPs. Some of the results they have obtained are reported in sections 3.3 a) and b).

3.3. Application of dynamic programming to some VRPs

DP has been used successfully to solve some VRPs to optimality or to obtain very sharp bounds on the value of their optimal solutions. Here are some examples.

a) *Capacitated VRP (CVRP)* - Chistofides et al. [13] present three formulations for this problem. The first one is provided by (26), subject to (27) and (28). Of the other two, one has been used to obtain a good lower bound on the optimum. It can be described as follows. Let $f_k(U, r)$ be the least cost of supplying a set U of customers, using only k vehicles, with the last customers of the corresponding k routes being among the customers $2, \ldots, r$ ($k \leq r \leq n$). Let $c(U, r)$ be the cost of the TSP solution through $U \cup \{1\}$ where the last customer (before the depot) is r. The recursion is then

$$f_k(U, r) = \begin{cases} \min\,[f_k(U, r-1), \min_{U^* \subset U} \{f_{k-1}(U - U^*, r-1) + c(U^*, r)\}] & (k, r > 1) \qquad (34) \\ c(U, r) & (k = 1) \end{cases}$$

subject to (27).

For this problem, the mapping function is based on q-routes, i.e.

$$g(U) = \sum_{i \in U} d_i. \qquad (35)$$

Recursion (31) then becomes

$$f_k(g(U), r) = \min\,[f_k(g(U), r-1), \\ \min_p \{f_{k-1}(g(U) - p, r-1) + \overline{c}(p, r)\}] \qquad (36)$$

subject to

$$g(N) - (\overline{m} - 1)D \leq p \leq \min\,(g(U), D) \qquad (37)$$

Using this relaxation, lower bounds on the optimal CVRP solutions were obtained for 10 problems containing from 10 to 25 cities. For the 10-city problem, the bound was in fact equal to the optimum. Otherwise the ratio «lower bound ÷ optimum» varied between 93.1% and 99.6%. More recently, Christofides [9] reported that CVRPs involving up to 50 nodes could be solved systematically by this approach (the largest size attained was 73). Problems containing up to 125 nodes were solved within 2% of the optimum in less than 15 minutes on a CYBER 855.

b) *TSP with time windows* (Christofides et al. [13]). Consider a TSP in which r_i time windows $[a_i^k, b_i^k]$ ($k = 1, \ldots, r_i$) are associated with each city i. Without loss of generality, it is assumed that the time windows are disjoint and ordered so that

$$b_i^{k-1} < a_i^k \qquad (k = 1, \ldots, r_i) \qquad (38)$$

$$b_i^0 = 0.$$

Also assume that there is a *processing time* δ_i at city i and that c_{ij} represents a *travel time* from i to j. Let t_i be the time of arrival at city i in the optimal solution. Then we seek the shortest Hamiltonian circuit through the n cities under the constraint

$$a_i^k \leqslant t_i \leqslant b_i^k \qquad \text{for some } k \in \{1, \ldots, r_i\}. \tag{39}$$

In this problem, the efficiency of the DP approach is largely dependent on the number of states which can be eliminated through feasibility considerations: tighter time windows will lead to the generation of fewer states. Here again the authors use state-space relaxation; the particular mapping $g(\cdot)$ which is used is not mentioned. The authors report results for 40 «moderately tight» problems ranging from 15 to 50 cities. The method only provides a lower bound on the optimum. For one particular 30-city problem, this bound attains 99.8% of the optimum.

c) *Dial-a-ride problems* - This class of problems can be described as follows. Consider n groups of customers requesting to be picked up at a given location and to be delivered to another location. In *many-to-many* problems, origins as well as destinations are all distinct points. In *many-to-one* problems, all delivery points coincide. What follows is valid for many-to-many problems. Let $\{1, \ldots, n\}$ be the set of origins and $\{n + 1, \ldots, 2n\}$, the set of destinations. There is a depot located at node 0. Travel times t_{ij} and distances c_{ij} between pairs of cities are given. In addition, there is a processing time δ_i at each node i. The problem consists of establishing optimal vehicle routes subject to some operating rules and constraints.

There are two main classes of operating rules. In the *static case*, customers request service and these requests are registered in a list. At a certain point T_0 in time, a vehicle becomes available and the task is to provide service to all customers in the list at the time; no new customers may be added to a route during its execution. In the *dynamic case* new customers become eligible for immediate inclusion in a route as they make their request.

The most common constraints are

(i) Vehicle capacities: at node i are associated d_i customers to be picked up (if $d_i > 0$) or delivered (if $d_i < 0$).
Assume that

$$\begin{aligned} d_i &> 0 & (i = 1, \ldots, n) \\ d_i &= -d_{i-n} & (i = n+1, \ldots, 2n). \end{aligned} \tag{40}$$

At most $\bar{m}$ vehicles of capacity D are used for the operation. At no time can vehicle capacity be exceeded.

(ii) Time windows: with each node i is associated a unique time window $[a_i, b_i]$. Vehicles may not leave i after b_i; they are however allowed to stay idle at the node if they arrive before a_i.

(iii) Maximum position shifts: these were described by Psaraftis [79] in the following terms:

«In any particular vehicle route we can identify the sequence of pickups and the sequence of deliveries, sequences which, in general, will be *merged* with one another. The position (1st, 2nd, etc.) that a particular customer holds in the sequence of pickups will not in general be the same as his First-Come-First-Served (FCFS) position in the initial list of customer requests, the difference of these two positions constituting the *pickup position shift* of that customer. For instance, a customer holding the 5th position in the initial list and being pickep up 3rd has a pickup position shift of $+2$ while this shift becomes -1 if the customer is picked up 6th. A *delivery position shift* can be similarly defined as the difference between the position of a customer in the sequence of deliveries and the FCFS position of that customer in the initial list of requests».

Then, for any customer, neither of these two position shifts many exceed a prescribed *maximum position shift*.

Several variants of the problem can then be defined. In [79], Psaraftis considers the static and the dynamic cases with capacity and maximum position shift constraints. The objective function is a linear combination of
 (i) route durations;
 (ii) customers waiting times (between T_0 and pick up times);
 (iii) customers riding times.

He proposes an $O(n^2 3^n)$ exact algorithm, an extension of the classical Held and Karp [47] DP algorithm for the TSP. Taking advantage of the fact that many infeasible states need not be defined, Psaraftis succeeds in solving optimally problems involving up to 9 customers in the static case. No computational results are reported in the dynamic case.

Psaraftis later described a similar algorigithm [80] for problems in which maximum position shift constraints are replaced by time windows. No computational results were reported, but the author stated that he did not expect the CPU times to be significantly different from those obtained in [79].

Desrosiers et al. [21] consider the static problem where the objective is to minimize the total distance travelled while respecting vehicle capacities and time windows. The method is based upon an earlier algorithm by the same authors for the shortest path problem with time windows [22 - 24]. As in the above methods, infeasible states are not defined and dominated states are discarded. Also, feasible states which, due to the presence of time windows, could not possibly be part of the solution, are eliminated. This explains, to a large extent, why this algorithm produces better results than reported in [79]: problems involving in to 40 requests (80 nodes) are solved in less than 6 seconds on the University of Montreal CYBER 173.

In a different paper [20], the same authors consider the multivehicle dial-a-ride problem. They use a decomposition approach which provides a suboptimal solution. Real-life problems involving up to 880 requests are solved.

4. Set partitioning formulations

Over the years, several integer linear programming formulations have been suggested for the VRP. Among these, set partitioning formulations cover a wide range of problems. Unfortunately, due to the large number of variables they contain, it will rarely be practicable to use them to derive optimal solutions. Two interesting exceptions are Agarwal's algorithm for the CVPR [1] and the algorithm developed by Desrosiers et al. [25] for school bus routing.

Balinski and Quandt [2] were among the first to suggest such a formulation. Consider all feasible routes j and let a_{ij} be a 0-1 coefficient taking the value 1 if and only if city i appears on route j. Let c_j^* be the optimal cost of route j and x_j be a 0-1 variable equal to 1 if and only if route j is used in the optimal solution. Then the problem is to

$$(\text{VRP1}) \qquad \text{minimize} \sum_j c_j^* x_j$$

subject to

$$\sum_j a_{ij} x_j = 1 \qquad (i \in N - \{1\}) \qquad (41)$$

$$x_j = 0, 1 \qquad (\text{for all } j). \qquad (42)$$

This formulation can easily be transformed to include the case where multiple passages through city i are allowed: it suffices to replace the equality sign of constraint i by a «$\geqslant$» sign. Similarly, the number of passages through city i may be set equal to a number greater than 1.

There are two major difficulties associated with this approach.

(i) The large number of binary variables which can run into the millions in most real-life cases. Only in extremely constrained problems (i.e. in problems with very few feasible solutions) will the number of variables be small enough to enable the problem to be solved directly.

(ii) The difficulty in computing the c_j^* values. For example, in the CVRP, each route j corresponds to a set of cities S_j satisfying

$$\sum_{i \in S_j} d_i \leqslant D. \qquad (43)$$

The value of c_j^* is then obtained by solving a TSP on S_j. As such, this formulation offers no obvious advantage over the general DP formulation.

However, if the number of variables is relatively small and if the objective is to minimize the number of vehicles, i.e. $c_j^* = 1$ for all j, the linear relaxation of (VRP1) often provides an integer solution [84]. If the solution (x^*) is fractional and gives a non-integral objective function, then the following cutting plane can be introduced

$$\sum_j x_j > \left\lceil \sum_j x_j^* \right\rceil. \tag{44}$$

Very few cuts are generally required to reach integrality [77].

Various authors have proposed column generation schemes to solve VRPs by the set partitioning approach. Rao and Zionts [82] consider the problem of establishing $\bar{m}$ vessel routes between ports. At each iteration, $\bar{m}$ shortest complete cycles (one for each vessel) are generated by means of the out-of-kilter algorithm [33] and the corresponding set partitioning problem is solved. New columns are introduced into the problem to replace old ones as long as they can reduce the value of the objective function. Unfortunately, no computational results are reported for this method.

Foster and Ryan [34] also suggest a column generation approach in which the routes are obtained by dynamic programming. These authors report that the method, although not run to optimlity, provides good routes when compared to those obtained by previous researchers [10, 36, 40, 88].

Agarwal [1] proposes an exact algorithm for the CVRP, based on the set partitioning formulation. Initially, only a limited number of columns are included in the master problem and the LP relaxation of (VRP1) is solved optimally. New columns j are gradually introduced into the problem as follows. Let $y = (y_2, \ldots, y_n)$ be the 0-1 vector of coefficients of the new column and let $(u_2, \ldots, u_n)$ be the vector of dual values associated with the current optimal solution of the master problem. Also, let $c^*(y)$ be the value of the optimal TSP solution associated with y. We seek the column of least reduced cost:

$$(\text{CG}) \qquad \text{minimize } c^*(y) - \sum_{i=2}^{n} u_i y_i$$

subject to

$$\sum_{i=2}^{n} d_i y_i \leq D \tag{45}$$

$$y_i = 0, 1 \qquad (i = 2, \ldots, n) \tag{46}$$

A lower bound on the optimal value of (CG) can be obtained by first considering a linear lower bound $\underline{c}^*(y) = \sum_{i=2}^{n} p_i y_i$ on $c^*(y)$ and by then solving the following knapsack problem:

$$(\text{KP}) \qquad \text{minimize } \sum_{i=2}^{n} (p_i - u_i) y_i$$

subject to (45) and (46).

The solution to (KP) is then embedded in a branch and bound scheme to solve (CG).

Agarwal employs various devices to estimate the dual values and to eliminate from consideration several candidate columns. At the end of the algorithm, the set covering problem (including the integrality conditions) need only be solved over a limited number of columns.

Using this approach, optimal results were obtained for 7 CVRPs (taken from [12]) involving between 15 and 25 cities, in times ranging from 6 to 156 seconds on an IBM 370/4381.

Orloff [77] discussed the use of column generation in the context of school bus scheduling and suggested some solution approaches. Desrosiers et al. [25] developed an efficient column generation algorithm for the same problem. In order to summarize their work, some definitions and notation will be required.

Consider a set of *trips* to be covered by buses and a set of *intertrips* corresponding to unproductive bus journeys between the trips. Every trip i is characterized by a place of origin, a destination, a duration, a cost and a time interval $[a_i, b_i]$ during which the trip must begin. The intertrip arc (i, j) goes from the end of trip i to the beginning of trip j. Its duration t_{ij} and its cost c_{ij} include respectively the duration and cost of trip i. A *route* is a sequence of trips and intertrips carried out by the same vehicle. The problem is to determine routes and schedules for all the trips so as to minimize the number of vehicles and travel costs for that number of vehicles, while respecting network and scheduling constraints.

The problem is formulated as (VRP1) where i represents a trip and j, a route. A column generation scheme is used and columns are provided by the solution of the following shortest path problem with time windows. First define the following notation.

P is the set of trips;

I is the set of intertrips arcs;

s is a source and t is a sink;

$P' = P \cup \{s, t\}$;

$A = P^2 \cup (\{s\} \times P) \cup (P \times \{t\})$;

$x_{ij} = 1$ if and only if (i, j) is used by a vehicle;

t_i is the starting time of trip i;

u_i is the ith dual variable of the linear relaxation of (VRP1).

Then the subproblem associated with the generation of a column is a shortest path problem with time windows:

(SPTW) $$\text{minimize} \sum_{(i,j) \in A} (c_{ij} - u_i) x_{ij}$$

subject to

$$\sum_{j \in P'} x_{ij} = \sum_{j \in P'} x_{ji} \quad (i \in P) \tag{47}$$

$$\sum_{j \in P} x_{sj} = \sum_{j \in P} x_{jt} = 1 \tag{48}$$

$$x_{ij} > 0 \Rightarrow t_i + t_{ij} \leq t_j \quad (i,j) \in I \tag{49}$$

$$a_i \leq t_i \leq b_i \quad (i \in P) \tag{50}$$

$$x_{ij} = 0,1 \quad ((i,j) \in A) \tag{51}$$

In this formulation, constraints (47) are flow conservation equations and constraint (48) expresses the fact that one vehicle leaves and enters the depot. Constraints (49) ensure that trip j can only be made immediately after trip i if this succession does not violate a time window constraint. It can be shown [25] that (SPTW) possesses an integer optimum even if constraints (51) are relaxed (i.e. the bounds on the variables are relaxed as well as the integrality conditions). Desrosiers et al. solve (SPTW) by means of a dynamic programming algorithm [22]. The authors suggest introducing several routes (columns) simultaneously into the master problem (VRP1), as opposed to only one at a time. The effect of this strategy is to reduce overall computation times by a factor of 2. The master problem is also solved by relaxing the integrality requirements and by introducing two cuts in order to eliminate solutions with a fractional cost or a fractional number of vehicles.

Desrosiers et al. report exact solutions for problems involving up to 151 trips. The running time for the largest problems varies between 94 seconds and 376 seconds on the University of Montreal CYBER 173 computer, in the case where time windows are relatively narrow. As expected, computation times increase as time windows widen.

5. Vehicle flow formulations

Most ILP algorithms for vehicle routing problems are based on vehicle flow formulations. These formulations use binary variables to indicate whether a vehicle travels between two given cities in the optimal solution. We distinguish two families of vehicle flow formulations: three-index formulations and two-index formulations. In the first case, three indices are attached to each flow variable: the origin, the destination and the vehicle making the trip; in the second case, the vehicle is not identified.

5.1. Three-index formulations

The following formulation is adapted from the work of Golden et al. [42] for asymmetrical VRPs involving at most $\bar{m}$ vehicles and r depots. In addition to the standard notation, define

D_k : the capacity of vehicle k

t_{ij}^k : the time taken by vehicle k to travel from i to j ($t_{ii} = \infty$)

δ_i^k : the time required for vehicle k to service city i ($\delta_i^k = 0$, $i \in R$)

L_k : the maximum allowed length of route k

c_{ij}^k : the cost of using vehicle k from i to j

$$x_{ij}^k = \begin{cases} 1 & \text{if vehicle } k \text{ travels directly from } i \text{ to } j \\ 0 & \text{otherwise.} \end{cases}$$

The VRP can be formulated as follows.

$$\text{(VRP2)} \quad \text{minimize} \quad \sum_{i=1}^{n} \sum_{j=1}^{n} \sum_{k=1}^{\bar{m}} c_{ij}^k x_{ij}^k$$

subject to

$$\sum_{i=1}^{n} \sum_{k=1}^{\bar{m}} x_{ij}^k = 1 \qquad (j = r+1, \ldots, n) \quad (52)$$

$$\sum_{j=1}^{n} \sum_{k=1}^{\bar{m}} x_{ij}^k = 1 \qquad (i = r+1, \ldots, n) \quad (53)$$

$$\sum_{i=1}^{n} x_{il}^k - \sum_{j=1}^{n} x_{lj}^k = 0 \qquad \begin{matrix}(k = 1, \ldots, \bar{m}; \\ l = 1, \ldots, n)\end{matrix} \quad (54)$$

$$\sum_{i=1}^{n} d_i \sum_{j=1}^{n} x_{ij}^k \leq D_k \qquad (k = 1, \ldots, \bar{m}) \quad (55)$$

$$\sum_{i=1}^{n} \delta_i^k \sum_{j=1}^{n} x_{ij}^k + \sum_{i=1}^{n} \sum_{j=1}^{n} t_{ij}^k x_{ij}^k \leq L_k \qquad (k = 1, \ldots, \bar{m}) \quad (56)$$

$$\sum_{i=1}^{r} \sum_{j=r+1}^{n} x_{ij}^{k} \leqslant 1 \qquad (k=1,\ldots,\bar{m}) \qquad (57)$$

$$\sum_{j=1}^{r} \sum_{i=r+1}^{n} x_{ij}^{k} \leqslant 1 \qquad (k=1,\ldots,\bar{m}) \qquad (58)$$

The solution may not contain subtours which do not include a depot. (59)

$$x_{ij}^{k} = 0, 1 \qquad (i,j=1,\ldots,n; k=1,\ldots,\bar{m}) \qquad (60)$$

In this formulation, constraints (52) and (53) specify that each city must be served exactly once by one and only one vehicle; constraints (54) ensure that every city is entered and left by the same vehicle; constraints (55) guarantee that vehicle capacities are never exceeded while constraints (56) are imposed in order to ensure that no vehicle route exceeds its time limit. Constraints (57) and (58) ensure that no more than $\bar{m}$ vehicles leave the deposts and there is at most one vehicle per depot.

Subtour elimination constraints specified by (59) may be one of the following:

$$\sum_{i \in S} \sum_{j \notin S} x_{ij}^{k} \geqslant 1 \qquad \begin{array}{l}(|S| \geqslant 1; R \subseteq S \subset N; \\ k=1,\ldots,\bar{m})\end{array} \qquad (59a)$$

$$\sum_{i \in S} \sum_{j \in S} x_{ij}^{k} \leqslant |S|-1 \qquad \begin{array}{l}(|S| \geqslant 1; S \subseteq N-R; \\ k=1,\ldots,\bar{m})\end{array} \qquad (59b)$$

$$y_i - y_j + n\, x_{ij}^{k} \leqslant n-1 \qquad \begin{array}{l}(k=1,\ldots,\bar{m}; \\ r+1 \leqslant i \neq j \leqslant n; \\ \text{for some } y_i, y_j \in \mathbb{R})\end{array} \qquad (59c)$$

Contraints (59a) and (59b) were first suggested by Dantzig et al. [18] for the TSP while constraints (59c) are those of Miller et al. [74].

It is relatively easy to include fixed vehicle costs f_k into the objective function by adding the term $\Sigma_{k=1}^{\bar{m}} f_k (\Sigma_{i=1}^{r} \Sigma_{j=r+1}^{n} x_{ij}^{k})$. It is also possible to impose the restriction that city i be visited within the time interval $[a_i, b_i]$ ($i = r+1, \ldots, n$). For this, define a variable t_i equal to the arrival time at city i and impose

$$t_i = 0 \qquad (i=1,\ldots,r) \qquad (61)$$

$$a_i \leqslant t_i \leqslant b_i \qquad (i=r+1,\ldots,n) \qquad (62)$$

$$\begin{cases} t_j \geq (t_i + \delta_i^k + t_{ij}^k) - (1 - x_{ij}^k) T \\ \\ t_j \leq (t_i + \delta_i^k + t_{ij}^k) + (1 - x_{ij}^k) T \end{cases} \quad \begin{array}{l}(i,j = 1, \ldots, n; \\ k = 1, \ldots, \overline{m})\end{array} \quad (63)$$

where T is an arbitrarily large number.

Fisher and Jaikumar [29] present a different three-index vehicle flow formulation for the single depot VRP with capacity restrictions, time windows and a fixed number $\overline{m}$ of vehicles. The authors have then developed an efficient algorithm based on this formulation. Although the algorithm seems to have been used to only provide a heuristic solution to the problem, as do some other approximate methods, it guarantees an optimal solution in a finite number of steps, if run to completion. The approach developed by Fisher and Jaikumar also possesses many other desirable features.

The notation is partly borrowed from (VRP2). Extra variables are introduced:

$$y_{ik} = \begin{cases} 1 & \text{if city } i \text{ is served by vehicle } k \\ 0 & \text{otherwise} \end{cases}$$

but travel costs and times are vehicle independent (although vehicle dependent parameters could easily be taken into account by the model and algorithm). There are no service times at the nodes. The formulation is

$$(\text{VRP3}) \qquad \text{minimize} \quad \sum_{i=1}^{n} \sum_{j=1}^{n} \sum_{k=1}^{\overline{m}} c_{ij} x_{ij}^k$$

subject to

$$\sum_{i=1}^{n} d_i y_{ik} \leq D_k \qquad (k = 1, \ldots, \overline{m}) \qquad (64)$$

$$\sum_{k=1}^{\overline{m}} y_{ik} = \begin{cases} \overline{m} & (i = 1) \\ 1 & (i = 2, \ldots, n) \end{cases} \qquad (65)$$

$$\sum_{i=1}^{n} x_{ij}^k = y_{jk} \qquad \begin{array}{l}(j = 1, \ldots, n; \\ k = 1, \ldots, \overline{m})\end{array} \qquad (66)$$

$$\sum_{j=1}^{n} x_{ij}^k = y_{ik} \qquad \begin{array}{l}(i = 1, \ldots, n; \\ k = 1, \ldots, \overline{m})\end{array} \qquad (67)$$

$$\sum_{i,j \in S} x_{ij}^k \leq |S| - 1 \qquad \begin{array}{l}(S \subset N; 2 \leq |S| \leq n - 1; \\ k = 1, \ldots, \overline{m})\end{array} \qquad (68)$$

$$\begin{cases} t_j \geq t_i + t_{ij} - (1-x_{ij}^k)\, T \\ \\ t_j \leq t_i + t_{ij} + (1-x_{ij}^k)\, T \end{cases} \quad \begin{matrix} (i,j = 1,\ldots,n; \\ k = 1,\ldots,\overline{m}) \end{matrix} \qquad (69)$$

$$a_i \leq t_i \leq b_i \qquad (i = 1,\ldots,n) \qquad (70)$$

$$y_{ik} = 0, 1 \qquad (i = 1,\ldots,n; k = 1,\ldots,\overline{m}) \qquad (71)$$

$$x_{ij}^k = 0, 1 \qquad (i,j = 1,\ldots,n; k = 1,\ldots,\overline{m}) \qquad (72)$$

All constraints included in this formulation are self-explanatory or have been described previously.

Essentially, two well known problems are contained in (VRP3):
(i) the generalized assignment problem (GAP) (constraints (64), (65), 71));
(ii) the TSP: when the y_{ik}'s are fixed to satisfy the GAP constraints, then for a given k, constraints (66) - (68) and (72) define a TSP for vehicle k.

The authors propose an algorithm based on Benders' decomposition [3]. The method iterates between solving a GAP master problem to assign cities to vehicles and solving a TSP with time windows (TSPTW) to determine the best route for each vehicle. The method has the advantage of producing a feasible solution, even when not run to completion. Also, since it repeatedly solves a GAP and TSPTW, it can benefit directly from any improvement in algorithms for these two problems. Moreover, the method has an intuitive appeal in that its two phases correspond to the two steps naturally followed by vehicle dispatchers.

In [30], Fisher and Jaikumar use their algorithm to derive approximate solutions to a number of problems involving capacity restrictions only and ranging from 50 to 199 cities. The results indicate that their method compares favourably with those of Christofides et al. [11], Clarke and Wright [16] and Gillet and Miller [40]. It is worth mentioning that Toth [85] lately improved the two-phase method described in [11] to produce a method which outperforms in most instances that of Fisher and Jaikumar [30].

5.2. Two-index formulations

Two-index formulations can be derived from three-index formulations by aggregating all x_{ij}^k variables into a single variable x_{ij} indicating whether or not a vehicle travels directly from i to j in the optimal solution. In symmetrical problems, x_{ij} indicates the number of times (0, 1 or 2) edge (i,j) $(i < j)$ is used in the optimal solution. More formally, x_{ij} is defined as

$$x_{ij} = \sum_{k=1}^{\overline{m}} x_{ij}^k \quad \text{or} \quad \sum_{k=1}^{m} x_{ij}^k. \qquad (73)$$

In two-index formulations, all vehicles are assumed to have the same costs and characteristics. These formulations were used for the exact solution of several types of VRPs by Laporte et al. [52 - 63, 76] and Fleischmann [31, 32]. The constraint relaxation algorithms developed by these authors belong to the same family as those used by several authors for the TSP and some of its extensions.

5.2.1. Symmetrical VRPs under capacity and distance restrictions

Two-index formulations have been mainly used for problems having a symmetrical distance matrix C. Consider first the single depot VRP under capacity and distance restrictions and assume C satisfies the triangle inequality. The following formulation is taken from [62]:

$$\text{(VRP4)} \qquad \text{minimize} \quad \sum_{i,j \in N} c_{ij} x_{ij}$$

subject to

$$\sum_{j \in N - \{1\}} x_{1j} = 2m \tag{74}$$

$$\sum_{i < k} x_{ik} + \sum_{j > k} x_{kj} = 2 \qquad (k \in N - \{1\}) \tag{75}$$

$$\sum_{i,j \in S} x_{ij} \leq |S| - V(S) \qquad (S \subseteq N - \{1\}, |S| \geq 3) \tag{76}$$

$$x_{ij} = \begin{cases} 0, 1, 2 & (i = 1, j \in N - \{1\}, \\ & c_{1j} \leq 1/2 \, L) \\ 0, 1 & (i, j \in N - \{1\}) \end{cases} \tag{77}$$

$$m \geq 1 \quad \text{and integer.} \tag{78}$$

In this formulation, L represents the maximum length of any route, m is the number of vehicles (a constant if it is fixed a priori or a variable) and $V(S)$ represents a lower bound on the number of vehicles required to visit the depot and all cities of S in the optimal solution. Its computation is developed later. Variables are only defined if each of the following three conditions holds:
 (i) $i < j$ (x_{ij} must be interpreted as x_{ji} whenever $i > j$);
 (ii) $d_i + d_j \leq D$;
 (iii) $c_{ij} + P(i) + P(j) \leq L$ where $P(i)$ is the length of the shortest path from 1 to i. (When C satisfies the triangle inequality, $P(i)$ is simply c_{1i}).

Constraints (74) and (75) specify the degree of each node while constraints (76) prohibit the formation of illegal subtours, i.e. subtours which are either
 (i) disconnected from the depot, or
 (ii) connected to the depot and having a total weight exceeding D, or

(iii) connected to the depot and having a total length exceeding L.

In [62], (VRP4) is solved by a constraint relaxation algorithm: integrality and subtour elimination constraints are first relaxed; integrality is obtained by branching on the variables while illegal subtours are eliminated as they are found to be violated (constraints (76) can be generated at non-integer solution). The algorithm can be summarized by the flow-chart depicted in Figure 1 ($\bar{z}$ represents the objective value of the current subproblem while z^* is the value of the incumbent). Most steps of the algorithm require no explanation in so far as they correspond to a standard branch and bound procedure. Howere, it is worth examining two steps in some detail.

(i) Subtour prevention constraints

Consider the solution at a given node h of the search tree: the solution contains

(i) sets of nodes $\{i_1, \ldots, i_u\}$ ($u > 1$) corresponding to chains $(i_1, \ldots, i_u)$ such that $1 \notin \{i_2, \ldots, i_{u-1}\}$ if $u > 2$ and for which *all* variables $x_{i_1 i_2}, x_{i_2 i_3}, \ldots, x_{i_{u-1} i_u}$ have been *fixed* at 1 and

(ii) nodes not belonging to such chains (we define for each such node i a singleton $\{i\}$).

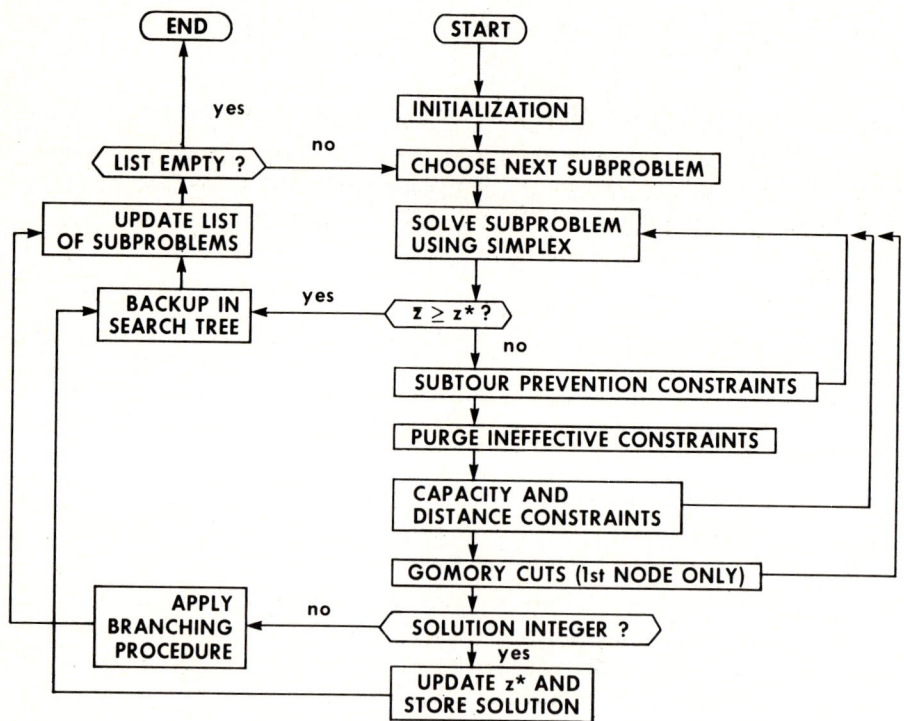

Fig. 1. Flow chart of the algorithm used for the solution fo (VRP4).

We refer to these sets of nodes S_k (corresponding to chains or single nodes) as *components*. Each S_k has an associated weight $w(S_k)$ defined as

$$w(S_k) = \sum_{i \in S_k} d_i \qquad (79)$$

and a length $l^h(S_k)$ defined as

$$l^h(S_k) = \begin{cases} \sum_{t=1}^{u-1} c_{i_t i_{t+1}} & \text{in the case of a chain} \\ & (i_1, \ldots, i_u) \\ 0 & \text{in the case of a node.} \end{cases} \qquad (80)$$

Consider a component S_r. If it corresponds to a chain, let p_r and q_r be the end nodes of that chain; if it corresponds to a node i, let $p_r = q_r = i$. In the first case, variable $x_{p_r q_r}$ can be *forced* to zero if
  (i) $p_r, q_r \in N - \{1\}$ or
  (ii) $p_r = 1$ and $l^h(S_r) + c_{1q_r} > L$.

Now consider to components S_r and S_s and let $i \in \{p_r, q_r\}, \{\overline{i}\} = \{p_r, q_r\} - \{i\}$, $j \in \{p_s, q_s\}, \{\overline{j}\} = \{p_s, q_s\} - \{j\}$. Variable x_{ij} can be forced to zero if
  (iii) $i, j \in N - \{1\}$ and $w(S_r) + w(S_s) > D$, or
  (iv) $i, j \in N - \{1\}$ and $P(\overline{i}) + P(\overline{j}) + l^h(S_r) + l^h(S_s) + c_{ij} > L$, or
  (v) $i = 1, j \in N - \{1\}$ and $P(\overline{j}) + l^h(S_s) + c_{1j} > L$.

At node h of the search tree, consider the graph $G' = (N - \{1\}, E', C)$ where E' is the set of all (undirected) edges (i, j) for which $i, j \in N - \{1\}$ and $x_{ij} > 0$. $N - \{1\}$ can be partitioned into K connected components $S_1, S_2, \ldots, S_K$ i.e. sets of nodes linked by «positive edges». Let $S_k^+ = S_k \cup \{1\}$. Each S_k^+ has a weight

$$w(S_k^+) = w(S_k) = \sum_{i \in S_k} d_i \qquad (81)$$

and a length

$$l^h(S_k^+) = \sum_{i,j \in S_k^+} c_{ij} x_{ij} \qquad (82)$$

It can be shown that the minimum number of vehicles required to visit all nodes of S_k^+ in the optimal solution is bounded below by

$$V^h(S_k) = \max\{\lceil w(S_k)/D \rceil, \lceil l^h(S_k^+)/L \rceil\}. \qquad (83)$$

Therefore, for each k, the following subtour elimination constraint can be imposed, whenever it is violated:

$$\sum_{i,j \in S_k} x_{ij} \leq |S_k| - V^h(S_k). \tag{84}$$

Recently, Gavish [37] suggested the use of a sharper bound $V_1^h(S_k)$ obtained by replacing in (83) $\lceil w(S_k)/D \rceil$ by the number of bins in the optimal solution of the bin packing problem [27] associated with S_k. Alternatively, one could use one of the known lower bounds for the bin packing problem [68].

In addition to imposing one type (84) constraint for each S_k, the authors also introduce a similar constraint for $S^* = S_1 \cup S_2 \cup \ldots \cup S_K$, whenever $K > 1$; this constraint is obtained by replacing S_k by S^* in (83) and (84). Computational experiments show that the presence of such a constraint helps to reduce the growth of the search tree.

Since constraints (84) are sometimes imposed when the current solution is fractional (see Figure 1), they may be satisfied at the time of their generation. In such cases, a violated constraint can sometimes be derived by means of a simple heuristic search [62].

It can be shown that if C does not satisfy the triangle inequality, $\lceil l^h(S_k^+)/L \rceil$ does not constitute a valid lower bound on the number of vehicles required to visit all nodes of S_k^+ in the optimal solution of subproblem h. In such cases, a valid constraint can be derived to eliminate subtours connected to the depot and which do not satisfy distance constraints. This constraint, which is derived in [62], can be expressed as

$$\sum_{j \in S - \{1\}} x_{1j} + 3 \sum_{\substack{i \in S - \{1\}, j \in \overline{S} \\ \text{or } i \in \overline{S}, j \in S - \{1\}}} x_{ij} \geq 4. \tag{85}$$

In (85), S is a subset of N satisfying
   (i) $1 \in S$;
   (ii) $|S| \geq 3$;
   (iii) the value of the TSP solution on S exceeds L. (This information is available when the constraint is generated).

Using the algorithm summarized by Figure 1 (or a slightly modified version for problems in which C did not satisfy the triangle inequality), Laporte et al. [62] have obtained exact solutions to problems involving up to 50 cities for problems in which C satisfied the triangle inequality and up to 60 cities for problems in which C did not possess this property.

(VRP4) can be stengthened by the inclusion of extra constraints such as *comb inequalities*. These constraints were first developed by Chvátal [15] for the TSP and later generalized by Grötschel and Padberg [44]. They constitute facets of the integer polytope associated with the TSP [45] and include subtour elimination constraints (6) as a special case. Their inclusion in a constraint relaxation algorithm for the TSP enabled Crowder and Padberg to solve exactly a 318-city problem [17].

Comb inequalities for the TSP were generalized for the VRP by Laporte and Nobert [60] and later modified by Laporte and Bourjolly [53]. In the case of the CVRP, they can be stated as follows. Let $W_l (l = 0, \ldots, k)$ be subsets of $N - \{1\}$ satisfying

$$|W_l - W_0| \geq 1 \qquad (l = 1, \ldots, k) \qquad (86)$$

$$|W_l \cap W_0| \geq 1 \qquad (l = 1, \ldots, k) \qquad (87)$$

$$|W_l \cap W_{l'}| = 1 \qquad (1 \leq l < l' \leq k). \qquad (88)$$

Then the following condition (comb inequality) holds for every feasible solution fo the CVRP:

$$\sum_{l=0}^{k} \sum_{i,j \in W_l} x_{ij} \leq \sum_{l=0}^{k} |W_l|$$

$$- \left\lceil \frac{1}{2} \sum_{l=1}^{k} [V_1(W_l) + V_1(W_l - W_0) + V_1(W_l \cap W_0)] \right\rceil \qquad (89)$$

where $V_1(S)$ is a lower bound on the optimal value of the bin packing solution associated with S.

It can be shown [53] that in general, comb inequalities defined by (89) do not constitute facets of the integer polytope associated with the CVRP.

5.2.2. Capacitated symmetrical road-TSP

Fleischmann [31, 32] uses the expression «road-TSP» to designate routing problems derived from a real-life (sparse) road network and in which multiple passages through the same city are allowed whenever they are economical. Such problems have also been referred to as complete cycle problems by other authors [73]. The two-index formulation used by Fleischmann belongs to the same family as those of Dantzig et al. [18], Miliotis [70 - 72], Land [51] and Crowder and Padberg [17]. It includes degree constraints, integrality constraints, connectivity constraints (preventing the occurence of disconnected components) and a new class of constraints called *3-star constraints*, generalizing comb inequalities. Fleischmann uses a pure cutting planes method and obtains exact solutions for road-TSPs involving up to 292 cities [32]. He applies a similar approach to the capacitated road-TSP (CRTSP) [31] by first considering the following relaxation of the problem:

$$\text{(VRP5)} \qquad \text{minimize} \sum_{i,j \in N} c_{ij} x_{ij}$$

subject to

$$\sum_{j=2}^{m} x_{1j} \geq 2m \tag{90}$$

$$\sum_{i<k} x_{ik} + \sum_{j>k} x_{kj} \geq 2 \text{ and even} \qquad (k \in N - \{1\}) \tag{91}$$

$$\sum_{\substack{i \in \underline{S}, j \in \overline{S} \\ \text{or } i \in \overline{S}, j \in S}} x_{ij} \geq 2 V(S) \qquad \begin{array}{l}(2 \leq |S| \leq n-2; \\ S \subset N - \{1\})\end{array} \tag{92}$$

$$x_{ij} \geq 0 \text{ and integer} \qquad (i < j) \tag{93}$$

$$m \geq 1 \text{ and integer.} \tag{94}$$

Here, $V(S)$ is defined as $\lceil \Sigma_{i \in S} d_i/D \rceil$, but could easily be replaced by the bin packing bound. (VRP5) is solved by means of a constraint relaxation algorithm of the type used for the road-TSP. 3-star constraints generalizing those used for the road-TSP (in the same way as comb inequalities used for the CVRP [60] generalize those of Grötschel and Padberg [44, 45]), are generated to eliminate fractional solutions not violating connectivity constraints (92). If the solution to (VRP5) is feasible for the CRTSP, it is then optimal; otherwise, it constitutes a valid lower bound on the value of the CRTSP solution. Fleischmann [31] gives an example of a solution satisfying (90) - (94), but infeasible for the CRTSP. Laporte [52] describes a procedure for generating constraints in order to eliminate such solutions.

Using this approach, Fleischmann succeeded in solving to optimality some CRTSPs ranging from 28 to 68 cities [31].

5.2.3. Capacitated location-routing symmetrical problems

Laporte et al. [61] used a two-index formulation to treat a family of CVRPs involving simultaneous depot location and routing. Consider $R \subseteq N$, a set of *potential depots*. The number P of such nodes used as depots in the optimal solution must lie between two prespecified bounds $\underline{P} \geq 1$ and $\overline{P} \leq |R|$. The cost of using node r as a depot is equal to g_r. There are m_r identical vehicles based at depot r, each with the same capacity D and a fixed cost f_r. To each node i of $N - R$, is associated a nonnegative requirement d_i ($\leq D$).

The problem consists of selecting depot sites (when $\underline{P} < |R|$), of determining how many vehicles are based at each selected depot and of establishing vehicle routes in such a way that

(i) each route starts and ends at the same depot;

(ii) all requirements are met exactly once by a vehicle (the same city i may be visited more than once if this saves distance but then, the requirement d_i of that city is satisfied only once by one vehicle);

(iii) the sum of all requirements satisfied by any vehicle does not exceed D;
(iv) $\underline{P} \leq P \leq \overline{P}$;
(v) for each node r used as a depot, the number of vehicles lies between two prespecified bounds $\underline{m}_r$ and $\overline{m}_r$: $1 \leq \underline{m}_r \leq m_r \leq \overline{m}_r$;
(vi) the total cost is minimized.

In addition to the notation already introduced, define

T : an arbitrarily large number

x_{ij} : a variable indicating the number of times edge (i, j) is used in the optimal solution. x_{ij} is not defined if $i \geq j$, if $i, j \in R$ or if $d_i + d_j > D$. x_{ij} must be interpreted as x_{ji} whenever $i > j$

y_r : a binary variable indicating whether node r is used as a depot ($y_r = 1$) or not ($y_r = 0$).

The problem can be formulated as follows:

$$\text{(VRP6)} \quad \text{minimize} \quad \sum_{i,j \in N} c_{ij} x_{ij} + \sum_{r \in R} (g_r y_r + f_r m_r)$$

subject to

$$\sum_{i<k} x_{ik} + \sum_{k<j} x_{kj} = 2 \qquad (k \in N - R) \qquad (95)$$

$$\sum_{i<r} x_{ir} + \sum_{r<j} x_{rj} = 2m_r \qquad (r \in R) \qquad (96)$$

$$\sum_{i,j \in S} x_{ij} \leq |S| - \left\lceil \frac{\sum_{k \in S} d_k}{D} \right\rceil \qquad (S \subseteq N - R, |S| \geq 3) \qquad (97)$$

$$x_{i_1 i_2} + 3 x_{i_2 i_3} + x_{i_3 i_4} \leq 4 \qquad (i_1, i_4 \in R; i_2, i_3 \in N - R) \qquad (98)$$

$$x_{i_1 i_2} + x_{i_{h-1} i_h} + \qquad\qquad (h \geq 5; i_1, i_h \in R;$$
$$\qquad\qquad\qquad\qquad\qquad\qquad i_2, \ldots, i_{h-1} \in N - R; \qquad (99)$$
$$+ 2 \sum_{i,j \in \{i_2, \ldots, i_{h-1}\}} x_{ij} \leq 2h - 5 \qquad d_{i_2} + \ldots + d_{i_{h-1}} \leq D)$$

$$y_r \leq m_r \leq T y_r \qquad (r \in R) \qquad (100)$$

$$\underline{m}_r \leq m_r \leq \overline{m}_r \qquad (r \in R) \qquad (101)$$

$$\underline{P} \leq \sum_{r \in R} y_r \leq \overline{P} \qquad (102)$$

$$y_r = 0, 1 \qquad (r \in R) \qquad (103)$$

$$x_{ij} = \begin{cases} 0, 1 & (i, j \in N - R) \\ 0, 1, 2 & (i \text{ or } j \in R). \end{cases} \qquad (104)$$

In this formulation, constraints (95) specify that each city not used as a depot must be serviced exactly once by a vehicle. Similarly, constraints (96) express the fact that m_r vehicles must leave and enter each city in R. Constraints (97) ensure that the solution does not contain illegal subtours, i.e. subtours disjoint from R or subtours having a total weight exceeding D. As previously, the bin packing bound could be substituted in the right hand side of (97). Constraints (98) and (99) are *chain barring constraints*. They ensure that each route starts and ends at the same depot. Their development is fairly lengthy and is fully provided in [61].

Using a constraint relaxation approach in which integrality constraints, subtour elimination constraints and chain barring constraints are first relaxed, the authors succeeded in solving to optimality

(i) problems without capacity restrictions involving up to 25 cities;

(ii) problems with capacity restrictions containing up to 20 cities.

It was observed that imposing large depot costs tended to produce easier problems (the locational aspect of the problem becoming less predominent); on the other hand, the size of vehicle costs did not seem to affect the behaviour of the algorithm one way or the other.

Other types of less sophisticated location-routing problems are also treated in [57, 63].

5.6.4. *Asymmetrical VRPs under capacity or distance restrictions*

Many of the concepts developed for symmetrical VRPs are also usable in the asymmetrical case. Two types of asymmetrical VRPs (the CVRP and the DVRP) are studied in [55] and [56]. It is assumed in each case that an upper bound $\overline{m}$ on the number of vehicles in known. Then the graph $G = (N, E, C)$ associated with the VRP is extended into a graph $G' = (N', E', C')$ as suggested by Lenstra and Rinnooy Kan [64] for example. First, $\overline{m} - 1$ artificial depots are introduced; let $N^* = \{1, n+1, n+2, \ldots, n+\overline{m}-1\}$ be the set of depots and let $N' = N \cup N^*$. $C' = (c'_{ij})$ is defined as in (9) except that $c_{ii} = \infty$ for all $i \in N'$. As explained in section 2, the nature of the problem varies according to λ.

The problem is formulated as (TSP) (see section 2.1) where N is replaced by N'. The solution of the TSP on G' can easily be transformed into the solution of a multiple TSP on G where $m \leq \overline{m}$:

(i) if $i \in N - \{1\}$ and $j \in N' - N$, replace (i, j) by $(i, 1)$;

(ii) if $i \in N' - N$ and $j \in N - \{1\}$, replace (i, j) by $(1, j)$;
(iii) if $i, j \in N' - N$, remove (i, j).

The transformed solution consists of at most $\bar{m}$ Hamiltonian circuits covering all nodes of G and having a unique common node at 1. This solution is feasible for the VRP if and only if

(i) the weight of each Hamiltonian circuit is less than or equal to D (in the CVRP);

(ii) the length of each Hamiltonian circuit is less than or equal to L (in the DVRP).

These two types of infeasibilities can be eliminated at the source, during the solution of the TSP of G'. This is done by extending or otherwise modifying the Carpaneto and Toth algorithm [6] for asymmetrical TSPs. Using this approach, exact solutions were obtained for problems containing up to 260 cities in the case of the CVRP and up to 100 cities in the case of the DVRP.

6. Commodity flow formulations

In addition to the x_{ij}^k (or x_{ij}) variables used in the CVRP formulations of the previous section, commodity flow formulations associate flow variables y_{ij}^l (or y_{ij}) with the arcs. These variables indicate how much of the demand destined for city l travels on arc (i, j). Such formulations were first proposed by Garvin et al. [35] in an oil delivery problem. Many variants were later developed and analyzed by Gavish and Graves [38, 39] for various TSP extensions. As far as the authors are aware, no exact VRP algorithm based on a commodity flow formulation has ever been constructed and tested. This section presents some of the commodity flow formulations set up for the CVRP. They are adapted from [37 - 39].

First consider CVRPs in which d_i represents the *demand* of city i and where the number $\bar{m}$ of vehicles is fixed. The following formulation is adapted from the the work of Gavish and Graves [38] (these authors also consider time limit constraints on the routes).

$$(\text{VRP7}) \qquad \text{minimize} \quad \sum_{i=1}^{n} \sum_{j=1}^{n} c_{ij} x_{ij}$$

subject to

$$\sum_{i=1}^{n} x_{ij} = 1 \qquad (j = 2, \ldots, n) \qquad (105)$$

$$\sum_{j=1}^{n} x_{ij} = 1 \qquad (i = 2, \ldots, n) \qquad (106)$$

$$\sum_{i=2}^{n} x_{i1} = \sum_{j=2}^{n} x_{1j} = \bar{m} \qquad (107)$$

$$\sum_{i=1}^{n} y_{ij}^l - \sum_{i=1}^{n} y_{ji}^l = \begin{cases} d_l \\ 0 \end{cases} \quad \begin{array}{l}(j=l; j=2,\ldots,n) \\ (j \neq l; j, l=2,\ldots,n)\end{array} \qquad (108)$$

$$\sum_{i=1}^{n} y_{ij}^l \leqslant D\, x_{ij} \qquad (i \neq j; i,j = ,\ldots,n) \qquad (109)$$

$$y_{ij}^l \geqslant 0 \qquad (i,j,l = 1,\ldots,n) \qquad (110)$$

$$x_{ij} = 0, 1 \qquad (i,j = 1,\ldots,n) \qquad (111)$$

In this formulation, constraints (105), (106) and (111) ensure that every city, except the depot is visited exactly once; constraints (107) and (111) specify that there must be $\bar{m}$ vehicles; constraints (108) are flow balancing constraints which guarantee that the demand of each node is satisfied; constraints (108) and (109) ensure that the solution will contain no illegal subtour; constraints (109) are imposed in order to ensure that vehicle capacities are never exceeded and that goods can only be moved through an arc if a vehicle uses this arc.

In the case where the number of vehicles is bounded by $\bar{m}$, constraints (107) can be replaced by

$$\sum_{i=2}^{n} x_{i1} = \sum_{j=2}^{n} x_{1j} \leqslant \bar{m} \qquad (107')$$

Similarly, if the number of vehicles is unspecified, constraints (107) become

$$\sum_{i=2}^{n} x_{i1} = \sum_{j=2}^{n} x_{1j} \qquad (107'')$$

As suggested by Magnanti [67], such a model can easily be transformed to deal with the case of heterogeneous fleets. This is done by defining variables x_{ij}^k indicating whether or not vehicle k travels directly from i to j. Constraints involving x_{ij} variables in (VRP7) must of course be modified accordingly.

In Gavish and Graves [38], the index l is removed and the problem becomes:

$$(\text{VRP8}) \qquad \text{minimize} \quad \sum_{i=1}^{n} \sum_{j=1}^{n} c_{ij}\, x_{ij}$$

subject to (105) - (107), (111) and

$$\sum_{i=1}^{n} y_{ij} - \sum_{i=1}^{n} y_{ji} = d_j \qquad (j=2,\ldots,n) \qquad (112)$$

$$y_{ij} \leq D\, x_{ij} \qquad (i,j=1,\ldots,n) \qquad (113)$$

$$y_{ij} \geq 0 \qquad (i,j=1,\ldots,n) \qquad (114)$$

In (VRP8), y_{ij} represents the quantity of goods travelling on arc (i,j).

Using different arguments, Magnanti [67] and Nobert [76] have proved that by applying Benders' decomposition to (VRP7) or (VRP8), one obtains the subtour elimination constraints vehicle flow model (VRP4) specialized to the CVRP.

In [37], Gavish presents some more constrained formulations.

$$\text{(VRP9)} \qquad \text{minimize} \quad \sum_{i=1}^{n}\sum_{j=1}^{n} c_{ij} x_{ij}$$

subject to (105) - (107), (110), (111) and

$$\sum_{i=1}^{n} y^l_{ij} - \sum_{i=1}^{n} y^l_{ji} = \begin{cases} d_l & (j=l;\, l=2,\ldots,n) \\ 0 & (j \neq l;\, j,l=2,\ldots,n) \\ -d_l & (j=1;\, l=2,\ldots,n) \end{cases} \qquad (115)$$

$$y^l_{ij} \leq d_l\, x_{ij} \qquad \begin{array}{l}(i,l=2,\ldots,n;\\ j=1,\ldots,n)\end{array} \qquad (116)$$

$$\sum_{i=2}^{n}\sum_{l=2}^{n} y^l_{ji} \leq D - d_j \qquad (j=1,\ldots,n;\, d_1=0) \qquad (117)$$

The elimination of index l yields a simpler model:

$$\text{(VRP10)} \qquad \text{minimize} \quad \sum_{i=1}^{n}\sum_{j=1}^{n} c_{ij} x_{ij}$$

subject to (105) - (107), (111), (112), (114) and

$$d_j x_{ij} \leq y_{ij} \leq (D - d_i) x_{ij} \qquad (i,j=1,\ldots,n) \qquad (118)$$

In order to establish comparisons between these formulations, consider LPi, the value of the LP relaxation of (VRPi). Gavish [37] shows that

(i) LP9 $\geq$ LP10   (This is a generalization of a similar result for the TSP proved in [87])

(ii) LP4 = LP10   (when $V(S) = \sum_{i \in S} d_i/D$ in (76))

(iii) $LP4 \geqslant LP9 \geqslant LP10$ (when $V(S) =$ bin packing bound in (76)).

These relationships and the equivalence between (VRP7) and (VRP4) suggest that it many be relatively efficient to use (VRP4) with the bin packing bound to solve the CVRP.

In [38], Gavish and Graves report some computational results relative to the bound associated with various relaxations of (VRP10) for CVRPs ranging from 10 to 30 cities. These results show the presence of fairly large gaps between the VRP optimum and the LP optimum when $n \geqslant 15$. Finally, we mention the existence of various commodity flow formulations for other types of VRPs. These were developed by Gavish and Graves [38, 39] for multi-depot VRPs, dial-a-ride problems and school bus problems.

7. Conclusion

Many person-years of research time have been spent on the development of solution methods for the VRP. Yet, despite all these efforts, most types of VRPs remain virtually unsolved: exact methods can only handle problems of relatively modest dimensions.

Of all methods considered in this survey, direct tree search algorithms offer, in our opinion, the least potential for growth. Dynamic programming algorithms, on the other hand do not appear to have reached their maximum potential; the state-space relaxation approach developed in [13] does not seem to have been exploited fully. Similarly, several ILP formulations have been provided, but few have led to the construction of exact algorithms. The rare cases that have been published would indicate that ILP works quite efficiently on some types of problems.

The most successful algorithms presented in this survey take the greatest possible advantage of the problem on hand. Tight problems cannot be handled in the same manner as loose problems and, in general, symmetrical and asymmetrical problems require different types of algorithms. One obvious tendency has been to move from general to particularized algorithms.

Looking ahead, it is safe to say that the study and development of exact algorithms for the vehicle routing problem will remain a major research topic in years to come and should yield new theoretical developments.

References

[1] Y.K. Agarwal, «Set Partitioning Approach to Vehicle Routing», *presented at the TIMS/ORSA Conference, Boston*, 1985.
[2] M. Baliński and R. Quandt, «On an Integer Program for a Delivery Problem», *Operations Research*, 12, 300 - 304, 1964.
[3] J.F. Benders, «Partitioning Procedures for Solving Mixed-Variables Programming Problems», *Numerische Mathematik* 4, 238 - 252, 1962.

[4] L.D. Bodin and B.L. Golden, «Classification in Vehicle Routing and Scheduling», *Networks* 11, 97 - 108, 1981.
[5] L.D. Bodin, B.L. Golden, A. Assad and M. Ball, «Routing and Scheduling of Vehicles and Crews, The State of the Art», *Computers & Operations Research* 10, 69 - 211, 1983.
[6] G. Carpaneto and P. Toth, «Some New Branching and Bounding Criteria for the Asymmetric Travelling Salesman Problem», *Management Science* 26, 736 - 743, 1980.
[7] N. Christofides, «The Vehicle Routing Problem», *RAIRO (recherche opérationnelle)* 10, 55 - 70, 1976.
[8] N. Christofides, «Vehicle Routing», in E.L. Lawler, J.K. Lenstra, A.H.G. Rinnooy Kan and D.B. Shmoys, eds., *The Traveling Salesman Problem: A Guided Tour of Combinatorial Optimization*, Wiley, 1985.
[9] N. Christofides, «Vehicle Scheduling and Routing», *presented at the 12th International Symposium of Mathematical Programming, Cambridge, Massachusetts*, 1985.
[10] N. Christofides and S. Eilon, «An Algorithm for the Vehicle Dispatching Problem», *Operational Research Quarterly* 20, 309 - 318, 1969.
[11] N. Christofides, A. Mingozzi and P. Toth, «The Vehicle Routing Problem», in N. Christofides, A. Mingozzi, P. Toth and C. Sandi, eds., *Combinatorial Optimization*, Wiley, 1979.
[12] N. Christofides, A. Mingozzi and P. Toth, «Exact Algorithms for the Vehicle Routing Problem, Based on Spanning Tree Shortest Path Relaxations», *Mathematical Programming* 20, 255 - 282, 1981.
[13] N. Christofides, A. Mingozzi and P. Toth, «Space State Relaxation Procedures for the Computation of Bounds to Routing Problems», *Networks* 11, 145 - 164, 1981.
[14] N. Christofides and M. Thornton, «A Shortest Path Algorithm for Generalized Weighted Matchings», *Report IC-OR-82/2, Imperial College of Science and Technology, London*, 1982.
[15] V. Chvátal, «Edmonds Polytopes and Weakly Hamiltonian Graphs», *Mathematical Programming* 5, 29 - 40, 1973.
[16] G. Clarke and J.W. Wright, «Scheduling of Vehicles from a Central Depot to a Number of Delivery Points», *Operations Research* 12, 568 - 581, 1964.
[17] H. Crowder and M.W. Padberg, «Solving Large-Scale Symmetric Travelling Salesman Problems to Optimality», *Management Science* 26, 495 - 509, 1980.
[18] G.B. Dantzig, D.R. Fulkerson and S.M. Johnson, «Solution of a Large Scale Travelling Salesman Problem», *Operations Research* 2, 393 - 410, 1954.
[19] G.B. Dantzig and J. Ramser, «The Truck Dispatching Problem», *Management Science* 6, 81 - 91, 1959.
[20] J. Desrosiers, Y. Dumas and F. Soumis, «The Multiple Vehicles Many to Many Routing Problem with Time Windows», *Cahiers du GERAD, G-84-13, Ecole des Hautes Commerciales de Montréal*, 1984.
[21] J. Desrosiers, Y. Dumas and F. Soumis, «A Dynamic Programming Method for the Large Scale Single Vehicle Dial-a-Ride with Time Windows», *American Journal of Mathematical and Management Sciences* (to appear).
[22] J. Desrosiers, P. Pelletier and F. Soumis, «Plus court chemin avec contraintes d'horaires», *RAIRO (recherche opérationnelle)* 17, 357 - 377, 1983.
[23] J. Desrosiers, M. Sauvé and F. Soumis, «Shortest Path with Time Windows Problem», *Publication 389, Centre de Recherche sur les Transports, Université de Montréal*, 1985.
[24] M. Desrochers and F. Soumis, «A Generalized Permanent Labelling Algorithms for the Shortest Path Problem with Time Windows», *Publication 394A, Centre de Recherche sue les Transports, Université de Montréal*, 1985.
[25] J. Desrosiers, F. Soumis and M. Desrochers, «Routing with Time Windows by Column Generation», *Networks* 14, 545 - 565, 1984.
[26] E.W. Dijkstra, «A Note on Two Problems in Connexion with Graphs», *Numerische Mathematik* 1, 269 - 271, 1959.
[27] S. Eilon and N. Christofides, «The Loading Problem», *Management Science* 17, 259 - 267, 1971.
[28] S. Eilon, C.D.T. Watson-Gandy and N. Christofides, *Distribution Management: Mathematical Modelling and Practical Analysis*, Griffin, 1971.
[29] M. Fisher and R. Jaikumar, «A Decomposition Algorithm for Large-Scale Vehicle Routing», *Working paper 78-11-05, Department of Decision Sciences, University of Pennsylvania*, 1978.
[30] M. Fischer and R. Jaikumar, «A Generalized Assignment Heuristic for Vehicle Routing», *Networks* 11, 109 - 124, 1981.
[31] B. Fleischmann, «Linear Programming Approaches to Travelling Salesman and Vehicle Scheduling Problems», *presented at the XI. International Symposium on Mathematical Programming, Bonn*, 1982.

[32] B. Fleischmann, «A Cutting Planes Procedure for the Travelling Salesman Problem on Road Networks», *Research Report QM-02-83, University of Hamburg*, 1983.
[33] L.R. Ford and D.R. Fulkerson, *Flows in Neworks*, Princeton University Press, 1962.
[34] B.A. Foster and D.M. Ryan, «An Integer Programming Approach to the Vehicle Scheduling Problem», *Operational Research Quarterly* 27, 367 - 384, 1976.
[35] W.M. Garvin, H.W. Crandall, J.B. John and R.A. Spellman, «Applications of Linear Programming in the Oil Industry», *Management Science* 3, 407 - 430, 1957.
[36] T.J. Gaskell, «Bases for Vehicle Fleet Scheduling», *Operational Research Quarterly* 19, 281 - 295, 1967.
[37] B. Gavish, «The Delivery Problem: New Cutting Planes Procedures», *presented at the TIMS XXVI Conference, Copenhagen*, 1984.
[38] B. Gavish and S.C. Graves, «The Travelling Salesman Problem and Related Problems», *Working Paper no. 7905, Graduate School of Management, University of Rochester*, 1979.
[39] B. Gavish and S.C. Graves, «Scheduling and Routing in Transportation and Distribution Systems: Formulations and New Relaxations», *Working Paper, Graduate School of Management, University of Rochester*, 1982.
[40] B.E. Gillet and L.R. Miller, «A Heuristic Algorithm for the Vehicle-Dispatch Problem», *Operations Research* 22, 341 - 349, 1974.
[41] B. Golden, L. Bodin, T. Doyle and W. Stewart, «Approximate Travelling Salesman Algorithms», *Operations Research* 28, 694 - 711, 1980.
[42] B. Golden, T.L. Magnanti and H.Q. Nguyen, «Implementing Vehicle Routing Algorithms», *Networks* 7, 113 - 148, 1977.
[43] B. Golden and W. Stewart, «Empirical Analysis of Heuristics», in E.L. Lawler, J.K. Lenstra, A.H.G. Rinnooy Kan and D.B. Shmoys, eds., *The Traveling Salesman Problem: A Guided Tour of Combinatorial Optimization*, Wiley, 1985.
[44] M. Grötschel and M.W. Padberg, «On the Symmetric Travelling Salesman Problem I: Inequalities», *Mathematical Programming* 16, 265 - 280, 1979.
[45] M. Grötschel and M.W. Padberg, «On the Symmetric Travelling Salesman Problem II: Lifting Theorems and Facets», *Mathematical Programming* 16, 281 - 302, 1979.
[46] R. Hays, «The Delivery Problem», *Management Science Research Report no. 106, Carnegie Institute of Technology*, 1967.
[47] M. Held and R.M. Karp, «A Dynamic Programming Approach to Sequencing Problems», *SIAM* 10, 196 - 210, 1962.
[48] D. Houck, J. Picard, M. Queyranne and R. Vemuganti, «The Travelling Salesman Problem as a Constrained Shortest Path Problem: Theory and Computational Experience», *Opsearch* 17, 93 - 109, 1980.
[49] A. Kolen, A.H.G. Rinnooy Kan and H. Trienekens, «Vehicle Routing with Time Windows», *Report 8433/0, Erasmus University, Rotterdam*, 1984.
[50] J.B. Kruskal, «On the Shortest Spanning Subtree of a Graph and the Travelling Salesman Problem», *Proceedings of the American Mathematical Society*, 2, 48 - 50, 1956.
[51] A.H. Land, «The Solution of Some 100-City Travelling Salesman Problems», *presented at the Tenth International Symposium on Mathematical Programming, Montreal*, 1979.
[52] G. Laporte, «An Integer Programming Approach to the Vehicle Scheduling Problem», *Cahiers du GERAD, G-82-10, Ecole des Hautes Etudes Commerciales de Montréal*, 1982.
[53] G. Laporte and J.-M. Bourjolly, «Some Further Results on k-Star Constraints and Comb Inequalities», *Cahiers du GERAD, G-84-17, Ecole des Hautes Etudes Commerciales de Montréal*, 1984.
[54] G. Laporte, M. Desrochers and Y. Nobert, «Two Exact Algorithms for the Distance Constrained Vehicle Routing Problem», *Networks* 14, 161 - 172, 1984.
[55] G. Laporte, H. Mercure and Y. Nobert, «An Exact Algorithm for the Asymmetrical Capacitated Vehicle Routing Problem», *Networks* (to appear).
[56] G. Laporte, T. Nguyen and Y. Nobert, «A Branch and Bound Algorithm for the Asymmetrical Distance Constrained Vehicle Routing Problem», *Mathematical Modelling* (to appear).
[57] G. Laporte and Y. Nobert, «An Exact Algorithm for Minimizing Routing and Operating Costs in Depot Location», *European Journal of Operational Research* 6, 224 - 226, 1981.
[58] G. Laporte and Y. Nobert, «A Branch and Bound Algorithm for the Capacitated Vehicle Routing Problem», *Operations Research Spektrum*, 5, 77 - 85, 1983.
[59] G. Laporte, Y. Nobert, «Algorithmes de relaxation de contraintes pour le problème du voyageur de commerce symétrique et ses extensions», *Annales des Sciences mathématiques du Québec*, VII, 109 - 137, 1983.

[60] G. Laporte and Y. Nobert, «Comb Inequalities for the Vehicle Routing Problem», *Methods of Operations Research* 51, 271 - 276, 1984.
[61] G. Laporte, Y. Nobert and D. Arpin, «An Exact Algorithm for Solving a Capacitated Location-Routing Problem», in *Location Decisions: Methodology and Applications*, J.C. Baltzer AG, Scientific Publishing Co. (to appear).
[62] G. Laporte, Y. Nobert and M. Desrochers, «Optimal Routing Under Capacity and Distance Restrictions», *Operations Research* 33, 1050 - 1073, 1985.
[63] G. Laporte, Y. Nobert and P. Pelletier, «Hamiltonian Location Problems», *European Journal of Operational Research* 12, 82 - 89, 1983.
[64] J.K. Lenstra and A.H.G. Rinnooy Kan, «Some Simple Applications of the Travelling Salesman Problem», *Operational Research Quarterly* 26, 717 - 734, 1975.
[65] J.K. Lenstra and A.H.G. Rinnooy Kan, «Complexity of Vehicle Routing and Scheduling Problems», *Networks* 11, 221 - 228, 1981.
[66] J.D.C. Little, K.G. Murty, D.W. Sweeney and C. Karel, «An Algorithm for the Travelling Salesman Problem», *Operations Research* 11, 972 - 989, 1963.
[67] T.L. Magnanti, «Combinatorial Optimization and Vehicle Fleet Planning: Perspectives and Prospects», *Networks* 11, 179 - 214, 1981.
[68] S. Martello and P. Toth, «Lower Bounds and Reduction Procedures for the Bin Packing Problem», *Working Paper 1-84, Progretto e Analisi di Algoritmi, University of Bologna*, 1984.
[69] S. Martello and P. Toth, «Linear Assignment Problems», this volume.
[70] P. Miliotis, «Combining Cutting Planes and Branch-and-Bound Methods to Solve Integer Programming Problems: Applications to the Travelling Salesman Problem and the 1-Matching Problem», *Ph.D. Thesis, University of London*, 1975.
[71] P. Miliotis, «Integer Programming Approaches to the Travelling Salesman Problem», *Mathematical Programming* 10, 367 - 378, 1976.
[72] P. Miliotis, «Using Cutting Planes to Solve the Symmetric Travelling Salesman Problem», *Mathematical Programming* 15, 177 - 188, 1978.
[73] P. Miliotis, G. Laporte and Y. Nobert, «Computational Comparison of Two Methods for Finding the Shortest Complete Cycle or Circuit in a Graph», *RAIRO (recherche opérationnelle)* 15, 233 - 239, 1981.
[74] C.E. Miller, A.W. Tucker and R.A. Zemlin, «Integer Programming Formulation of Travelling Salesman Problems», *Journal of ACM* 7, 326 - 329, 1969.
[75] R. Mole, «A Survey of Local Delivery Vehicle Routing Methodology», *Journal of the Operational Research Society* 30, 245 - 252, 1979.
[76] Y. Nobert, «Construction d'algorithmes optimaux pour des extensions au problème du voyageur de commerce», *Thèse de doctorat, Département d'Informatique et de Recherche Opérationnelle, Université de Montréal*, 1982.
[77] C. Orloff, «Route Constrained Fleet Scheduling», *Transportation Science* 10, 149 - 168, 1976.
[78] J.F. Pierce, «Direct Search Algorithms for the Truck-Dispatching Problem», *Transportation Research* 3, 1 - 42, 1969.
[79] H.N. Psaraftis, «A Dynamic Programming Solution to the Single Vehicle Many-to-Many Immediate Request Dial-a-Ride Problem», *Transportation Science* 14, 130 - 154, 1980.
[80] H.N. Psaraftis, «An Exact Algorithm for the Single Vehicle Many-to-Many Dial-a-Ride Problem with Time Windows», *Transportation Science* 17, 351 - 360, 1983.
[81] H.N. Psaraftis, «k-Interchange Procedures for Local Search in a Precedence Constrained Routing Problem», *European Journal of Operational Research* 13, 391 - 402, 1983.
[82] M.R. Rao and S. Zionts, «Allocation of Transportation Units to Alternative Trips - A Column Generation Scheme with Out-of-Kilter Subproblems», *Operations Research* 16, 52 - 63, 1968.
[83] R.W. Simpson and P. Kivestu, «Network Models in Transportation: a Bibliography», *Report 76-11, Center for Transportation Studies, Massachusetts Institute of Technology*, 1976.
[84] C. Toregas and C. ReVelle, «Location Under Time or Distance Constraints», *Papers of the Regional Science Association*, 28, 133 - 143, 1972.
[85] P. Toth, «Heuristic Algorithms for the Vehicle Routing Problem», *presented at the Workshop on Routing Problems, Hamburg*, 1984.
[86] C.D.T. Watson-Gandy and L.R. Foulds, «The Vehicle Scheduling Problem: a Survey», *New Zealand Operational Research* 9, 73 - 92, 1981.
[87] R.T. Wong, «Integer Programming Formulations of the Travelling Salesman Problem», *Proceedings of the IEEE International Conference on Circuits and Computers*, 149 - 152, 1980.
[88] A. Wren and A. Holliday, «Computer Scheduling of Vehicles from One or More Depots to a Number of Delivery Points», *Operational Research Quarterly* 23, 333 - 344, 1972.

Gilbert Laporte
Ecole des Hautes Etudes Commerciales
5255 avenue Decelles
Montréal H3T 1V6
Canada

Yves Nobert
Départment des Sciences administratives
Université du Québec à Montréal
1495 rue St. Denis
Montréal H3C 3P8
Canada

THE STEINER PROBLEM IN GRAPHS*

Nelson MACULAN

1. Introduction

The purpose of this paper is to review some formulations and some procedures that have been suggested for the solution of the Steiner Problem in graphs.

In section 1 we present the definition and some results associated with the classical Steiner Problem also known as the Euclidean Steiner Problem (ESP). The Steiner Problem in graphs is considered in section 2, where the Steiner Problem in an undirected graph (SPUG) and the Steiner Problem in a directed graph (SPDG) are defined. In this same section we show the transformation of a SPUG in a SPDG. The Lawler algorithm for the SPUG is presented in section 4. In section 5, we present four integer programming formulations of the Steiner Problem in graphs, compare the linear relaxations of three of these formulations and consider an application of Benders method to solve one of these formulations illustrated by an example. In section 6 solution methods based on these formulations are presented and some computational results are described. The last section discusses applications and conclusions.

2. Euclidean Steiner Problem

Our historical sketch about the Euclidean Steiner Problem (ESP) is based on the works written by Courant and Robbins [10] and Kuhn [28].

The problem posed by Fermat early in the 17th century was stated as follows: given three points in the plane, find a fourth point such that the sum of its distances to the three given points is a minimum. This problem may have travelled to Italy with Mersenne, see Hiriart-Urruty [25]; it is known that before 1640 Torricelli had solved the problem.

He asserted that the circles circumscribing the equilateral triangles constructed on the sides of and outside the given triangle intersect in the point (Torricelli point) that is sought. In Cavalieri's «Exercitationes Geometricae» of 1647, it is

---

*This work was partially supported by Conselho Nacional de Desenvolvimento Cientifico e Tecnologico, CNPq 300195-83 and by FINEP.

shown that the sides of the given triangle subtend angles of 120° from the Torricelli point. F. Heinen, apparently, is the first to prove in 1834 that, for a triangle in which one angle is greater or equal to 120°, the vertex of this angle is the minimizing point.

The General Fermat Problem refers to the problem of minimizing the weighted sum of distances from n given points in the plane. Let there be given n distinct points $P_i \equiv (x_i, y_i)$ in the plane and n possible weights l_i, where $i = 1, 2, \ldots, n$. For $P \equiv (X, y)$ let $d_i(P) = \sqrt{(x - x_i)^2 + (y - y_i)^2}$, the General Fermat Problem asks for a point P that minimizes $f(P) = \Sigma_{i=1}^{n} l_i d_i(P)$, see [28].

Jacob Steiner, the famous representative of Geometry at the University of Berlin in the early 19th century wrote on the subject. The Fermat Problem has been popularized in [10] under the name of the Steiner Problem. But the Euclidean Steiner Problem considered here is to find a tree which spans n given points in the Euclidean plane with minimum length. A minimum tree which spans these n points, called Steiner Tree, may contain nodes other (Steiner points) than the points which are spanned.

Other related problems with the Euclidean Steiner Problem are considered in Melzak [35].

We will depict some examples of the Euclidean Steiner Problem for $n = 3$, $n = 4, n = 5$ and $n = 6$ in figure 2.1.

For the regular polygons when $n \geqslant 6$, there is no Steiner point and the total length of the Steiner Tree is equal to $(n - 1)l$, where l is the lenght of the side.

We have presented some easy examples; in Cockayne [9] it is presented an algorithm for the Steiner Problem in the Euclidean plane (ESP) and some extensions for any metric space and in Gilbert and Pollak [19] it is presented the conditions which simplify the task of constructing a Steiner minimal tree.

3. The Steiner Problem in graphs

In this section we will consider the Steiner Problem in an undirected graph (SPUG) and the Steiner Problem in a directed graphs (SPDG).

3.1. The Steiner Problem in an undirected graph (SPUG)

The Steiner Problem in an undirected graph is the problem of connecting together at minimum length a set of vertices (nodes) in an undirected graph. This problem is derived from the Euclidean Steiner Problem, Hakimi [23] and Dreyfus and Wagner [13] were the first to study it.

Let $G_u = (S, E)$ be a connected undirected graph, where S is the set of vertices (nodes), $|S| < \infty$, and E the set of edges. With each edge $a = [i, j] \in E$ $(i \neq j, i \in S$ and $j \in S)$ is associated a positive number $l_a \equiv l_{ij}$ (length of a). Let X, Y be a partition of S, i.e., $X \cup Y = S$ and $X \cap Y = \phi$.

The problem we are considering is that of computing a subset Z of E such that:

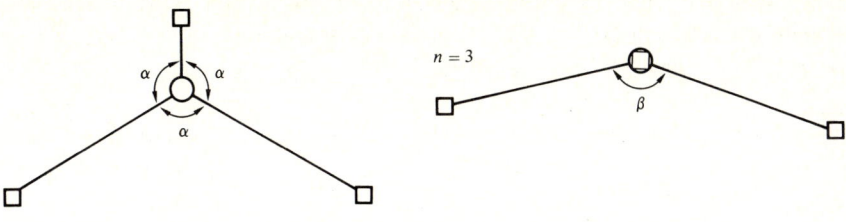
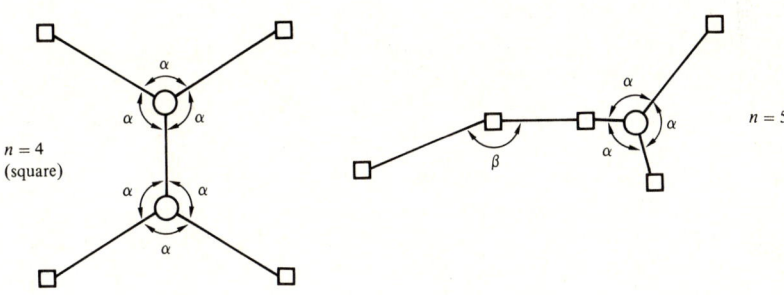
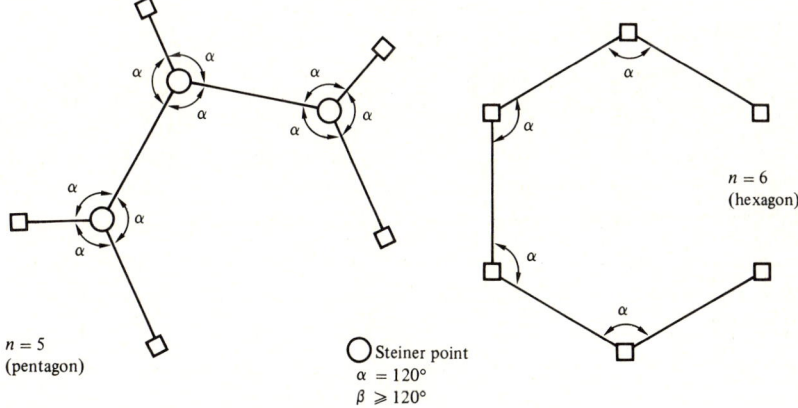

Fig. 2.1.

(i) all nodes of X are connected by chains composed only of edges in Z,
(ii) $\Sigma_{a \in Z} l_a = \text{minimum} \{\Sigma_{a \in Z' \subseteq E} l_a | Z' \text{ satisfies (i)}\}$.
Y is the set of optional vertices (Steiner points).

It is very easy to verify that the subset Z forms a tree because $l_a > 0$, $a \in E$.

When $|X| = 2$, the Steiner problem in a graph is reduced to the «shortest path problem», for which we know very efficient algorithms, see Dijkstra [12]. When $Y = \phi$, i.e., $X = S$, we have the «minimum-cost spanning tree problem», see Kruskal [27] and Prim [40]. For both cases we have polynomial time algorithms in $|S|$ for solving the problem. But when $2 < |X| < |S|$, no polynomial time algorithm to solve the Steiner problem in an undirected graph is likely to exist, since Karp [26] showed that this problem is NP-complete. We must say that the Steiner problem in an undirected graph is NP-hard under the optimization formulation.

3.2. The Steiner Problem in a directed graph (SPDG)

We suppose that Nastansky, Selkow and Stewart [37] is the first reference to consider the directed form of the Steiner problem in graphs. This problem has been considered lately by Arpin, Maculan and Nguyen [3], Beasley [4], Claus and Maculan [7], Maculan [30] and Wong [47].

Let $G_d = (S, A)$ be a strong by connected directed graph, where S is the set of vertices (nodes). A is the set of arcs: $A \subset S \times S$, with each arc $a = (i, j) \in A$ is associated a positive number $l_a \equiv l_{ij}$ (length of a).

Let $\{s\}$, S_0, S_f be a partition of S, where s is called the root. The Steiner Problem in a directed graph is to find directed paths from this root to all the vertices of S_0 of minimum total cost. S_f is the set of optional vertices (Steiner points).

When $|S_0| = 1$, i.e., the solution of the SPDG is the «Shortest path problem» from s to the only vertex of S_0, we have very good polynomial time algorithms to solve this case, see [12] and Gondran and Minoux [20]. When $S_f = \phi$, i.e., we wish to find the shortest s-directed spanning tree of G_d, in this case we have the intersection of two matroids, and then polynomial time algorithms to solve this problem, see [20]. When $1 < |S_0| < |S| - 1$, we have a «difficult» problem, NP-hard, see [47]. We can observe that the solution of the SPDG will be a minimum cost directed subtree whose root is s.

A SPUG can be transformed in a SPDG as follows. Let $G_u = (S, E)$ be an undirected graph, X and Y a partition of S, where X is the set of points to be spanned and Y the set of Steiner points; $l_a = l_{ij} > 0$ associated with the edge $a = [i, j] \in E$. We construct a directed graph $G_d = (S, A)$ where (i, j) and $(j, i) \in A$ if $[i, j] \in E$, then take any vertex belonging to X, for example $s \in X$, and form the new partition of $S : \{s\}$, $S_0 = X - \{s\}$ and $S_f = Y$. With the arcs (i, j) and (j, i) we associate the length $l_{ij} = l_a$. Then the solution of SPDG in G_d gives the solution of SPUG in G_u, see [7] and [47].

4. Lawler algorithm for the SPUG

We wish to solve the SPUG in $G_u = (S, E)$. The first algorithm for this problem was presented in [13] and requires time proportional to

$$|S|^3/2 + |S|^2(2^{|X|-1} - |X| - 1) + |S|(3^{|X|-1} - 2^{|X|} + 3)/2,$$

although no computational results were given, it is clear that this algorithm is only computationally effective if $|X|$ is small.

In this section we will present an algorithm proposed by Lawler ([29], pp. 291 - 294) which requires time proportional to $|X|^2 \, 2^{|Y|}$. It is based the idea that we can solve a minimum spanning tree problem for each of several possible choice of Steiner points.

Theorem ([29], pp. 291 and [19]). *Suppose the edge lengths l_{ij} of a graph, satisfy the metric requirement, i.e., they are nonnegative and $l_{ij} \leq l_{ik} + l_{kj}$ for all $i, j, k \in S$. Then for any $|X|$ points to be spanned, there exists a Steiner tree in the graph which contains no more than $|X| - 2$ Steiner points.*

Proof. Let p, $0 \leq p \leq |Y|$, denote the number of Steiner points in a minimal tree. Let q denote the mean number of tree edges incident to a Steiner point, and r denote the mean number of tree edges incident to the $|X|$ points to be spanned. The number of edges in the tree is $|X| + p - 1 = (qp + r|X|)/2$ but because of the metric condition, $q \geq 3$ and $r \geq 1$. It follows that

$$|X| + p - 1 \geq (3p + |X|)/2 \quad \text{and} \quad p \leq |X| - 2. \qquad \square$$

Lawler algorithm ([29], pp. 292 - 294)

Step 1 (shortest path computation). If the edges lengths do not satisfy the metric condition, compute shortest paths between all pairs of vertices, see Floyd [15] and replace the edge lengths with shortest path lengths, adding edges to the graph where necessary.

Step 2 (minimum spanning tree computation). For each possible subset $K \subseteq Y$, such that $0 \leq |K| \leq |X| - 2$, solve a minimum spanning tree of a graph $G'_u = (S', E')$ where $S' = X \cup K$ and $E' \subseteq E$.

Step 3 (construction of Steiner tree). Select the least lengthly spanning tree from those computed in step 2, and transform it into a tree of the original graph, i.e., replace each edge of the spanning tree with the edges of the shortest path between the vertices in question.

The algorithm requires the solution of a minimal spanning tree problem on no more that $|X| + |X| - 2 = 2|X| - 2$ vertices for each of σ choices of Steiner points, where

$$\sigma = \sum_{i=0}^{|X|-2} \binom{|Y|}{i} \leq 2^{|Y|}.$$

It follows that the overall computational complexity is no more than $O(|X|^2 2^{|Y|})$. This does not include the shortest path computation of step 1 which is $O(|S|^3)$.

It is clear that the Lawler algorithm is computationally effective if $|Y|$ is small.

5. Integer programming formulations for the SPUG and SPDG

As we have seen in section 3 a SPUG can be transformed in a SPDG. In this section we will present three formulations for the SPDG in $G_d = (S, A)$ as zero--one integer programming and one for the SPUG in $G_u = (S, E)$.

We consider that $I(i)$ is the set of all vertices $j \in S$ such that $(j, i) \in A$ and $0(i)$ the set of all vertices $j \in S$ such that $(i, j) \in A$. Let $C_e = (X_e, \overline{X}_e) \subseteq A$ be a cutset in G_d such that $X_e \cup \overline{X}_e = S$, $X_e \cap \overline{X}_e = \phi$, $s \in X_e$ and $S_0 \cap \overline{X}_e \neq \phi$ where $(i, j) \in C_e$ if and only if $i \in X_e$ and $j \in \overline{X}_e$. If we consider all the possible cuts: $C_1, C_2, \ldots, C_p$, where $p = O(2^{n-1})$.

Let $y_{ij} = 1$ if arc $(i, j) \in A$ is in the solution and $y_{ij} = 0$ otherwise. We will consider for all the formumations $I(s) = \phi$.

5.1. First formulation

We consider a network flow problem in which the root s offers $|S_0|$ units of flow and each $k \in S_0$ demands one unit of flow. Let x_{ij} be the amount of flow on arc (i, j). This kind of formulation is presented in location problems and in the traveling salesman problem see Gavish and Graves [17] and Nobert [38], and for the SPDG was presented in [3]:

(P_1) minimize $\sum_{(i,j) \in A} l_{ij} y_{ij}$

subject to

$$\sum_{j \in 0(i)} x_{ij} - \sum_{j \in I(i)} x_{ji} = 0 \qquad i \in S_f$$

$$\sum_{j \in 0(k)} x_{kj} - \sum_{j \in I(k)} x_{jk} = -1 \qquad k \in S_0$$

$$0 \leq x_{ij} \leq |S_0| y_{ij} \qquad (i, j) \in A$$

$$y_{ij} \in \{0, 1\} \qquad (i, j) \in A$$

5.2. Second formulation

This formulation was presented in [47] and in [7] for the SPDG and in [4] for the SPUG. We consider a network synthesis problem with nonsimultaneous single-commodity flow requirements. Let x_{ij}^k be the amount of commodity k between s and $k \in S_0$ on arc (i, j). The root s offers one unit of each commodity $k \in S_0$:

$$(P_2) \qquad \text{minimize} \qquad \sum_{(i,j) \in A} l_{ij} y_{ij} \tag{5.1}$$

subject to

$$\sum_{j \in O(i)} x_{ij}^k - \sum_{j \in I(i)} x_{ji}^k = 0 \qquad j \in S - \{s, k\}, \quad k \in S_0 \tag{5.2}$$

$$\sum_{j \in O(k)} x_{kj}^k - \sum_{j \in I(k)} x_{jk}^k = -1 \qquad k \in S_0 \tag{5.3}$$

$$0 \leq x_{ij}^k \leq y_{ij} \quad (i, j) \in A, \quad k \in S_0 \tag{5.4}$$

$$y_{ij} \in \{0, 1\}, \quad (i, j) \in A. \tag{5.5}$$

5.3. Third formulation

This integer programming formulation was given by Aneja [2] for the SPDG and in [47] and in [3] for the SPDG:

$$(P_3) \qquad \text{minimize} \qquad \sum_{(i,j) \in A} l_{ij} y_{ij} \tag{5.6}$$

subject to

$$\sum_{(i,j) \in C_q} y_{ij} \geq 1 \qquad q = 1, 2, \ldots, p \tag{5.7}$$

$$y_{ij} \in \{0, 1\} \quad (i, j) \in A. \tag{5.8}$$

When we replace $y_{ij} \in \{0, 1\}$ by $0 \leq y_{ij} \leq 1$ in P_1, P_2 and P_3 we will have three linear programming relaxations for the SPDG, LP_1, LP_2 and LP_3. Let $v(\,.\,)$ denote the optimal solution value for problem $(\,.\,)$. We are interested in obtaining lower bounds for the SPDG optimal solution value. Using the results stated in [47] and in [3] we have the following proposition.

Proposition 5.1. $v(LP_1) \leq v(LP_2) = v(LP_3)$.

Proof. First we show that $v(LP_1) \leq v(LP_2)$. For the LP_2 formulation we can write

$$\sum_{k \in S_0} \left(\sum_{j \in O(i)} x_{ij}^k - \sum_{j \in I(i)} x_{ji}^k \right) = 0 \qquad i \in S_f$$

$$\sum_{k \in S_0} \left(\sum_{j \in O(i)} x_{ij}^k - \sum_{j \in I(i)} x_{ji}^k \right) = -1 \qquad i \in S_0$$

and

$$0 \leq \sum_{k \in S_0} x_{ij}^k \leq |S_0| y_{ij} \qquad (i,j) \in A.$$

Then

$$\sum_{j \in O(i)} \left(\sum_{k \in S_0} x_{ij}^k \right) - \sum_{j \in I(i)} \left(\sum_{k \in S_0} x_{ij}^k \right) = 0 \qquad i \in S_f$$

$$\sum_{j \in O(i)} \left(\sum_{k \in S_0} x_{ij}^k \right) - \sum_{j \in I(i)} \left(\sum_{k \in S_0} x_{ij}^k \right) = -1 \qquad i \in S_0,$$

define $x_{ij} = \Sigma_{k \in S_0} x_{ij}^k$ then LP_1 can be stated as a linear aggregation of the LP_2, then $v(LP_1) \leq v(LP_2)$.

We show now that $v(LP_2) = v(LP_3)$.

(i) We prove that $v(LP_2) \leq v(LP_3)$. Let $\bar{y}_{ij}$ be a feasible solution of LP_3, i.e., $\Sigma_{(i,j) \in C_q} \bar{y}_{ij} \geq 1$, $q = 1, 2, \ldots, p$ and $0 \leq \bar{y}_{ij} \leq 1$, $(i,j) \in A$. If we consider $\bar{y}_{ij}$ the capacity of arc (i,j) in LP_2 then it must be possible to flow one unit of commodity k from s to $k \in S_0$: by the max flow-min cut theorem, see [20] the total capacity of any cut separating nodes s and k must be at least one; $\bar{y}_{ij}$ will be a feasible solution of LP_2, then $v(LP_2) \leq v(LP_3)$.

(ii) We will prove that $v(LP_2) \geq v(LP_3)$. Let $(\bar{x}_{ij}^k, \bar{y}_{ij})$ be a feasible solution of LP_2, then for all $k \in S_0$ the flow problem associated gives $\Sigma_{(i,j) \in C_q} \bar{y}_{ij} \geq 1$, $q = 1, 2, \ldots, p$, $\bar{y}_{ij}$ must be a feasible solution of LP_3 then $v(LP_2) \geq v(LP_3)$. □

Let $v(P)$ be the optimal solution value for the SPDG, then $v(P) = v(P_1) = v(P_2) = v(P_3)$. Using this notation we can observe that $v(LP_1) \leq v(LP_2) = v(LP_3) \leq v(P)$, that is $v(LP_2)$ or $v(LP_3)$ are a better lower bound for $v(P)$ than $v(LP_1)$.

5.4. Application of Benders Decompostion Method to Problem P_2

We will apply Benders [6] decomposition method to the formulation given by problem P_2 as it was suggested by Guyard [21] and in [3]. Given $y_{ij} = \bar{y}_{ij} \in \{0, 1\}$, $(i, j) \in A$, the dual program of LP_2 is stated as follows:

(DP_2) maximize $\sum_{k \in S_0} \left(-u_k^k - \sum_{(i,j) \in A} \bar{y}_{ij} w_{ij}^k \right)$

subject to

$$u_i^k - u_j^k - w_{ij}^k \leq 0 \quad (i, j) \in A, i \neq s, j \neq s, \; k \in S_0$$
$$-u_j^k - w_{sj}^k \leq 0 \quad j \in 0(s), k \in S_0$$
$$w_{ij}^k \geq 0 \quad (i, j) \in A, k \in S_0$$

where u_i^k are the dual variables associated with the flow constraints (5.2) and (5.3), and w_{ij}^k associated with $-x_{ij}^k \geq -\bar{y}_{ij}$.

We consider the following problem:

(DP_2^k) maximize $-u_k^k - \sum_{(i,j) \in A} \bar{y}_{ij} w_{ij}^k$

subject to

$$u_i^k - u_j^k - w_{ij}^k \leq 0 \quad (i, j) \in A, i \neq s, \; j \neq s$$
$$-u_j^k - w_{sj}^k \leq 0 \quad j \in 0(s)$$
$$w_j^k \geq 0 \quad (i, j) \in A.$$

The solution of DP_2 can be obtained by solving separetely all the DP_2^k problems for $k \in S_0$. If $v(DP_2^k) < \infty$ for $k \in S_0$ then $v(DP_2) = \Sigma_{k \in S_0} v(DP_2^k)$. If DP_2^k is unbounded for at least one $k \in S_0$, then DP_2 is also unbounded.

The DP_2^k constraints define a polyhedral convex cone, then the master problem of the Benders decomposition will be:

minimize z

subject to $z \geq \sum_{(i,j) \in A} l_{ij} y_{ij}$

$$r_k^h[-e_k, -y] \leq 0 \quad k \in S_0, \; h = 1, 2, \ldots, t$$
$$y_{ij} \in \{0, 1\} \quad (i, j) \in A$$

where $e_k = (00 \ldots 010 \ldots 0)^T$
$\quad\quad\quad\quad\quad\quad\uparrow$
$\quad\quad\quad$ associated with node k

$y = (y_{a_1} y_{a_2} \ldots y_{a_m})^T \quad a_i \in A, \; m = |A|$

and r_k^h is an extreme ray of the polyhedral cone we are considering, t is the total number of these extreme rays.

If DP_2^k is unbounded for one $k \in S_0$ then the capacities $\bar{y}_{ij}$ of arcs (i, j) define a network where it is not possible to flow one unit of the commodity k from s to k.

Let S_1, S_2 be a partition of S, such that $s \in S_1$ and $g \in S_1$ if the node g can receive one unit of flow from s when we consider $\bar{y}_{ij}$ as the capacity of arc (i, j). If $k \in S_0$ and $k \in S_2$ and then DP_2^k is unbounded, if not $v(DP_2^k) = 0$. When DP_2^k is unbounded, i.e., $k \in S_2$, we can generate an extreme ray of the polyhedral convex cone defined by the constraints of DP_2^k, see [17], as follows:

$$u_i^k = \begin{cases} 0 & \text{if } i \in S_1 - \{s\} \\ -1 & \text{if } i \in S_2 \end{cases}$$

$$w_{ij}^k = |u_i^k - u_j^k|, \quad (i, j) \in A, \quad i \neq s, \quad j \neq s$$

$$w_{sj}^k = -u_j^k, \quad j \in 0(s) \quad \text{then the Benders constraint will be}$$

$$-u_k^k - \sum_{(i,j) \in A} y_{ij} w_{ij}^k \leq 0,$$

if we replace in this last inequality u_k^k and w_{ij}^k by their values then

$$1 - \sum_{(i,j) \in (S_1, S_2)} y_{ij} \leq 0$$

$$\sum_{(i,j) \in (S_1, S_2)} y_{ij} \geq 1 \tag{5.9}$$

where (S_1, S_2) is a cut set in G_d.

If we introduce all these constraints (5.9) in the master program of the Benders decomposition method we have the same formulation given by P_3.

We observe that (5.9) is the subtour-breaking constraint proposed in Dantzig, Fulkerson and Johnson [11] for the traveling salesman problem.

If we applied Benders decomposition method to the formulation given by problem P_1 we would obtain the same master problem, i.e., P_3. We must observe that if we apply Benders decompostion method to LP_2, we will have LP_3 as the master problem of this decomposition method, but if we apply the method to LP_1 we will not obtain LP_3 as the master problem, see [17].

If we wish to solve P_2 using Benders decomposition method, we do not have to solve the DP_2^k problem ($k \in S_0$) for generating the Benders constraints; we propose the following procedure: given $\bar{y}_{ij} \in \{0, 1\}$ solution of a relaxation of the master problem, we solve the maximum flow problem from s to each $k \in S_0$. If for all $k \in S_0$ this flow value is greater or equal to one, then $\bar{y}_{ij}$ will be an optimal solution, otherwise we will have a minimum cut value equal to zero for at least one k. This cut (S_1, S_2) will generate a new Benders constraint which is introduced in the master problem. Our procedure is close to that proposed in [2] in solving P_3.

Example

$G_d = (S, A)$, $s \equiv 1$, $S = \{1\} \cup S_0 \cup S_f$, $S_0 = \{4, 5, 8\}$ and $S_f = \{2, 3, 6, 7\}$. This graph is depicted in figure 5.1, where the values of l_{ij} are on the arcs (i, j).

We start the master problem introducing the constraints associated with the vertices belonging to $\{1\} \cup S_0$, see figure 5.2: the cut sets $(\{1\}, S - \{1\})$, $(S - \{4\}, \{4\})$, $(S - \{5\}, \{5\})$, $(S - \{8\}, \{8\})$:

(a) $y_{12} + y_{13} \geq 1$,   (b) $y_{24} + y_{34} \geq 1$,
(c) $y_{25} + y_{45} \geq 1$ and   (d) $y_{38} + y_{68} \geq 1$.

The solution of the master problem, see figure 5.6, associated with (a), (b), (c) and (d) is $\bar{y}_{13} = \bar{y}_{24} = \bar{y}_{45} = \bar{y}_{68} = 1$ and the other $\bar{y}_{ij} = 0$. We will consider this solution $(\bar{y}_{ij})$ as the capacities of the arcs (i, j), see figure 5.3 where $\bar{y}_{ij}$ is written on arc (i, j).

For the network depicted in figure 5.3 we will compute the maximum flow from 1 to 4, from 1 to 5 and from 1 to 8. From 1 to 4 the maximum flow is equal to zero and the min-cut is $(S_1, S_2) = \{(1, 2), (3, 2), (3, 4), (3.6), (3.8)\}$, which generate the constraint (e) $y_{12} + y_{32} + y_{34} + y_{36} + y_{38} \geq 1$. We have this same solution for the other two cases ($k = 5$ and $k = 8$) we introduce (e) in the master program, see figure 5.6, then the new solution will be:

$\bar{y}_{13} = \bar{y}_{24} = \bar{y}_{32} = \bar{y}_{45} = \bar{y}_{68} = 1$ and the other $\bar{y}_{ij} = 0$, the new network is depicted in figure 5.4.

From 1 to 4 and from 1 to 5 the maximum flow is equal to one, i.e., DP_2^4

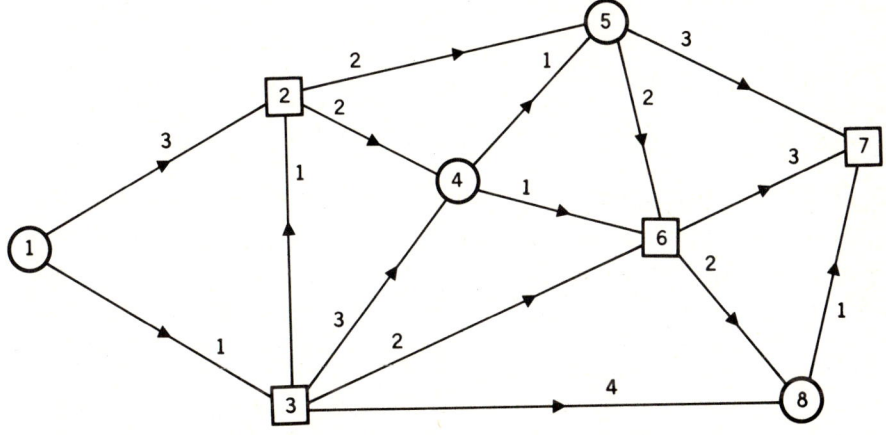

Fig. 5.1.

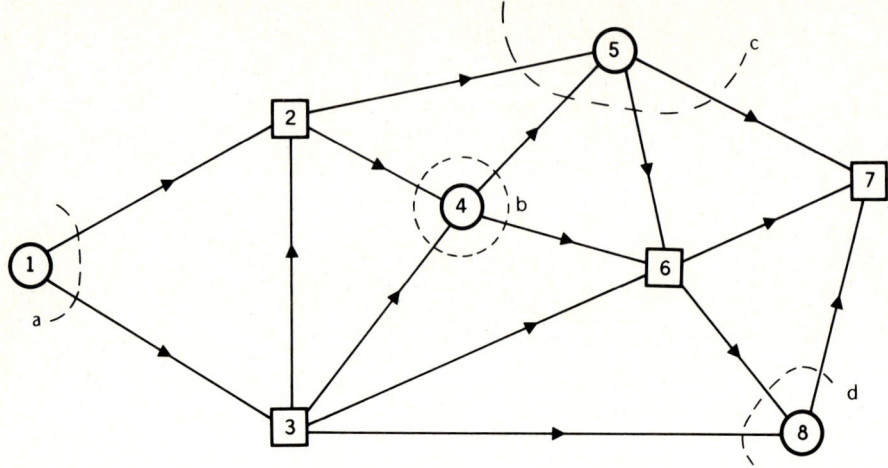

Fig. 5.2.

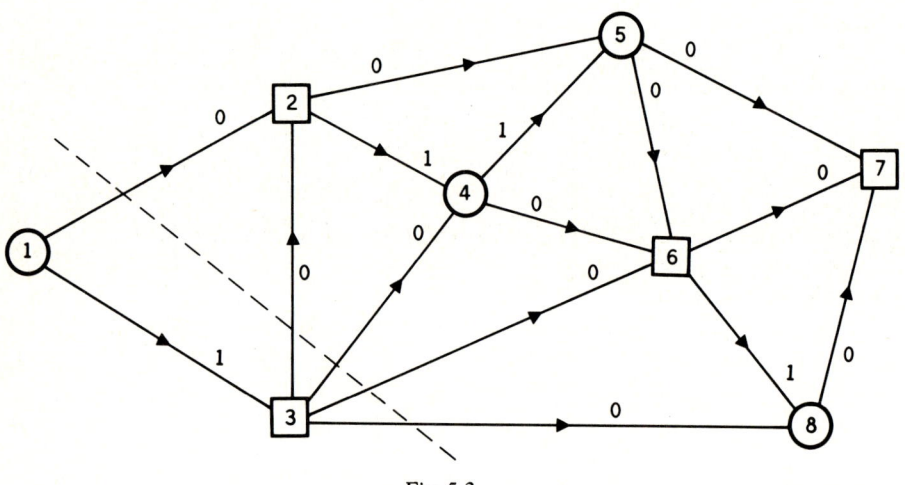

Fig. 5.3.

and DP_2^5 have finite optimum solution. From 1 to 8 the maximum flow is equal to zero and the min cut $(S_1, S_2) = \{(3, 6), (3, 8), (4, 6), (5, 6), (5, 7)\}$ which generate the constraint (f) $y_{36} + y_{38} + y_{46} + y_{56} + y_{57} \geqslant 1$. We introduce (f) in the master program, see figure 5.6, then the new solution will be $\bar{y}_{13} = \bar{y}_{24} = \bar{y}_{32} = \bar{y}_{45} = \bar{y}_{68} = \bar{y}_{46} = 1$, the other $\bar{y}_{ij} = 0$.

This last solution gives a new network for which the maximum flows from 1 to 4, from 1 to 5 and from 1 to 8 are equal to 1: then we find the optimal solution for the example we are considering. The Steiner deirected tree or Steiner arborescence will be depicted in figure 5.5, whre the vertices 2,3 and 6 are the

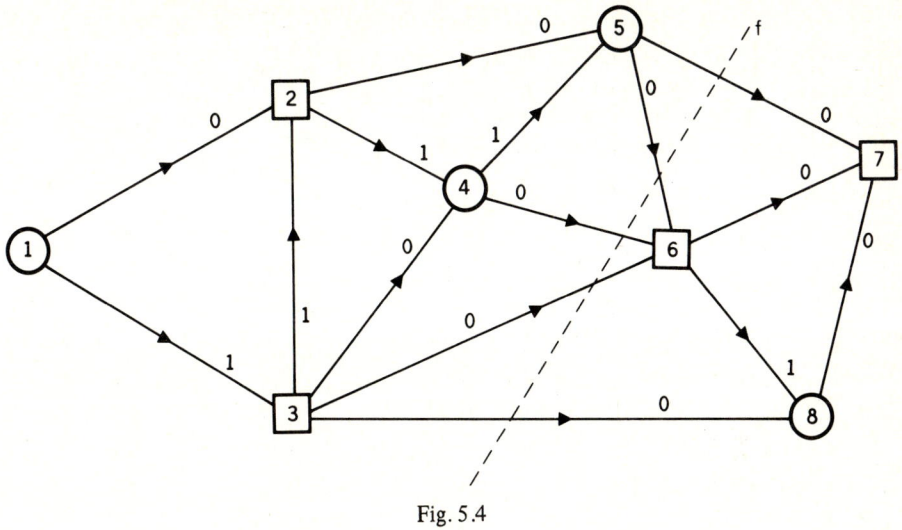

Fig. 5.4

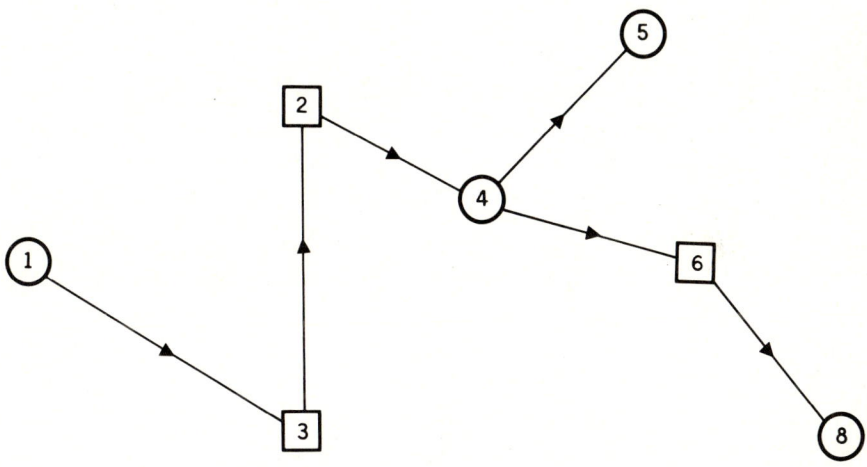

Fig. 5.5.

Steiner points.

Remark 1. In section 3 we have seen that $G_d = (S, A)$ when $S_f = \phi$ the SPUG is the shortest s-directed spanning tree problem. For this case the LP_3 formulation has all the vertices of the polyhedron formed by (5.7) and $0 \leq y_{ij} \leq 1, (i, j) \in A$, are integer, see Edmonds [14] and Schrijver [44]. For LP_1 we can have noninteger vertices of the polyhedron formed by its constraints, We have showed, see Maculan [32], that all the vertices of the polyhedron formed by the constraints of LP_2: (5.2), (5.3), (5.4) and $0 \leq y_{ij} \leq 1, (i, j) \in A$, are also integer, when $S_f = \phi$. A

Master Problem for the Benders Decomposition Method

arcs (i,j)	(1,2)	(1,3)	(2,4)	(2,5)	(3,2)	(3,4)	(3,6)	(3,8)	(4,5)	(4,6)	(5,6)	(5,7)	(6,7)	(6,8)	(8,7)		
lengths	3	1	2	2	1	3	2	4	1	1	2	3	3	2	1		
	1	1														≥ 1	(a)
			1			1										≥ 1	(b)
				1					1							≥ 1	(c)
							1							1		≥ 1	(d)
	1				1	1	1	1								≥ 1	(e)
							1	1		1	1	1				≥ 1	(f)

Fig. 5.6.

different proof has been given by Minoux and Guyard [36].

This remark gives us a good motivation for solving LP_2 or LP_3 when $S_f \neq \phi$ getting either an optimal integer solution or a good lower bound of $v(P_2)$.

We will present an example of $G_d = (S, A)$ when $S_f \neq \phi$ for which we have a non integer vertex.

Let $G_d = (S, A)$ where $S = \{1, 2, 3, 4, 5, 6, 7\}$ and $s \equiv 1$, $S_0 = \{5, 6, 7\}$, $S_f = \{2, 3, 4\}$, and $l_{ij} = 1$, $(i, j) \in A$, where the set of arcs A can be seen in the figure 5.7 below.

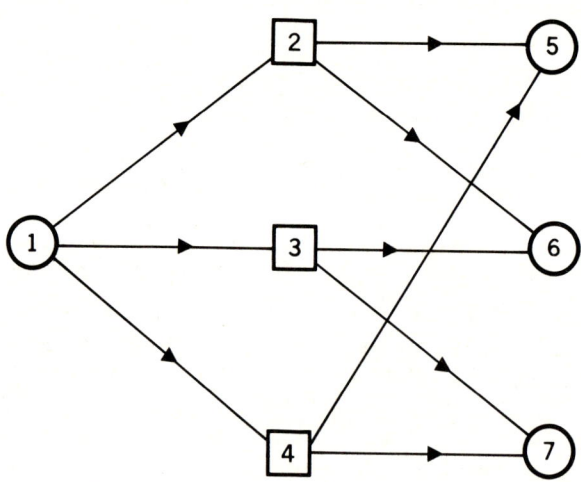

Fig. 5.7.

It is easy to verify that $y_{ij} = 1/2$, for all $(i,j) \in A$ is a vertex of LP_3 and LP_2. In LP_2 and LP_3 we have not the total unimodularity property.

5.5. Fourth Formulation

This formulation was presented by Beasley [5] and is based on an algorithm for the shortest spanning tree problem.

Let $G_u = (S, E)$ be a connected undirected graph as we have defined in section 3. Given this graph we will introduce one vertex and some edges as follows:
- one adds an artificial vertex (vertex 0) to the graph G_u,
- for each vertex $i \in Y$ add an edge $[0, i]$ of cost zero,
- for one vertex belonging to X, say 1, i.e., $1 \in X$, add an edge $[0, 1]$ of cost zero.

We will find in the resulting graph the shortest spanning tree subject to the additional restriction: any vertex $i \in Y$ connected by edge $[0, i]$ to vertex 0 must have degree one.

If we define this resulting graph as $G_u^0 = (S_0, E_0)$, where $S_0 = S \cup \{0\}$, $E_0 = E \cup \{[0,i] \mid i \in Y \text{ or } i = 1\}$, we can write this formulation as follows:

(P_4) minimize $\sum_{[i,j] \in E_0} l_{ij} y_{ij}$

subject to

(y_{ij}) forms a spanning tree on (S_0, E_0),

$y_{0i} + y_{pq} \leq 1 \quad [p,q] \in T_i, \quad i \in Y$

$y_{ij} \in \{0, 1\}, \quad [i,j] \in E_0,$

where $T_i = \{[p,q] \in E \mid p = i \text{ or } q = i\}$.

The solution of P_4 will be of the form shown in figure 5.8.

6. Solving the integer programming formulations P₂ and P₄

We will consider the solution of P_2 and P_4. The solution of P_3 has been considered in the example of section 5, Aneja has proposed in [2] a solution of P_3 associated with SPUG. We do not consider the solution of P_1 because the lower bound presented by its linear programming relaxation is not very good, but Palma-Pacheco [39] has considered this formulation for developing a heuristic method to solve the SPDG.

6.1. Solving the P₂ formulation

As P_2 is an integer programming formulation of SPDG an enumerative method (branch and bound) can be used; in [47], [3] and [4] one has proposed branch and

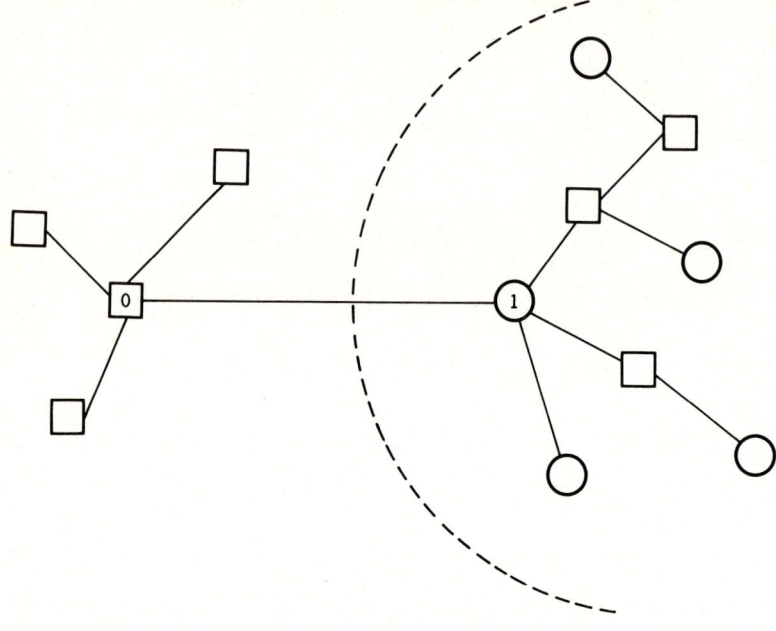

Fig. 5.8.

bound algorithms to solve this integer programming problem.

For each node of a branch and bound procedure we can consider two problems.

(i) the solution of LP_2 associated with this node,
(ii) problem reduction.

We will consider now the solution of LP_2.

6.1.1. A dual ascent algorithm for solving LP_2

A dual ascent approach for SPDG was presented by Wong [47]. Let the dual of LP_2 be as follows:

(DLP_2) maximize $\sum_{k \in S_0} u_k^k$

subject to

$$u_j^k - u_i^k - w_{ij}^k \leq 0 \qquad (i,j) \in A, \quad i \neq s, \quad k \in S_0$$

$$u_j^k - w_{sj}^k \leq 0 \qquad j \in 0(s), \quad k \in S_0$$

$$\sum_{k \in S_0} w_{ij}^k \leq l_{ij} \qquad (i,j) \in A$$

$$w_{ij}^k \geq 0 \qquad (i,j) \in A, \quad k \in S_0,$$

where the dual variable u_i^k is associated with the conservation of flow equation for commodity k at vertex i.

An auxiliary directed graph $G_d' = (S, A')$ is defined, where $A' \subseteq A$.

If vertex i is connected to vertex j (there is a directed path from i to j in G_d') but not vice-versa, Wong (1984) says that vertex i *dangles* from node j. A strong connected component T is also a *root component* if T contains a member of S_0 but no member of $S_0 \cup \{s\}$ dangles from a member of T. Let $C(k)$ be the set of vertices that are connected to vertex k. The cut set of k, $CS(k)$, is a set of arcs whose members (i,j) satisfy: $(i,j) \in A$, $(i,j) \notin A'$, $j \in C(k)$ and $i \in \overline{C}(k)$.

We present now an ascent procedure for computing a (hopefully) near optimal solution of DLP_2 proposed by Wong [47].

Step 0. (Initialize)

$u_i^k := 0, \quad k \in S_0, \quad i \in S$
$w_{ij}^k := 0, \quad k \in S_0, \quad (i,j) \in A$
$R(i,j) := l_{ij} - \Sigma_{k \in S_0} w_{ij}^k, \quad (i,j) \in A.$

From the auxiliary graph $G_d' = (S, A')$ with $A' = \phi$ (initially all vertices are strongly connected components and all members of S_0 are root components).

Step 1. Select a root component T. If there are no root components then *stop*.

Step 2

Select a vertex $k \in S_0 \cap T$
Let $R(i^*, j^*) = \text{minimum}\{R(i,j) \mid (i,j) \in CS(k)\}$;
For each node $h \in C(k)$,
$\quad u_h^k := u_h^k + R(i^*, j^*)$
For each $(i,j) \in CS(k)$,
$\quad w_{ij}^k := w_{ij}^k + R(i^*, j^*)$
and
$\quad R(i,j) := R(i,j) - R(i^*, j^*).$

Step 3

Update the auxiliary graph by setting
$A' := A' \cup \{(i^*, j^*)\}$
Go back to Step 1.

As we have seen in section 5 this ascent algorithm wields lower bounds for the SPDG and we can use it to help find feasible solutions (and upper bounds) for the SPDG.

On the auxiliary graph G'_d obtained when the ascent algorithm terminates we compute the shortest s-directed spanning tree. For that, Wong suggests in [47] to use this ascent algorithm and a procedure for recovering the optimal tree from this auxiliary graph G'_d. This optimal directed tree is a upper bound solution to the SPDG. When this upper bound is equal to $v(DLP_2)$, we have found an optimal solution of the SPDG.

6.1.2. A V.U.B. (Variable upper bounds) simplex method for solving LP_2

A V.U.B. simplex method for solving LP_2 has been presented in [3]. In the constraints $x_{ij}^k \leq y_{ij}$, $(i,j) \in A$ and $k \in S_0$, the variables y_{ij} will be considered as V.U.B. For developing this V.U.B. simplex method we have considered the works written by Schrage in [42] and [43] and Todd [46]. We start with a feasible basic solution for LP_2 (we introduce the slack variables) associated with a s-directed spanning tree of G_d. We have used a s-directed spanning tree of G_d as the union of $|S_0 \cup S_f|$ shortests paths from the vertex s to each vertex in $S_0 \cup S_f$.

6.1.3. Lagrangean relaxations for P_2

Two Lagrangean relaxations have been presented in [4] for P_2, these relaxations have the integrality property, see Geoffrion [18], then we are considering LP_2. Beasley has presented in [4] the P_2 formulation for the SPUG as follows:

$$\text{minimize} \quad \sum_{[i,j] \in E} l_{ij} y_{ij} \tag{6.1}$$

subject to

$$x_{ij}^k \leq y_{ij} \quad k \in S_0, \quad [i,j] \in E \tag{6.2}$$

there exists an elementary path (x_{ij}^k) from s to $k \in S_0$ (6.3)

$$x_{ij}^k \in \{0, 1\}, \quad k \in S_0, \quad [i,j] \in E \tag{6.4}$$

$$y_{ij} \in \{0, 1\}, \quad [i,j] \in E \tag{6.5}$$

$$d_1 \leq \sum_{[i,j] \in E} y_{ij} \leq d_2 \tag{6.6}$$

where s is a vertex in X and $S_0 = X - \{s\}$ as we have stated in section 3, $d_1 = |S_0|$ and $d_2 = |S| - 1$. We observe that (6.6) is redundant. In his paper [4], Beasley has proposed to choose the root vertex s corresponding to

$$\sum_{k \in S} \text{dist}(s, k) \geq \sum_{k \in S} \text{dist}(i, k) \text{ for all } i \in S_0, \text{ where dist}(i, j)$$

is the length of the shortest path connecting i to j.

Lower bound 1

The first Lagrangean relaxation considered by Beasley [4] can be stated as follow.

For $q \geqslant 0$, i.e., $q = (q_{ij}^k)$ where $q_{ij}^k \geqslant 0$,

$$L_1(q) = \text{minimum} \sum_{[i,j] \in E} \left(l_{ij} - \sum_{k \in S_0} q_{ij}^k \right) y_{ij} + \sum_{k \in S_0} \sum_{[i,j] \in E} q_{ij}^k x_{ij}^k$$

subject to (6.3), (6.4), (6.5) and (6.6).

This Lagrangean relaxation problem can be easily solved because this optimization problem can be decomposed into a number of separate problems. The solution consists of:

(i) the d_1 y_{ij} with the smallest Lagrangean objective function coefficients $l_{ij} - \Sigma_{k \in S_0} q_{ij}^k$ and at most $d_2 - d_1$ of the remaining y_{ij} providing their Lagrangean coefficients are nonpositive will be equal to one;

(ii) $|S_0|$ shortest elementary path calculations-finding the shortest path from vertex s to vertex $k \in S_0$.

We wish to compute the best lower bound then

$$L_1(q^*) = \underset{q \geqslant 0}{\text{maximum}}\, L_1(q) \qquad (6.7)$$

Lower bound 2

The P_2 formulation for the SPUG has been written by Beasley in [4] as follows:

$$\text{minimize} \quad \sum_{[i,j] \in E} l_{ij} y_{ij} \qquad (6.8)$$

subject to

$$x_{ij}^k + x_{ji}^k \leqslant y_{ij} \quad k \in S_0, \quad [i,j] \in E \qquad (6.9)$$

$$d_1 \leqslant \sum_{[i,j] \in E} y_{ij} \leqslant d_2 \qquad (6.10)$$

$$\sum_{(s,j) \in A} x_{sj}^k \geqslant 1 \quad k \in S_0 \qquad (6.11)$$

$$\sum_{(i,k) \in A} x_{ik}^k \geqslant 1 \quad k \in S_0 \qquad (6.12)$$

$$\sum_{j \in 0(i)} x_{ij}^k - \sum_{j \in I(i)} x_{ji}^k \geq 0 \quad k \in S_0, \quad i \in S - \{s, k\} \tag{6.13}$$

$$x_{ij}^k \in \{0, 1\}, \quad k \in S_0, \quad (i,j) \in A \tag{6.14}$$

$$y_{ij} \in \{0, 1\}, \quad [i,j] \in E \tag{6.15}$$

This formulation can be stated using the ideas we developed in section 3. The second relaxation considered in [4] can be stated as follows:

For $t \geq 0$, i.e., $t = (t_i^k)$, $t_i^k \geq 0$,

$$L_2(t) = \text{minimum} \sum_{[i,j] \in E} l_{ij} y_{ij} + \sum_{(i,j) \in A} \sum_{k \in S_0} c_{ij}^k x_{ij}^k + \sum_{k \in S_0} (t_s^k + t_k^k)$$

subject to (6.9), (6.10), (6.14) and (6.15),

where
$$c_{ij}^k = t_i^k - t_j^k \quad i \neq k, \quad j = k$$
$$= -t_i^k + t_j^k \quad i \neq k, \quad j \neq s, k$$
$$= 0 \text{ otherwise.}$$

To compute $L_2(t)$, for $t \geq 0$, we can consider

$$b_{ij} = l_{ij} + \sum_{k \in S_0} \text{minimum } \{0, c_{ij}^k, c_{ji}^k\};$$

then

$$L_2(t) = \text{minimum} \sum_{[i,j] \in E} b_{ij} y_{ij} + \sum_{k \in S_0} (t_s^k + t_k^k)$$

subject to (6.10) and (6.15).

The solution of this last optimization problem consists of d_1 y_{ij} with the smallest b_{ij} values and then at most $d_2 - d_1$ of the remaining y_{ij} provided their Lagrangean coefficients b_{ij} are nonpositive.

We compute the best lower bound

$$L_2(t^*) = \underset{t \geq 0}{\text{maximum}} \; L_2(t) \tag{6.16}$$

Proposition 6.1. $v(LP_2) = L_1(q^*) = L_2(t^*)$.

Proof. Both Lagrangean relaxations have the integrality property, see [18] and Maculan, Campello and Lopes [31]. □

Solving (6.7) and (6.16)

The lower bound 1 requires $|E||S_0|$ Lagrange multipliers and the lower bound 2 requires $|S||S_0|$ Lagrange multipliers, but $|S|<|E|$. Hence lower bound 2 geives a tight bound as lower bound 1 but with fewer Lagrange multipliers.

Subgradient procedure to solve the dual problems (6.7) and (6.16).

Step 0. Set initial values of the multipliers of

$q_{ij}^k := 0 \quad [i,j] \in E, \quad k \in S_0$
$t_i^k := 0 \quad i \in S, \quad k \in S_0$.

Step 1: solve $L_1(q)$ and $L_2(t)$ and let the solutions be $\bar{y}_{ij}, \bar{x}_{ij}^k$.

Step 2. If the Lagrangean solution $\bar{y}_{ij}$ is a feasible solution to P_2, then update the upper bound (Z_{UB}). Update the maximum lower bound found $(Z_{max}): Z_{max} := $ $= \max\{Z_{max}, L_1(q)\}$ for the lower bound 1, or $Z_{max} := \max\{Z_{max}, L_2(q)\}$ for the lower bound 2.

Step 3. *Stop* if $Z_{UB} = Z_{max}$ since then Z_{UB} is the optimal solution, else go to step 4.

Step 4. Calculate the subgradients
For the lower bound 1

$$g_{ij}^k = \bar{x}_{ij}^k - \bar{y}_{ij}, \quad [i,j] \in E, \quad k \in S_0$$

For the lower bound 2

$$h_s^k = 1 - \sum_{i \in 0(s)} \bar{x}_{si}^k, \quad k \in S_0$$

$$h_k^k = 1 - \sum_{i \in I(k)} \bar{x}_{ik}^k, \quad k \in S_0$$

$$h_i^k = \sum_{j \in I(i)} \bar{x}_{ji}^k - \sum_{j \in 0(i)} \bar{x}_{ij}^k, \quad k \in S_0, \quad i \in S - \{s, k\}.$$

Step 5. Define a step size d by

$$d = \lambda [Z_{UB} - Z_{max}] / \sum_{k \in S_0} \sum_{[i,j] \in E} (g_{ij}^k)^2$$

$$d = \lambda [Z_{UB} - Z_{max}]/\sum_{i \in S}\sum_{k \in S_0} (h_i^k)^2$$

where $0 \leq \lambda \leq 2$ and update the Lagrange multipliers by

$$q_{ij}^k := \text{maximum}\{0, q_{ij}^k + d\, g_{ij}^k\}, \quad [i,j] \in E, \quad k \in S_0,$$

$$t_i^k := \text{maximum}\{0, t_i^k + d\, h_i^k\}, \quad i \in S, \quad k \in S_0.$$

Step 6. Go to step 1 to solve the Lagrangean problems, $L_1(q)$ or $L_2(t)$, with this new set of multipliers unless sufficient subgradient interations have been performed, in which case *stop*.

In calculating a value of λ, Beasley [4] followed the approach of Held, Wolfe and Crowder [24] in letting $\lambda = 2$ for $|S|$ iterations, then successively halving both λ and the number of iterations until the number of iteration reached a threshold value of 5, λ was then halved every 5 iterations. The subgradient procedure was stopped when λ fell bellow 0.05.

Now we consider the problem reduction.

When we are solving P_2 by an enumerative method it is very important to reduce the size of the problem (in terms of vertices and/or edges). Beasley [4] proposed some tests to reduce the size of the problem: least cost, degree tests, nearest vertex, reachability, penalties on number of edges, edge penalties and components.

In our branch and bound procedure, see [30] we have used the classical ideals developed in Forrest, Hirst and Tomlin [16].

6.2. Solving the P_4 formulation

The integer programming formulation P_4 of SPUD can also be solved by enumerative methods solving in each tree node of a branch and bound methods a Lagrangean relaxation of P_4. Beasley [5] proposed a Lagrangean relaxation to P_4 as follows:

For $t \geq 0$, i.e., $t = (t_{ipq})$, $t_{ipq} \geq 0$ we have

$$L_3(t) = \text{minimum} \sum_{[i,j] \in E_0} c_{ij} y_{ij} - \sum_{i \in Y}\sum_{[p,q] \in T_i} t_{ipq}$$

subject to

(y_{ij}) forms a spanning tree on (S_0, E_0)

$y_{ij} \in \{0, 1\}, (i,j) \in E_0,$

where

$$c_{ij} = \sum_{[p,q] \in T_j} t_{jpq} \quad \text{if} \quad i = 0 \quad \text{and} \quad j \in Y,$$

$$= l_{ij} + t_{iij} \quad \text{if} \quad [i,j] \in E \quad \text{and} \quad i \in Y \quad \text{and} \quad j \in X$$

$$= l_{ij} + t_{jij} \quad \text{if} \quad [i,j] \in E \quad \text{and} \quad i \in Y \quad \text{and} \quad j \in Y$$

$$= l_{ij} + t_{iij} + t_{jij} \quad \text{if} \quad [i,j] \in E \quad \text{and} \quad i \in Y \quad \text{and} \quad j \in Y$$

$$= l_{ij} \quad \text{otherwise.}$$

We note here that $c_{ij} \geq 0$, for all $(i,j) \in E_0$.

To compute $L_3(t)$ we have to solve a shortest spanning tree problem. This problem can be easily solved by the Kruskal [27] algorithm or Prim [40] algorithm.

The best lower bound for this relaxation is

$$L_3(t^*) = \underset{t \geq 0}{\text{maximum}} \; L_3(t). \tag{6.17}$$

We can verify that $v(LP_4) = L_3(t^*)$. Beasley proposed in [5] a subgradient procedure to solve (6.17). As upper bound Beasley [5] has used the solution of a heuristic algorithm to get a «good» initial feasible solution proposed by Takahashi and Matsuyama [45].

6.3. Computational Results

Using the dual ascent approach for STDG, Wong [47] has reported his computational experience for $|S| = 40$ and $|S| = 60$; $|A| = 120, |A| = 160, |A| = 180$, $|A| = 240$; $|S_0| = |S|/2$, the graphs considered were symetric, 24 test problems were solved and for only two of them at the end of the dual ascent approach an optimal solution could not be found. An IBM 3033 computer was used and the average CPU time (sec) has been observed in the $[0.3783, 0.5875]$ interval. Guyard [22] reports similar results.

We have solved 80 test problems using the V.U.B. simplex algorithm, see [3] to solve LP_2 for $10 \leq |S| \leq 40$, $20 \leq |A| \leq 50$, $5 \leq |S_0| \leq 8$ using a CDC-CYBER 835 computer, the average CPU time (sec) has been observed in the $[0.857, 8.910]$ interval. For all these problems we have got integer optimal solutions for LP_2.

In [4] it has been reported computational results for the SPUG when 18 test problems are solved for $50 \leq |S| \leq 100$, $63 \leq |E| \leq 200$, $9 \leq |X| \leq 50$. Most of these test problems have positive duality gap, that is, LP_2 does not solve in these cases the SPUG. Using a CDC-7600 computer the CPU time (sec) has been observed in the $[1.0, 124.4]$ interval. Three test problems did not finish after 250 seconds.

In [5] it has been reported computational results for the P_4 formulation of SPUG, 12 test problems are treated $|S| = 500$, $625 \leq |E| \leq 2500$, $5 \leq |X| \leq 250$

using a CRAY-1S computer, the CPU time (sec) has been observed in the [16.55, 1016,22] interval. One test problem did not finish after 1200 seconds.

7. Applications and conclusions

The uncapacitated plant location problem can be formulate as an SPDG, see [47]. Let J be the set of customers and I be the set of potential plant sites. The transportation cost of servicing customer $j \in J$ at plant site $i \in I$ is c_{ij} and the fixed cost of locating a plant at site i is f_i. The plant location problem is to locate the plants and assign the customers to these plants so that the total fixed cost plus total customer transportation cost is minimized. This problem can be stated as follows:

$$\text{minimize} \quad \sum_{i \in I} \sum_{j \in J} c_{ij} x_{ij} + \sum_{i \in I} f_i y_i$$

subject to

$$\sum_{i \in I} x_{ij} = 1, \quad j \in J$$

$$x_{ij} \leq y_i, \quad (i, j) \in I \times J$$

$$x_{ij} \text{ and } y_i \in \{0, 1\}, \quad (i, j) \in I \times J.$$

If we define a directed graph $G_d = (S, A)$ where $S = \{s\} \cup S_0 \cup S_f$, $S_0 = J$, $S_f = I$ and $A = \{(s, i) \mid i \in I\} \cup \{(i, j) \mid (i, j) \in I \times J\}$ we associate the cost l_{gh} with each $(g, h) \in A$ as follows

$$l_{gh} = \begin{cases} f_i & \text{if} \quad g = s \quad \text{and} \quad h = i \in I \\ c_{ij} & \text{if} \quad g = i \in I \quad \text{and} \quad h = j \in J. \end{cases}$$

If we solve the SPDG in G_d we will obtain the solution of the plant location problem considered. A variety of problems associated with the network design problem can be viewed as generalizations of the SPDG, see Magnanti and Wong [33].

The SPUG can be used to solve a problem in molecular evolution, see Shore, Foulds and Gibbons [41].

As conclusions we can propose a research area associated with heuristic approaches for solving the SPDG when we have very large systems associated with the SPDG to be solved, for example $|S| \geq 500$ and $|A| \geq 3000$. For this size of problems Palma-Pacheco [39] has developed a heuristic method using two phases:

phase 1: we apply the dual ascent approach, see Wong [47], and we get a lower bound and an upper bound solution.

phase 2: we try to imporve the upper bound solution obtained in phase 1 using the P_1 formulation.

Computational results about this method are shown in [39], where some Steiner Problems in directed graph with $|S| = 900, |A| = 4500, 90 \leq |S_0| \leq 800$ are treated in an IBM 4341 computer, the CPU time (sec) has been observed in the [3.463, 66.616] interval.

This approach can be used to get «good» solutions when we have to solve very large SPDG.

We did not consider in this paper the Rectilinear Steiner Tree Problem, where a minimal rectilinear Steiner tree for a set S of points in the plane is a tree which interconnects S using horizontal and vertical lines, see Aho, Garey and Hwang [1] and Matos [34].

The idea of the P_2 formulation can be used to treat the Travelling Salesman Problem, see Claus [8], no computation results are reported. We proposed to solve the Travelling Salesman Problem using this integer programming formulation described in [8] and in [30].

Aknowlodgement

The author wishes to thank Jacques Desrosiers, Pierre Hansen, Silvano Martello and Michel Minoux for their helpful advice and suggestions.

References

[1] A.V. Aho, M.R. Garey and F.K. Hwang, «Rectilinear Steiner Trees: Efficient Special-Case Algorithms», *Networks* 7, 37 - 58, 1977.

[2] Y.P. Aneja, «An Integer Linear Programming Approach to the Steiner Problem in Graphs», *Networks* 10, 167 - 178, 1980.

[3] D. Arpin, N. Maculan and S. Nguyen, «Le Problème de Steiner sur un Graphe Orienté: Formulations et Relaxations», *Publication 315, Centre de Recherche sur les Transports, Université de Montréal*, 1983.

[4] J.E. Beasley, «An Algorithm for the Steiner Problem in Graphs», *Networks* 14, 147 - 159, 1984.

[5] J.E. Beasley, «An SST-Based Algorithm for the Steiner Problem in Graphs», *Department of Management Science, Imperial College, London*, 1985.

[6] J.F. Benders, «Partitioning Procedures for Solving Mixed Variables Programming Problems», *Numerische Mathematik* 4, 238 - 252, 1962.

[7] A. Claus and N. Maculan, «Une Nouvelle Formulation du Problème de Steiner sur un Graphe», *publication 280, Centre de Recerche sur les Transports, Université de Montréal*, 1983.

[8] A. Claus, «A New Formulation for the Travelling Salesman Problem», *SIAM Journal Algebraic and Discrete Methods* 5, 21 - 25, 1984.

[9] E.J. Cockayne, «On the Steiner Problem», *Canadian Mathematical Bulletin* 10, 431 - 450, 1967.

[10] R. Courant and H. Robbins, *What is Mathematics?*, Oxford University Press, NY, 1941.

[11] G.C. Dantzig, D.R. Fulkerson and S.M. Johnson, «Solution of a Large Scale Traveling Salesman Problem», *Operations Research* 2, 393 - 410, 1954.

[12] E.W. Dijkstra, «A Note on Two Problems in Connection with Graphs», *Numerische Mathematik* 1, 269 - 271, 1959.

[13] S.E. Dreyfus and R.A. Wagner, «The Steiner Problem in Graphs», *Networks* 1, 195 - 207, 1972.

[14] J. Edmonds, «Optimum Branchings», *Journal of Research of the National Bureau of Standards*, 71B, 4, 233 - 240, 1967.

[15] R.W. Floyd, «Algorithm 97: Shortest Path», *Communication of ACM* 5, 345, 1962.

[16] J.J.H. Forrest, J.P.H. Hirst and J.A. Tomlin, «Practical Solution of Large Mixed Integer Programming Problem with UMPIRE», *Management Science* 20, 736 - 773, 1974.

[17] R. Gavish and S.C. Graves, «Scheduling and Routing in Transportation Systems: Formulations and New Relaxations», *Graduate School of Management, University of Rochester,* N.Y., 1982.

[18] A.M. Geoffrion, «Lagrangean Relaxation for Integer Programming», *Mathematical Programming Study* 2, 82 - 114, 1974.

[19] E.N. Gilbert and Pollak, «Steiner Minimal Trees», *SIAM Journal on Applied Mathematics* 16, 1 - 29, 1968.

[20] M. Gondran and M. Minoux, *Graphs and Algorithms*, John Wiley, N.Y., 1984.

[21] L. Guyard, «Applications de la Programmation Linéaire aux Problèmes de Flots Non Simultanés dans un Graphe», *mémoire de fin d'études, CNAM, Paris*, 1983.

[22] L. Guyard, «Le Problème de l'Arbre de Steiner: Modélisation par Programmation Linéaire et Résolution par des Techniques de Décomposition (Application à un Modèle de Bases de Données Relationnelles)», *Docteur-Ingénieur dissertation, Ecole Nationale Supérieure des Télécommunications, Paris*, 1985.

[23] S.L. Hakimi, «Steiner's Problem in Graphs and its Implications», *Networks* 1, 113 - 133, 1971.

[24] M. Held, P. Wolfe and H.P. Crowder, «Validation of Subgradient Optimization», *Mathematical Programming*, 6, 62 - 88, 1974.

[25] J.B. Hiriart-Urruty, «Flâneries Mathématiques», presented at the *Symposium of Convexity and Duality in Optimization, Rijksuniversiteit, Groningen*, 1984.

[26] R.M. Karp, «Reducibility among Combinatorial Problems», in R.E. Miller and J.W. Thatcher, eds., *Complexity of Computer Computations*, Plenun Press, N.Y., 1972.

[27] J.B. Kruskal Jr., «On the Shortest Spanning Subtree of a Graph and the Traveling Salesman Problem», *Proceedings of the American Mathematical Society* 7, 48 - 50, 1956.

[28] H.W. Kuhn, «On a Pair of Dual Nonlinear Programs», in J. Abadie, ed., *Nonlinear Programming*, North-Holland, Amsterdam, 1967.

[29] E.L. Lawler, *Combinatorial Optimization: Networks and Matroids*, Holt, Rihehart and Winston, N.Y., 1976.

[30] N. Maculan, «O Problema de Steiner em Grafos Orientados», *Proceedings of the Latin American Conference on Operations Research and Systems Engineering*, Buenos Aires, 1984.

[31] N. Maculan, R.E. Campello and L. Lopes, «Relaxaçao Lagrangeana em Programaçao Inteira», *Mini Cursos, VII Brazilian Symposium of Applied Mathematics*, Campinas, 293 - 316, 1984.

[32] N. Maculan, «A New Linear Programming Formulation for the Shortest s-Directed Spanning Tree Problem», *Technical report ES 54-85, Systems Engineering and Computer Science, COPPE, Federal University of Rio de Janeiro*, presented at the *12th Symposium on Mathematical Programming, Boston*, 1985.

[33] T.L. Magnanti and R.T. Wong, «Network Design and Transportation Planning: Models and Algorithms», *Transportation Science* 18, 1 - 55, 1984.

[34] R.R.L. Matos, «Rectilinear Arborescence and Rectilinear Steiner Tree Problems», *Ph.D. dissertation, Faculty of Science and Engineering, University of Birmingham*, 1980.

[35] Z.A. Melzak, «On the Problem of Steiner», *Canadian Mathematical Bulletin* 4, 143 - 148, 1961.

[36] M. Minoux and L. Guyard, «A Nonsimultaneous Network Flow Polytope having Arborescences as Extreme Points», (to appear).

[37] L. Nastansky, S.M. Selkow and N.F. Stewart, «Cost-Minimal Trees in Directed Acyclic Graphs», *Zeitschrift für Operations Research* 18, 59 - 67, 1974.

[38] Y. Nobert, «Construction d'Algorithmes Optimaux pour des Extensions du Problème de Voyageur de Commerce», *Ph.D. Dissertation, publication 297, Centre de Recherche sur les Transports, Université de Montréal*, 1982.

[39] O.I. Palma-Pacheco, «Contribuiçao para a Resoluçao do Problema de Steiner num Grafo Direcionado: un Método Heuristico», *Ph.D. Dissertation, Systems Engineering and Computer Sciences, COPPE, Federal University of Rio de Janeiro*, 1985.

[40] R.C. Prim, «Shortest Connection Networks and Some Generalizations», *Bell System Technical Journal* 36, 1389 - 1401, 1957.

[41] M.L. Shore, L.R. Foulds and P.B. Gibbons, «An Algorithm for the Steiner Problem in Graphs», *Networks* 12, 323 - 333, 1982.

[42] L. Schrage, «Implicit Representation of Variable Upper Bounds in Linear Programming», *Mathematical Programming Study* 4, 118 - 132, 1975.

[43] L. Scharge, «Implicit Representation of Generalized Variable Upper Bounds in Linear Programming», *Mathematical Programming* 14, 11 - 20, 1978.
[44] A. Schrijver, «Min-Max Results in Combinatorial Optimization», in A. Bachem, M. Grötschel and B. Korte, eds., *Mathematical Programming, the State of the Art*, Springer Verlag, 1983.
[45] H. Takahashi and A. Matsuyama, «An Approximate Solution for the Steiner Problem in Graphs», *Mathematica Japonica* 24, 573 - 577, 1980.
[46] M.J. Tood, «An Implementation of the Simplex Method for Linear Programming Problems with Variable Upper Bounds», *Mathematical Programming* 23, 34 - 49, 1982.
[47] R.T. Wong, «A Dual Ascent Approach to Steiner Tree Problems on a Directed Graph», *Mathematical Programming* 28, 271 - 287, 1984.

Nelson Maculan
COPPE and Instituto de Matemàtica
Federal University of Rio de Janeiro
P.O. Box 68501
21945 Rio de Janeiro
Brazil

ALGORITHMS FOR KNAPSACK PROBLEMS

Silvano MARTELLO and Paolo TOTH

1. Introduction

A great variety of practical problems can be represented by a set of entities, each having an associated value, from which one or more subsets has to be selected in such a way that the sum of the values of the selected entities is maximized, and some predefined conditions are respected. The most common condition is obtained by also associating a size to each entity and establishing that the sum of the entity sizes in each subset does not exceed some prefixed bound. These problems are generally called *knapsack problems*, since they recall the situation of a hitch-hiker having to fill up his knapsack by selecting from among various possible objects those which will give him the maximum comfort.

In the present survey we will adopt the following terminology. The entities will be called *items* and their number will be indicated by n. The value and size associated with the j-th item will be called *profit* and *weight*, respectively, and denoted by p_j and w_j ($j = 1, \ldots, n$).

The majority of problems considered in this survey are *single knapsack problems*, where one container must be filled with an optimal subset of items. The *capacity* of such a container will be denoted by c. In Section 4 we consider the more general case where m containers, of capacities c_i ($i = 1, \ldots, m$), are available (*multiple knapsack problems*).

We will suppose that profits, weights and capacities are positive integers.

With the exception of very particular cases – see for example Hirshberg and Wong [30] –, knapsack problems are NP-hard in the Garey and Johnson [22] sense, although they can generally be solved in pseudo-polynomial time by dynamic programming. We will review both exact and approximate algorithms for the solution of the most important types of knapsack problems.

Exact solutions are generally obtained through branch-and-bound algorithms or dynamic programming, or a combination of the two approaches. The performance of these algorithms will be evaluated by computational experiments with randomly generated test problems.

Approximate algorithms are of the greedy type or are based on scaling. They will generally be evaluated on the basis of their worst-case performance. The worst-case performance of a heuristic algorithm h is defined as follows. Given any instance P of the considered problem, let $z(P)$ be the value of the optimal solution and $h(P)$ the value of the solution found by h. The *worst-case performance ratio* of h is then the greatest $r(h)$ value such that $r(h) \leq h(P)/z(P)$ for all possible instances P.

Knapsack problems have been intensively studied both because they arise as subproblems in various integer programming problems and may represent many practical situations. The most typical applications are in capital budgeting and industrial production. Various capital budgeting models have been studied by Weingartner [77, 78], Weingartner and Ness [79], Cord [14] and Kaplan [41]. Among industrial applications, the classical studies an *Cargo Loading Problems* (Bellman and Dreyfus [7]) and on *Cutting Stock Problems* (Gilmore and Gomory [25, 26, 27, 28]) are worth mentioning. More detailed reviews of applications can be found in Salkin [69] and Salkin and de Kluyver [70].

Almost all books on Integer Programming or Combinational Optimization contain a chapter on knapsack problems. In particular mention is made of those of Hu [32], Garfinkel and Nemhauser [23], Salkin [69], Martello and Toth [53], Syslo, Deo and Kowalik [72].

2. 0-1 knapsack problem

2.1. The problem

The *0-1 Knapsack Problem (KP)* can be mathematically formulated through the following integer linear program

$$\text{maximize} \quad z = \sum_{j=1}^{n} p_j x_j \quad (2.1)$$

$$\text{subject to} \quad \sum_{j=1}^{n} w_j x_j \leq c \quad (2.2)$$

$$x_j = 0 \text{ or } 1 \quad (j = 1, \ldots, n)$$

where x_j is 1 if the j-th item has been selected, otherwise 0. Since profits and weights are positive, it will be supposed, without loss of generality, that

$$\sum_{j=1}^{n} w_j > c \quad (2.3)$$

$$w_j \leq c \quad (j = 1, \ldots, n) \quad (2.4)$$

KP is NP-hard, as can easily be verified by considering that the Subset-Sum Problem, which will be proved to be NP-hard in Section 3.1, is a special case of KP.

KP is a well-known problem and several exact and heuristic algorithms have been proposed for its solution. The exact algorithms can be subdivided into two classes: *branch and bound* methods (Kolesar [43], Greenberg and Hegerich [29], Horowitz and Sahni [31], Fayard and Plateau [19, 20], Ahrens and Finke [2], Barr and Ross [6], Nauss [65], Martello and Toth [48, 52, 61], Zoltners [81], Suhl [71], Balas and Zemel [5]) and *dynamic programming* procedures (Horowitz and Sahni [31], Arhens and Finke [2], Toth [75]). The performance of both classes of algorithms largely depends on the size of the problems to be solved. This can generally be decreased by applying *reduction procedures* (Ingargiola and Korsh [36], Toth [74], Dembo and Hammer [16], Fayard and Plateau [20]) so as to fix the value of as many variables as possible. One of the essential ingredients of the implicit enumeration algorithms and the reduction procedures is the quality of the *upper bounds* used in the computation. The most effective upper bounds for KP will be presented in Section 2.2. The following sections will summarize the reduction procedures (Section 2.3), the branch and bound methods for small size (Section 2.4) and large size (Section 2.5) problems and the dynamic programming approaches (Section 2.6). An experimental analysis of these exact algorithms is given in Section 2.7.

The main heuristic schemes are presented in Section 2.8.

2.2. Relaxations and upper bounds

For computation of the upper bounds it is assumed that the items are ordered according to nonincreasing values of the profit per unit weight, that is, so that

$$p_j/w_j \geq p_{j+1}/w_{j+1} \quad \text{for } j = 1, \ldots, n-1.$$

If this is not the case, the sorting of the items can be performed in $O(n \log n)$ time through any efficient sorting procedure (see, for instance, Aho, Hopcroft and Ullman [1]).

Let the *critical item s* be defined as

$$s = \min\left\{ j : \sum_{i=1}^{j} w_i > c \right\}$$

(because of (2.3) and (2.4), we have $1 < s \leq n$).

Given a problem instance P, we denote with $z(P)$ the value of any optimal solution of P. If P can be formulated as an integer linear program, we denote with $\bar{P}$ the *linear programming relaxation* of P.

The linear programming relaxation of KP, also called the *continuous knapsack problem*, is defined by

(KP̄)   (2.1), (2.2) and

$$0 \leq x_j \leq 1 \quad (j = 1, \ldots, n).$$

The optimal solution $\bar{x}$ of $\overline{KP}$ can easily be obtained in the following way (Dantzig [15])

$$\bar{x}_j = 1 \quad \text{for} \quad j = 1, \ldots, s-1$$

$$\bar{x}_j = 0 \quad \text{for} \quad j = s+1, \ldots, n$$

$$\bar{x}_s = \bar{c}/w_s,$$

where $\bar{c} = c - \sum_{j=1}^{s-1} w_j.$

The optimal solution value of $\overline{\text{KP}}$ is given by

$$z(\overline{\text{KP}}) = \sum_{j=1}^{s-1} p_j + \bar{c}\, p_s/w_s.$$

Because of the integrality of p_j and x_j, a valid upper bound for KP is:

$$u_1 = [z(\overline{\text{KP}})] = \sum_{j=1}^{s-1} p_j + [\bar{c}\, p_s/w_s],$$

where $[r]$ is the largest integer not greater than r.

The computational complexity of $\overline{\text{KP}}$, and hence of the Dantzig bound u_1, is clearly $O(n)$ if we assume that the items are already sorted, otherwise $O(n \log n)$.

An improved upper bound has been proposed by Martello and Toth [48] according to the following consideration: since, in KP, x_s cannot assume a fractional value, the optimal solution of KP can be obtained from the corresponding continuous solution $\bar{x}$ of $\overline{\text{KP}}$ either without inserting the s-th item (i.e. by setting $\bar{x}_s = 0$) or by inserting it (i.e. by setting $\bar{x}_s = 1$). In the former case, the solution value cannot exceed

$$b_1 = \sum_{j=1}^{s-1} p_j + [\bar{c}\, p_{s+1}/w_{s+1}],$$

which corresponds to the case of filling the residual capacity $\bar{c}$ with items having the best possible value of p_j/w_j (i.e. p_{s+1}/w_{s+1}). In the latter, since it is necessary to remove at least one of the first $s-1$ items, the best solution value is given by

$$b_2 = \sum_{j=1}^{s-1} p_j + [p_s - (w_s - \bar{c})\, p_{s-1}/w_{s-1}],$$

where it has been supposed that the item to be removed has exactly the minimum necessary value of w_j (i.e. $w_s - \bar{c}$) and the worst possible value of p_j/w_j (i.e. p_{s-1}/w_{s-1}). A valid upper bound for KP is so given by

$$u_2 = \max\{b_1, b_2\}.$$

It can be proved (see [48]) that $u_2 \leq u_1$.

The consideration on which the Martello-Toth bound is based can be further exploited to compute more restrictive upper bounds than u_2. This can be achieved by replacing the values b_1 and b_2 with stronger values, say b_3 and b_4, which take more carefully into account the exclusion and inclusion of the s-th item. Hudson [33] has proposed computing b_4 through the continuous relaxation of KP with the additional constraint $x_s = 1$, that is

$$b_4 = [z(\overline{\text{KP}} \text{ with } x_s = 1)].$$

Fayard and Plateau [20] and, independently, Villela and Bornstein [76], have proposed computing b_3 as well through continuous relaxation of KP, by imposing constraint $x_s = 0$, that is,

$$b_3 = [z(\overline{\text{KP}} \text{ with } x_s = 0)].$$

The corresponding bound is

$$u_3 = \max\{b_3, b_4\}.$$

It can easily be proved that, since $b_3 \leq b_1$ and $b_4 \leq b_2$, we have $u_3 \leq u_2$.

Bound u_3 can also be seen as the result of the application of Dantzig's bound at the two terminal nodes of a decision tree having the root node corresponding to KP and two descendent nodes, say N1 and N2, corresponding to the exclusion and inclusion of the s-th item. Clearly, the maximum among the upper bounds corresponding to all the terminal nodes of a decision tree represents a valid upper bound for the original problem corresponding to the root node. So, if b_3 and b_4 are the Dantzig bounds corresponding respectively to nodes N1 and N2, u_3 represents a valid upper bound for KP.

An improved bound u_4 can be obtained by considering decision trees having more than two terminal nodes; this approach has been proposed by Martello and Toth in [61]. If, for example, nodes N1 and N2 have each two descendent nodes, corresponding to the exclusion and inclusion of the $(s + 1)$-th item, a valid upper bound is given by

$$\bar{u}_4 = \max\{b_5, b_6, b_7, b_8\},$$

where $b_5 = [z(\overline{\text{KP}} \text{ with } x_s = x_{s+1} = 0)]$, $b_6 = [z(\overline{\text{KP}} \text{ with } x_s = 0 \text{ and } x_{s+1} = 1)]$, $b_7 = [z(\overline{\text{KP}} \text{ with } x_s = 1 \text{ and } x_{s+1} = 0)]$, $b_8 = [z(\overline{\text{KP}} \text{ with } x_s = x_{s+1} = 1)]$. Since $\max\{b_5, b_6\} \leq b_3$ and $\max\{b_7, b_8\} \leq b_4$, then $\bar{u}_4 \leq u_3$.

A different approach to the computation of upper bounds, proposed by Müller-Merbach [64], is based on *Lagrangean Relaxation*. Given a non-negative multiplier λ, the Lagrangean relaxation of KP is defined as

$$(\text{LR}_\lambda) \quad \text{maximize} \quad \tilde{z} = \sum_{j=1}^n p_j x_j + \lambda \left(c - \sum_{j=1}^n w_j x_j \right) \quad (2.5)$$

$$\text{subject to} \quad x_j = 0 \text{ or } 1 \quad (j = 1, \ldots, n).$$

The objective function (2.5) can be restated as

$$\text{maximize } \tilde{z} = \sum_{j=1}^{n} \tilde{p}_j x_j + \lambda c$$

where $\quad \tilde{p}_j = p_j - \lambda w_j \quad (j = 1, \ldots, n).$ (2.6)

The optimal solution of LR_λ can easily be obtained as $\tilde{x}_j = 1$ if $\tilde{p}_j > 0$, $\tilde{x}_j = 0$ or 1 if $\tilde{p}_j = 0$, $\tilde{x}_j = 0$ if $\tilde{p}_j < 0$.

Note that $z(\overline{LR}_\lambda) = z(LR_\lambda)$ for any $\lambda \geq 0$.

It can easily be verified that for $\lambda = \tilde{\lambda} = p_s/w_s$, $z(LR_\lambda)$ assumes its minimum value. In fact, $\tilde{p}_j \geq 0$ for $j = 1, \ldots, s$ and $\tilde{p}_j \leq 0$ for $j = s+1, \ldots, n$; from which:

$$z(LR_{\tilde{\lambda}}) = z(\overline{LR}_{\tilde{\lambda}}) = z(\overline{KP}) \geq z(KP);$$
$$\tilde{x}_j = \bar{x}_j \text{ for } j = 1, \ldots, s-1, s+1, \ldots, n.$$

It is noted that $|\tilde{p}_j|$ represents the decrease of $z(LR_{\tilde{\lambda}})$ corresponding to the change of the j-th variable from $\tilde{x}_j$ to $1 - \tilde{x}_j$. Other properties of the Lagrangean relaxation for KP have been studied by Maculan [45].

In order to obtain an integer solution to KP from the continuous one, either the fractional variable $\bar{x}_s$ alone has to be set to 0 or at least one of the other variables, say $\bar{x}_j$, has to change its value (from 1 to 0 or from 0 to 1). If the latter change takes place, for any $j \neq s$ we have

$$z(KP \text{ with } x_j = 1 - \tilde{x}_j) \leq [z(LR_{\tilde{\lambda}} \text{ with } x_j = 1 - \tilde{x}_j)] =$$
$$= [z(LR_{\tilde{\lambda}}) - |\tilde{p}_j|] = [z(\overline{KP}) - |\tilde{p}_j|].$$

Hence the Müller-Merbach upper bound for KP is

$$u_5 = \max\{z(\overline{KP}) - \bar{c} p_s/w_s, [z(\overline{KP}) - \min\{|\tilde{p}_j| : j = 1, \ldots, n; j \neq s\}]\}.$$

It can easily be seen that $u_5 \leq u_1$.

An improved bound with respect to both u_3 and u_5 has recently been proposed by Dudzinski and Walukiewicz [17]. Consider any feasible solution $\hat{x}$ to KP, of value $\hat{z}$ and satisfying

$$\sum_{i=1}^{n} w_i \hat{x}_i + w_j > c \quad \text{for any } j \text{ such that } \hat{x}_j = 0.$$

If $\hat{x}$ is not optimal for KP, then an optimal solution can be obtained only if at least two variables, say $\hat{x}_k$ and $\hat{x}_h$ with $\hat{x}_k = 0$ and $\hat{x}_h = 1$, change their values to $\hat{x}_k = 1$ and $\hat{x}_h = 0$. The following upper bound can thus be derived:

$$u_6 = \max\{\hat{z}, [\min\{z(\overline{KP} \text{ with } x_s=1), z(\overline{KP}) - \min\{\tilde{p}_j : j=1,\ldots,s-1\}\}],$$
$$[\min\{z(\overline{KP} \text{ with } x_s=0), z(\overline{KP}) + \max\{\tilde{p}_j : j \in \hat{N}\}\}]\}$$

where $\hat{N} = \{j : \hat{x}_j = 0, j = 1, \ldots, n, j \neq s\}$. It can be proved that $u_6 \leq \min\{u_3, u_5\}$.

Further improvements on u_6, leading to more time-consuming bounding procedures, have been presented in [17].

Since u_6 can be less, equal, or greater than u_4, an improved upper bound for KP is

$$u_7 = \min\{u_4, u_6\}.$$

Although all the above bounds have the same computational complexity, that is $O(n)$ if the items are already sorted, their average performances are quite different, as can easily be seen from the corresponding definitions. When implementing a branch-and-bound algorithm, the bound to be computed at each node of the branch-decision tree must be selected by considering that u_1, u_2, u_3 and u_4 can easily be computed through parametric techniques (the value of s generally changes slightly from one node to the next), while u_5 and u_6 need examination of value $\tilde{p}_j$ for all currently unfixed variables x_j.

Most of the algorithms for KP require initial sorting of items so that $p_j/w_j \geq$ $\geq p_{j+1}/w_{j+1}$ ($j = 1, \ldots, n-1$). For algorithms not requiring this sorting, all upper bounds can still be computed in $O(n)$ time since the continuous solution is completely identified by the value of s, which can be computed in $O(n)$ time through a partitioning technique, as shown in detail is Section 2.5.

2.3. Reduction procedures

The number of binary variables (x_j) and the value of c can be decreased by applying reduction procedures ([36], [74], [16], [20]) which fix the optimal value of as many variables as possible. These procedures partition set $N = \{1, 2, \ldots, n\}$ into three subsets:

$$J1 = \{j \in N : x_j = 1 \text{ in an optimal solution to KP}\};$$

$$J0 = \{j \in N : x_j = 0 \text{ in an optimal solution to KP}\};$$

$$F = N - (J1 \cup J0).$$

The original KP can now be transformed into the reduced form

$$\text{maximize} \quad z = \sum_{j \in F} p_j x_j + \hat{p}$$

$$\text{subject to} \quad \sum_{j \in F} w_j x_j \leq \hat{c}$$

$$x_j = 0 \text{ or } 1 \quad (j \in F)$$

where $\hat{p} = \sum_{j \in J1} p_j$, $\hat{c} = c - \sum_{j \in J1} w_j$.

Subsets $J1$ and $J0$ are obtained by considering the implications corresponding to setting a variable x_j to 0 or to 1. If setting a variable x_j to a given value a ($a = 0$ or 1) implies an infeasible solution to KP or a solution not better than an existing one, variable x_j can be fixed to $1-a$, since this is the only choice which can lead to feasible or improved solutions. The implications of what occurs by setting a variable to a given value can be derived either through dominance relations or the evaluation of upper bounds. For KP, the most efficient reduction procedures are based on the latter approach (the former will be considered in Section 4.3 for reduction of the Multiple Knapsack Problem).

Let $\hat{x}$ be any feasible solution to KP and $\hat{z}$ the corresponding value ($\hat{z}$ represents a lower bound on the optimal solution value z^* to KP). Moreover, for $j \in N$, let z_j (resp. $\bar{z}_j$) be the upper bound corresponding to KP with the additional constraint $x_j = 1$ (resp. $x_j = 0$). Then we have

$$J1 = \{j \in N : \bar{z}_j \leq \hat{z}\},$$
$$J0 = \{j \in N : z_j \leq \hat{z}\}.$$

The effectiveness and computational efficiency of the reduction procedures clearly depend on the techniques used in computing $\hat{z}, z_j$ and $\bar{z}_j$.

The upper bounds z_j and $\bar{z}_j$ can be computed through any of the approaches of Section 2.2. In particular, bound u_1 has been used in [36] and [20], bound u_2 in [74]. Since the computation of bounds z_j (resp. $\bar{z}_j$) requires $O(n)$ time to find the critical item s_j (resp. $\bar{s}_j$) in the corresponding continuous solution, the overall complexity of the reduction procedures is $O(n^2)$. However, if bounds u_1, u_2, u_3 or u_4 are used, the average computing times can be greatly decreased since, as mentioned in Section 2.2, these bounds can easily be computed through parametric techniques. A further decrease of the average time can be obtained by considering that it is useless to compute z_j (resp. $\bar{z}_j$) if $x_j = 1$ (resp. $x_j = 0$) in the solution corresponding to the upper bound u computed for the original KP, since, in these cases, we have $z_j = u$ (resp. $\bar{z}_j = u$). For instance, if u is given by u_1, it is useless to compute $\bar{z}_j$ for $j < s$ and z_j for $j > s$. In addition, the worst-case time complexity of the reduction procedures can be decreased to $O(n \log n)$ as follows.

Procedure REDUCTION

sort the items so that $p_j/w_j \geq p_{j+1}/w_{j+1}$ ($j = 1, \ldots, n-1$);
for $j = 1$ to n do compute $t_j = \Sigma_{i=1}^{j} w_i$;
find, through binary search, s such that $t_{s-1} \leq c < t_s$;
for $j = 1$ to s do compute $\bar{z}_j$ by finding the corresponding value $\bar{s}_j$ such that
$t_{\bar{s}_j - 1} \leq c + w_j < t_{\bar{s}_j}$;
for $j = s$ to n do compute z_j by finding the corresponding value $\bar{s}_j$ such that
$t_{\bar{s}_j - 1} \leq c - w_j < t_{\bar{s}_j}$.

The time complexity is $O(n \log n)$ for the first step, $O(n)$ for the second, $O(\log n)$

for the third, and $O(n \log n)$ for the last two steps.

A faster, but less effective, procedure of time complexity $O(n)$, has been derived in [16] by considering that value $|\tilde{p}_j|$, defined by (2.6), represents the decrease of $z(\overline{KP})$ corresponding to the change of the j-th variable from $\bar{x}_j$ to $1 - \bar{x}_j$. The upper bounds z_j and $\bar{z}_j$ can thus be given by $[z(\overline{KP}) - |\tilde{p}|]$. It is easy to prove that the values z_j and $\bar{z}_j$ obtained in this way are not lower than any bound previously considered, so this reduction procedure is generally less effective than the previous ones.

The cardinality of subset F can be further decreased by imposing conditions (2.3) and (2.4) on the reduced problem (i.e. by setting $J0 = J0 \cup \{j \in F : w_j > \hat{c}\}$ and, if $\Sigma_{j \in F} w_j \leq \hat{c}$, $J1 = J1 \cup F$) and by repeating the reduction procedure for the items in F until no further variable is fixed (insertion of an item into $J1$ or $J0$ can decrease the value of the upper bounds z_j and $\bar{z}_j$ for $j \in F$).

The lower bound $\hat{z}$ can be computed by using any heuristic procedure for KP (see Section 2.8), for instance the *greedy solution* as proposed in [16] and [20], or by taking the best of the integer solutions obtained as a by-product from the computation of the upper bounds z_j and $\bar{z}_j$ ($j \in N$), as proposed in [36] and [74].

2.4. Branch-and-bound algorithms

Many branch-and-bound methods have been proposed in the last few decades for the exact solution of KP ([43], [29], [31], [19], [2], [6], [65], [48], [52], [81], [71], [5], [20], [61]). The algorithms differ in the upper bounding technique (see Section 2.2), branching strategy, the application of dominance or infeasibility criteria at the nodes of the branch-decision tree, and the parametrization used in computing the upper bounds.

The most efficient methods apply a depth-first binary branching scheme which, at each node, selects a not yet fixed item, say the i-th, having the maximum ratio profit/weight, and generates two descendent nodes by fixing x_i, respectively, to 1 and 0. The search continues from the node associated with the insertion of the i-th item in the current solution ($x_i = 1$), i.e. from the node having the maximum value of the corresponding upper bound. In the following, we will describe two of the most efficient branch-and-bound methods for KP, that is, those proposed in [31] and [48].

The algorithm of Horowitz and Sahni [31] starts by sorting the items according to decreasing values of profit per unit weight and then by taking the continuous solution of KP (see Section 2.2) with $x_s = 0$ as first current solution. A *forward move* consists of inserting the largest set of new consecutive items into the current solution under condition (2.2). A *backtracking move* consists of removing the last inserted item from the current solution. Whenever a forward move is exhausted, the upper bound u_1 corresponding to the current solution is computed and compared with the best solution so far in order to check wether further forward moves could lead to a better one: if so, a new forward move is performed, otherwise a backtracking follows. When the last item has been considered, the current solution is complete and possible updating of the best solution so far occurs. The algorithm stops when no further backtracking can be performed.

The Martello and Toth algorithm [48] differs from the above method in the following main respects.

   i. Upper bound u_2 is used instead of u_1.

   ii. The forward move associated with the selection of the j-th item is split into two phases: *building of a new current solution* and *saving of the current solution*. In the first phase the largest set N_j of consecutive items which can be inserted into the current solution starting from the j-th, is defined and the upper bound u_j corresponding to the insertion of the j-th item computed. If u_j is not greater than the value of the best solution so far, a backtracking move immediately follows. Otherwise, the second phase, that is, insertion of the items of set N_j into the current solution, is performed only if the value of such a new solution does not represent the maximum which can be obtained by inserting the j-th item. If this is not the case, the best solution so far is changed but the current solution is not updated, so that useless backtrackings on the items in N_j are avoided.

   iii. A particular forward procedure, based on dominance criteria, is performed whenever, before a backtracking move on the i-th item, the residual capacity $\hat{c}$ does not allow insertion into the current solution of any item following the i-th. The procedure is based on the following consideration: the current solution could be improved only if the i-th item is replaced by an item having greater profit and a weight small enough to allow its insertion, or by at least two items having global weight not greater than $w_i + \hat{c}$. By this approach it is generally possible to eliminate most of the useless nodes generated at the lowest levels of the decision tree.

   iv. The upper bounds associated with the nodes of the decision-tree are computed through a parametric technique based on the storing of information related to the current solution. Suppose, in fact, that the current solution has been built by inserting the items from the j-th to the r-th: then, when trying to build a new solution starting from the i-th item ($j < i < r$), if no insertion or removal occurred for the items preceding the j-th, it is possible to insert at least the items from the i-th to the r-th into the new current solution.

Detailed description of the algorithm follows.

Algorithm MT1

1. [*initialization*]
    sort the items so that $p_j/w_j \geq p_{j+1}/w_{j+1}$ ($j = 1, \ldots, n-1$);
    find $r = \max\{j : \Sigma_{i=1}^{j} w_i \leq c\}$ and compute $p^* = \Sigma_{j=1}^{r} p_j$, $w^* = \Sigma_{j=1}^{r} w_j$;

for $j = 1$ **to** n **do** compute $m_j = \min \{w_i : i = j+1, \ldots, n\}$ and set $x_j = 0$;
set $p_{n+1} = 0$, $w_{n+1} = \infty$, $m_n = \infty$, $z = 0$, $z^* = 0$, $\hat{r} = n$, $\hat{c} = c$;
set $u = u_2 = p^* + \max \{[(c - w^*)p_{r+2}/w_{r+2}], [p_{r+1} - (w_{r+1} - (c - w^*))p_r/w_r]\}$;
set $j = 1$ and go to Step 3.

2. [*building of a new current solution*]
 while $w_j > \hat{c}$ **do**
    if $z^* \geq z + [\hat{c} p_{j+1}/w_{j+1}]$ **then** go to Step 5;
        else set $j = j + 1$;
 find $r = \max \{i : \bar{w}_j + \Sigma_{k=\bar{r}_j}^{i} w_k \leq \hat{c}\}$ (if $\bar{w}_j + w_{\bar{r}_j} > \hat{c}$, set $r = \bar{r}_j - 1$);
 compute $p^* = \bar{p}_j + \Sigma_{i=\bar{r}_j}^{r} p_i$, $w^* = \bar{w}_j + \Sigma_{i=\bar{r}_j}^{r} w_i$;
 if $w^* = \hat{c}$ **or** $r = n$
  then if $z^* < z + p^*$ **then** set $z^* = z + p^*$;
        for $i = 1$ **to** $j - 1$ **do** set $x_i^* = x_i$;
        for $i = j$ **to** r **do** set $x_i^* = 1$;
        for $i = r + 1$ **to** n **do** set $x_i^* = 0$;
        if $z^* = u$ **then** stop;
   go to Step 5;
  else if $z^* < z + p^* + [(\hat{c} - w^*)p_{r+2}/w_{r+2}]$
    then go to Step 3;
    else if $z^* \geq z + p^* + [p_{r+1} - (w_{r+1} - (\hat{c} - w^*))p_r/w_r]$ **then** go to Step 5.

3. [*saving of the current solution*]
 set $\hat{c} = \hat{c} - w^*$, $z = z + p^*$; **for** $i = j$ **to** r **do** set $x_i = 1$;
 set $\bar{w}_j = w^*$, $\bar{p}_j = p^*$, $\bar{r}_j = r + 1$;
 for $i = j + 1$ **to** r **do** set $\bar{w}_i = \bar{w}_{i-1} - w_{i-1}$, $\bar{p}_i = \bar{p}_{i-1} - p_{i-1}$, $\bar{r}_i = r + 1$;
 for $i = r + 1$ **to** $\hat{r}$ **do** set $\bar{w}_i = 0$, $\bar{p}_i = 0$, $\bar{r}_i = i$;
 set $\hat{r} = r$, $j = r + 2$;
 if $\hat{c} \geq m_{j-1}$ **then** go to Step 2.

4. [*updating of the best solution so far*]
 if $z^* < z$ **then** set $z^* = z$;
       for $i = 1$ **to** n **do** set $x_i^* = x_i$;
       if $z^* = u$ **then** stop.

5. [*backtracking*]
 find $i = \max \{k : x_k = 1, k < j\}$; **if** no such i exists **then** stop;
 set $\hat{c} = \hat{c} + w_i$, $z = z - p_i$, $x_i = 0$;
 if $\hat{c} - w_i \geq m_i$ **then** set $j = i + 1$ and go to Step 2;
        else set $j = i$, $h = i$.

6. [*replacement of the i-th item with the h-th*]
 set $h = h + 1$;
 if $z^* \geq z + [\hat{c} p_h/w_h]$ **then** go to Step 5;
 if $w_h = w_i$ **then** repeat Step 6;

if $w_h > w_i$ then if $w_h > \hat{c}$ or $z^* > z + p_h$ then repeat Step 6;
    set $z^* = z + p_h$;
    for $k = 1$ to n do set $x_k^* = x_k$;
    set $x_h^* = 1$;
    if $z^* = u$ then stop;
    set $i = h$ and repeat Step 6;
  else if $\hat{c} - w_h < m_h$ then repeat Step 6;
    set $\hat{c} = \hat{c} - w_h$, $z = z + p_h$, $x_h = 1, j = h + 1$;
    set $\bar{w}_h = w_h$, $\bar{p}_h = p_h$, $\bar{r}_h = h + 1$;
    for $k = h + 1$ to $\hat{r}$ do set $\bar{w}_k = 0, \bar{p}_k = 0, \bar{r}_k = k$;
    set $\hat{r} = h$ and go to Step 2.

2.5. Large-size problems

As will be shown in Section 2.7, many instances of KP can be solved by branch-and-bound algorithms for very high values of n. For such problems, the preliminary sorting of the items requires, on average, a comparatively high computing time (for example when $n \geq 2000$ the sorting time is about 90 percent of the total time required by the algorithms of Section 2.4). In the present section we review some algorithms which do not require preliminary sorting of all the items.

The first algorithm of this kind was presented by Balas and Zemel [5] and is based on the so-called «core problem». Suppose the items are sorted so that $p_j/w_j \geq p_{j+1}/w_{j+1}$ for $j = 1, \ldots, n-1$, and, for an optimal solution (x^*), let

$$j_1 = \min\{j : x_j^* = 0\}, \quad j_2 = \max\{j : x_j^* = 1\},$$

$$I = \{j_1, \ldots, j_2\};$$

the *core problem* is then defined as

$$\text{maximize} \quad \tilde{z} = \sum_{j \in I} p_j x_j$$

$$\text{sibject to} \quad \sum_{j \in I} w_j x_j \leq c - \sum_{j \in \{i : p_i/w_i > p_{j_1}/w_{j_1}\}} w_j$$

$$x_j = 0 \text{ or } 1 \quad \text{for } j \in I.$$

In general, for large problems, the size of the core is a very small fraction of n. Hence, if we knew «a priori» the values of j_1 and j_2, we could easily solve the complete problem by setting $x_j^* = 1$ for all j such that $p_j/w_j > p_{j_1}/w_{j_1}$, $x_j^* = 0$ for all j such that $p_j/w_j < p_{j_2}/w_{j_2}$ and by solving the core problem through any branch-and-bound algorithm (so that only the items in I would have to be sorted). j_1 and j_2 cannot be «a priori» identified, but a good approximation of the core problem can be obtained if we consider that, in most cases, given the *critical* *item* s, which identifies the solution of the continuous relaxation of KP (see Section 2.2), we have $j_1 \geq s - \vartheta/2$ and $j_2 \leq s + \vartheta/2$ for some $\vartheta \ll n$. Balas and Zemel have proposed the following procedure, having a prefixed value ϑ, to identify item s, the continuous solution and set I of the items in the approximate core problem, without preliminary sorting.

Procedure CORE

set $J0 = J1 = \emptyset$, $F = \{1, 2, \ldots, n\}$, $\hat{c} = c$;
while $|F| > \vartheta$ do
  let λ be the median of the first 3 ratios p_j/w_j in F;
  partition F into: $N^> = \{j \in F : p_j/w_j > \lambda\}$,
                      $N^< = \{j \in F : p_j/w_j < \lambda\}$,
                      $N^= = \{j \in F : p_j/w_j = \lambda\}$;
set $c_1 = \Sigma_{j \in N^>} w_j$, $c_2 = c_1 + \Sigma_{j \in N^=} w_j$;
if $c_1 < \hat{c} \leq c_2$
  then ($\lambda = p_s/w_s$, so) the continuous solution is obtained by setting $\bar{x}_j = 1$ for $j \in J1 \cup N^>$, $x_j = 0$ for $j \in J0 \cup N^<$ and then filling the residual capacity $\hat{c} - c_1$ with items in $N^=$ in any order (thus identifying item s);
  sort the items in F according to decreasing p_j/w_j ratios and define I as a (sorted) subset of F such that $|I| = \vartheta$ and s is contained in the middle third of I; stop;
  else if $c_1 \geq \hat{c}$ then set $J0 = J0 \cup N^< \cup N^=$, $F = N^>$;
          else set $J1 = J1 \cup N^> \cup N^=$, $F = N^<$, $\hat{c} = \hat{c} - c_2$;
set $I = F$;
sort I according to decreasing p_j/w_j ratios and determine the continuous solution by setting $\bar{x}_j = 1$ for $j \in J1$, $\bar{x}_j = 0$ for $j \in J0$ and then solving the continuous relaxation of the core problem by Dantzig's method.

The Balas-Zemel method also makes use of two procedures, H and R. Procedure H is a heuristic, including dominance relations among the items, which finds an approximate solution for a given core problem I (let z_h be the corresponding value of the heuristic solution of KP). Procedure R is a reduction algorithm (see Section 2.3) which fixes as many variables as possible by applying the Dembo-Hammer test and then the Ingargiola-Korsh test, modified so as to compute an approximate continuous solution when the items are not sorted.

The overall algorithm can be outlined as follows. Let $N = \{1, 2, \ldots, n\}$.

Algorithm BZ

apply procedure CORE and let z_c be the value of the continuous solution to KP;
apply procedures H and R to the items in I;
if $z_h = [z_c]$ then (an optimal solution has been found, so) stop;

apply procedure R to the items in $N-I$ and let $J1$ and $J0$ be those subsets of N whose corresponding variables are respectively fixed to 1 and 0;
define a new core problem $I = N - J1 - J0$;
if $|I| > \gamma$ (γ being a threshold value)
 then apply procedures H and R to the items in I and define the new sets $J1$, $J0$ and $I = N - J1 - J0$;
  if $z_h = [z_c]$ then stop;
sort the items in I and exactly solve the core problem through the enumerative Zoltners algortithm [81].

Two algorithms to solve large 0-1 knapsack problems without sorting all the items, have been derived from Balas and Zemel's [5] basic idea. That of Fayard and Plateau [20], presented together with an efficient Fortran implementation, can be briefly described as follows.

Algorithm FP

find the continuous solution of the problem through a procedure similar to CORE and let this solution be defined by s, $J1$ and z_c, with
$z_c = \Sigma_{j \in J1} p_j + (c - \Sigma_{j \in J1} w_j) p_s/w_s$;
find a heuristic solution of value $z_h = \Sigma_{j \in J1} p_j + g$, where g is obtained by applying the greedy algorithm (without sorting) to the items in $\{1, 2, \ldots, n\} - J1$;
if $z_h = [z_c]$ then stop;
apply the reduction procedure of Dembo and Hammer [16] and let F be the set of items whose corresponding variables are not fixed;
sort the items in F according to increasing values of $|\tilde{p}_j| = |p_j - w_j p_s/w_s|$ and exactly solve the corresponding problem through a particular enumerative technique.

Martello and Toth [61] have recently proposed another algorithm − based on the core problem − and the corresponding Fortran implementation. The method can be outlined as follows.

Algorithm MT2

find the set I of items in the core problem through a procedure similar to CORE and let $J1$ be the set of items whose corresponding variables are temporarily fixed to 1;
if $|I| \leq n/\alpha$ (α a prefixed parameter)
 then sort the items in I according to decreasing p_j/w_j ratios;
  find an exact solution (of value $\bar{z}$) to the core problem through a modified version of algorithm MT1 of Section 2.4, which also gives upper bound u_4 on KP (see Section 2.2);
  if $\Sigma_{j \in J1} p_j + \bar{z} = u_4$ then stop;
  apply a modified version of the Toth algorithm [74] to reduce the complete problem without sorting, and let F be the set of items whose correspond-

ing variables are not fixed;
   if $F \subseteq I$ **then** $(\Sigma_{j \in J_1} p_j + \bar{z}$ is an optimal solution value to KP, so) **stop**;
   sort the items in F according to decreasing p_j/w_j ratios;
else sort all the items according to p_j/w_j ratios;
   apply the reduction algorithm of Toth and define the corresponding set F;
exactly solve the reduced problem through algorithm MT1.

2.6. Dynamic programming algorithms

For each integer m $(m = 1, 2, \ldots, n)$ and for each integer b $(b = 0, 1, \ldots, c)$ let

$$f_m(b) = \max \left\{ \sum_{j=1}^{m} p_j x_j : \sum_{j=1}^{m} w_j x_j \leq b, x_j = 0 \text{ or } 1 \text{ for } j = 1, \ldots, m \right\},$$

$X_m(b) =$ set of items inserted in the optimal solution corresponding to $f_m(b)$ (i.e. $f_m(b) = \Sigma_{j \in X_m(b)} p_j$).

From the above definitions it follows that

$$f_1(b) = 0, \quad X_1(b) = \emptyset \quad \text{for} \quad 0 \leq b < w_1,$$

$$f_1(b) = p_1, \quad X_1(b) = \{1\} \quad \text{for} \quad w_1 \leq b \leq c.$$

The *backward recursive equations* for the m-th stage $(m = 2, \ldots, n)$ and for $b = 0, \ldots, c$ are defined as

$$f_m(b) = f_{m-1}(b), X_m(b) = X_{m-1}(b)$$
$$\text{if } b < w_m \text{ or } f_{m-1}(b) \geq f_{m-1}(b - w_m) + p_m;$$
otherwise $\quad f_m(b) = f_{m-1}(b - w_m) + p_m, X_m(b) = X_{m-1}(b - w_m) \cup \{m\}.$

The optimal solution for KP is given by

$$z = f_n(c),$$
$$x_j = 1 \text{ if } j \in X_n(c), \text{ otherwise } x_j = 0.$$

The time complexity of the dynamic programming algorithms based on the above recursions is clearly $O(nc)$ (pseudo-polynomial algorithms).

An improvement over the basic scheme can be obtained through the following considerations. Each triplet (b, g, Y) with $g = f_m(b)$ and $Y = X_m(b)$ represents a *state* at stage m. It is generally possible to obtain a considerable reduction of the number of states considered at a given stage by eliminating all the *dominated states*, that is, the states (b, g, Y) for which at least one state $(\bar{b}, \bar{g}, \bar{Y})$, having $\bar{b} < b$ and $\bar{g} \geq g$, exists at the same stage. The states at stage m $(m = 2, \ldots, n)$ can thus be defined by considering only the undominated states at stage $m - 1$ and applying *forward recursive equations* (see [31], [2], [75]). This approach requires no specific sorting of the items. However, its average efficiency increases if the items are sorted so that $p_j/w_j \geq p_{j+1}/w_{j+1}$ for $j = 1, \ldots, n-1$, since the number of undominated states at each stage decreases. It can be seen that the maximum number of states at stage m is given by min $\{2^m - 1, \Sigma_{j=1}^m w_j, c\}$.

Horowitz and Sahni [31] have proposed an algorithm based on splitting of the original problem into two subproblems having, respectively, $q = [n/2]$ and $r = n - q$ items. For each subproblem, a list containing all the undominated states relative to the last stage (respectively the q-th and the r-th), is computed. Then the two lists are merged to find the optimal solution of the original problem. The main feature of this approach is given by the property of having, in the worst case, two lists, each of $(2^q - 1)$ states, instead of a list of $(2^n - 1)$ states as required by the original problem.

Ahrens and Finke [2] have independently obtained an algorithm where the above technique is combined with a branch and bound procedure in order to reduce the storage requirements. This method works well for problems having small values of n and very large values of w_j and c.

Toth [75] has proposed an algorithm based on the computation of upper bounds for fathoming the states not leading to optimal solutions and on the elimination of the unutilized states, that is, of the states which will no longer be considered in the following stages. In addition, the method performs a combination of forward and backward recursive equations taking into account the number of undominated states and the value of c.

2.7. Computational results

In this section we analyze the computational performance of the algorithms of the previous sections on sets of randomly generated test problems. Since the difficulty of such problems is greatly affected by the correlations between profits and weights, we consider three uniformly randomly generated data sets, with different degrees of correlation:

uncorrelated : $1 \leq w_j \leq \bar{w}, \quad 1 \leq p_j \leq \bar{p}$;

weakly correlated : $1 \leq w_j \leq \bar{w}, \quad w_j - r \leq p_j \leq w_j + r$;

strongly correlated: $1 \leq w_j \leq \bar{w}, \quad p_j = w_j + r$.

For each data set we consider two values of the capacity: $c = 2\bar{w}$ and $c = 0.5 \Sigma_{j=1}^n w_j$. In the first case the optimal solution contains very few items, while in the second it contains about half of them.

We give separate tables for small-size problems ($n = 50, 100, 200$) and large-size problems ($n = 500, 1000, 2000, 5000, 10000$).

We compare the Fortran IV implementations of the following algorithms:

HS	= Horowitz-Sahni [31] (Section 2.4);
MT1	= Martello-Toth [48, 52] (Section 2.4);
R + HS	= HS with reduction (Section 2.3);
R + MT1	= MT1 with reduction;

R + DPT = Toth's dynamic programming [75] (Section 2.6) with reduction;
BZ = Balas-Zemel [5] (Section 2.5);
FP = Fayard-Plateau [20] (Section 2.5);
MT2 = Martello-Toth [61] (Section 2.5).

MT1, FP and MT2 are published codes. HS, R and DPT have been coded by us. For BZ we give the computing times presented by the authors [5], which have been obtained by choosing $\vartheta = 25$ and $\gamma = 50$.

All runs have been executed on a CDC-Cyber 730. For each data set, value of c and value of n, the tables give the average running time, expressed in seconds, computed over 20 problem instances. Since Balas and Zemel give times obtained an a CDC-6600, which we verified to be about two times faster than the CDC-Cyber 730 on this kind of problems, the times given in the tables for BZ are those in [5] multiplied by 2.

Code FP includes its own sorting procedure. The sortings needed by HS, MT1, DPT and MT2 have been obtained through Fortran subroutine SORTZV of the CERN library. For $n = 50, 100, 200, 500, 1000, 2000, 5000, 10000$ this subroutine requires respectively 0.008, 0.018, 0.041, 0.114, 0.250, 0.529, 1.416, 3.010 seconds on a CDC-Cyber 730. All times given in the tables include the corresponding sorting time.

Tables 2.1, 2.2 and 2.3 compare algorithms HS, R + HS, MT1, R + MT1, FP and R + DPT on small-size problems (we do not give the times of MT2, which are almost equal to those of R + MT1). For all data sets, $\overline{w} = 100, \overline{p} = 100, r = 10$. Table 2.1 refers to uncorrelated problems, Table 2.2 to weakly correlated problems. All algorithms solved the problems very quickly with the exception of HS and, for weakly correlated problems, R + DPT. MT1 is only slightly improved by previous application of R, contrary to what happens for HS. Table 2.3 refers to strongly correlated problems. Because of the high times generally involved, a time limit of 500 seconds has been assigned to each algorithm for solution of the 60 problems generated for each value of c. The dynamic programming approach

Table 2.1
Uncorrelated problems: $1 \leq w_j \leq 100, 1 \leq p_j \leq 100$. CDC-Cyber 730 seconds. Average times over 20 problems.

c	n	HS	R + HS	MT1	R + MT1	FP	R + DPT
200	50	0.022	0.013	0.015	0.012	0.013	0.013
	100	0.039	0.024	0.026	0.025	0.018	0.029
	200	0.081	0.050	0.051	0.050	0.032	0.055
$0.5 \sum_{j=1}^{n} w_j$	50	0.031	0.016	0.016	0.013	0.013	0.020
	100	0.075	0.028	0.030	0.026	0.021	0.043
	200	0.237	0.065	0.068	0.057	0.053	0.090

Table 2.2
Weakly correlated problems: $1 \leq w_j \leq 100$, $w_j - 10 \leq p_j \leq w_j + 10$. CDC-Cyber 730 seconds. Average times over 20 problems.

c	n	HS	R + HS	MT1	R + MT1	FP	R + DPT
200	50	0.031	0.018	0.017	0.014	0.016	0.022
	100	0.049	0.029	0.032	0.024	0.023	0.041
	200	0.091	0.052	0.055	0.048	0.030	0.066
$0.5 \sum_{j=1}^{n} w_j$	50	0.038	0.025	0.022	0.020	0.021	0.071
	100	0.079	0.042	0.040	0.031	0.039	0.158
	200	0.185	0.070	0.069	0.055	0.057	0.223

Table 2.3
Strongly correlated problems $1 \leq w_j \leq 100$, $p_j = w_j + 10$. CDC-Cyber 730 seconds. Average times over 20 porblems.

c	n	HS	R + HS	MT1	R + MT1	FP	R + DPT
200	50	0.165	0.101	0.028	0.025	0.047	0.041
	100	1.035	0.392	0.052	0.047	0.096	0.070
	200	3.584	2.785	0.367	0.311	0.928	0.111
$0.5 \sum_{j=1}^{n} w_j$	50	time limit	time limit	4.870	4.019	17.895	0.370
	100	–	–	time limit	time limit	time limit	1.409
	200	–	–	–	–	–	3.936

appears clearly superior to all branch-and-bound algorithms (among which MT1 has the best performance).

For large-size problems we do not consider strongly correlated problems, because of the impractical times involved. Tables 2.4 and 2.5 compare algorithms MT1, BZ, FP and MT2. Dynamic programming is not considered because of excessive memory requirements, HS because of clear inferiority. The best algorithms are FP and MT2 for uncorrelated problems, BZ and MT2 for weakly correlated problems.

MT1, which is not designed for large problems, is generally the worst algorithm. However, about 90 percent of its time is spent for sorting, so its use can be convenient when several KP's are to be solved for the same item set and different values of c. A situation of this kind arises for multiple knapsack problems, as will be seen in Section 4.

The core memory requirements of codes MT1, FP and MT2 are, respectively, $8n$, $7n$ and $8n$ words. MT1 consists of 180 statements, FP of 600, and MT2 of 1000.

Table 2.4
Uncorrelated problems. $1 \leq w_j \leq 1000$, $1 \leq p_j \leq 1000$, $c = 0.5 \sum_{j=1}^{n} w_j$. CDC-Cyber 730 seconds. Average times over 20 problems.

n	MT1	BZ	FP	MT2
500	0.199	–	0.104	0.157
1000	0.381	0.372	0.188	0.258
2000	0.787	0.606	0.358	0.462
5000	1.993	0.958	1.745	0.982
10000	4.265	1.514	7.661	1.979

Table 2.5
Weakly correlated problems. $1 \leq w_j \leq 1000$, $w_j - 100 \leq p_j \leq w_j + 100$, $c = 0.5 \sum_{j=1}^{n} w_j$. CDC-Cyber 730 seconds. Average times over 20 problems.

n	MT1	BZ	FP	MT2
500	0.367	–	0.185	0.209
1000	0.663	0.588	0.271	0.293
2000	1.080	0.586	0.404	0.491
5000	2.188	0.744	1.782	0.771
10000	3.856	1.018	19.481	1.608

2.8. Heuristic algorithms

In this section we review some approximation schemes to determine an approximate solution $\hat{X}$ of value $\hat{z}$ to KP. An *approximation scheme* is a parametric algorithm which allows one to obtain any *worst-case performance ratio* (see Section 1) by prefixing the value of a parameter.

The first approximation scheme for KP was proposed by Sahni [68] and makes use of a greedy procedure, which finds a lower bound solution by filling, in order of nonincreasing p_j/w_j ratios that part of c which is left vacant after the items of a given set M have been put into the knapsack. Given $M \subset \{1, \ldots, n\}$ and assuming that the items are sorted so that $p_j/w_j \geq p_{j+1}/w_{j+1}$ for $j = 1, \ldots, n-1$, the procedure is:

Procedure $g(M)$
let $g = \sum_{j \in M} p_j$, $c' = c - \sum_{j \in M} w_j$, $X = M$;
for $j = 1$ to n do
    if $j \notin M$ and $w_j \leq c'$ then set $g = g + p_j$, $c' = c' - w_j$, $X = X \cup \{j\}$.

The Sahni scheme, MT^k, where k is a nonnegative integer parameter, is

Algorithm S^k

Sort the items so that $p_j/w_j \geq p_{j+1}/w_{j+1}$ for $j = 1, \ldots, n-1$;
set $\hat{z} = 0$;
for each $M \subset \{1, 2, \ldots, n\}$ such that $|M| \leq k$ and $\Sigma_{j \in M} w_j \leq c$ **do**
 if $g(M) > \hat{z}$ **then** set $\hat{z} = g(M)$, $\hat{X} = M \cup X$.

Since the time complexity of procedure g is $O(n)$, algorithm S^k runs in $O(n^{k+1})$ time. The space complexity is clearly $O(n)$. Sahni has proved that the worst-case performance ratio of S^k is $r(S^k) = k/(k+1)$. Hence the Sahni algorithm is a *polynomial approximation scheme*, in the sense that any worst-case performance ratio can be obtained in a time bounded by a polynomial. However, the degree of the polynomial increases with k, so the time complexity of the algorithm is exponential in the inverse of the *worst-case relative error* $\epsilon = 1 - r$.

Ibarra and Kim [35] have obtained a *fully polynomial approximation scheme*, i.e. a parametric algorithm which allows one to obtain any worst-case relative error (note that imposing ϵ is equivalent to imposing r) in polynomial time and space, and such that the time and space complexities grow polynomially also with the inverse of ϵ. The basic ideas in the Ibarra-Kim algorithm are: (a) to separate items, according to profits, into a class of «large» items and one of «small» items; (b) to solve the problem for the large items only, with profits scaled by a suitable scale factor δ, through dynamic programming. The dynamic programming list is stored in a table T of length $[(3/\epsilon)^2] + 1$; $T(k) = \emptyset$ or is of the form $(L(k), P(k), W(k))$, where $L(k)$ is a subset of $\{1, \ldots, n\}$, $P(k) = \Sigma_{j \in L(k)} p_j$, $W(k) = \Sigma_{j \in L(k)} w_j$ and $k = \Sigma_{j \in L(k)} \bar{p}_j$ with $\bar{p}_j = [p_j/\delta]$.

Algorithm IK^ϵ

Sort the items so that $p_j/w_j \geq p_{j+1}/w_{j+1}$ for $j = 1, \ldots, n-1$;
find the largest r such that $\bar{w} = \Sigma_{j=1}^{r} w_j \leq c$;
if $\bar{w} = c$ **then** set $\hat{z} = \Sigma_{j=1}^{r} p_j$, $\hat{X} = \{1, \ldots, r\}$ and **stop**;
set $\tilde{z} = \Sigma_{j=1}^{r+1} p_j$ (note that $\tilde{z}/2 \leq z < \tilde{z}$) and define the scale factor $\delta = \tilde{z}(\epsilon/3)^2$;
set $S = \emptyset$, $T(0) = (\emptyset, 0, 0)$, $q = [\tilde{z}/\delta] = [(3/\epsilon)^2]$;
for $i = 1$ **to** q **do** set $T(i) = \emptyset$;
for $j = 1$ **to** n **do**
 if $p_j \leq \epsilon \tilde{z}/3$ **then** (the item is «small», so) set $S = S \cup \{j\}$, preserving the
   nonincreasing profit/weight order;
   else define the scaled profit $\bar{p}_j = [p_j/\delta]$;
    for $i = q - \bar{p}_j$ **to** 0 **step** -1 **do**
     if $T(i) \neq \emptyset$ and $W(i) + w_j \leq c$ **then**
      if $T(i + \bar{p}_j) = \emptyset$ or $W(i + \bar{p}_j) > W(i) + w_j$ **then**
       set $T(i + \bar{p}_j) = (L(i) \cup \{j\}, P(i) + p_j, W(i) + w_j)$;
set $\hat{z} = 0$;
for $i = 1$ **to** q **do**

if $T(i) \neq \emptyset$ then set $\bar{z} = P(i) + \Sigma_{j \in A}\, p_j$, where A is obtained by filling the residual capacity $c - W(i)$ with items in S, in order;
if $\bar{z} > \hat{z}$ then set $\hat{z} = \bar{z}, \hat{X} = L(i) \cup A$.

The time complexity of IK^ϵ is $O(n \log n + \frac{1}{\epsilon^4} \log(\frac{1}{\epsilon}))$, the space complexity $O(n + \frac{1}{\epsilon^3})$.

A fully polynomial approximation scheme for KP, based on the same ideas contained in the Ibarra-Kim algorithm, was independently found by Babat [4].

The Ibarra-Kim scheme has been modified by Lawler [44] to obtain time complexity $O(n \log(\frac{1}{\epsilon}) + \frac{1}{\epsilon^4})$ and space complexity $O(n + \frac{1}{\epsilon^3})$, and by Magazine and Oguz [47] to obtain time complexity $O(n^2 \log n/\epsilon)$ and space complexity $O(n/\epsilon)$.

Note that the core memory requirements of the fully polynomial approximation schemes depend on ϵ and can become impractical for small values of this parameter. On the contrary, the space complexity of Sahni's polynomial approximation scheme is $O(n)$, independently of r.

3. Subset-sum problem

3.1. The problem

The *Subset-Sum Problem (SSP)* is a particular case of the 0-1 Knapsack Problem where $p_j = w_j$ for all j. It can thus be formulated as

$$\text{maximize } z = \sum_{j=1}^{n} w_j x_j$$

$$\text{subject to } \sum_{j=1}^{n} w_j x_j \leq c,$$

$$x_j = 0 \text{ or } 1 \quad (j = 1, \ldots, n).$$

SSP is also called the *Value-Independent Knapsack Problem* or *Stickstacking Problem*. We will suppose, without loss of generality, that $w_j \leq c$ $(j = 1, \ldots, n)$ and $\Sigma_{j=1}^{n} w_j > c$.

SSP is NP-complete, as we can see by polynomial transformation from *Partition*, which is the following basic NP-complete problem (see Karp [42]):

Instance: finite set A, size $s(a) \in Z^+$ for each $a \in A$, positive integer b.
Question: is there a subset $A' \subseteq A$ such that the sum of the sizes of elements in A' is exactly b?

Given any instance of the Partition Problem, we can define the following instance of SSP:

$$n = |A|;$$

$$w_j = s(a_j) \ (j = 1, \ldots, n), \text{ where } A = \{a_1, a_2, \ldots, a_n\};$$

$$c = b.$$

Let z be the solution value of this SSP and (x^*) the corresponding solution vector. Clearly the answer to the Partition Problem instance is «yes» if and only if $z = c$, so that $A' = \{a_j \in A : x_j^* = 1\}$.

SSP can obviously be solved by any algorithm for the 0-1 Knapsack Problem (some computational results have been given by Martello and Toth [59]). It can easily be verified, however, that all the upper bounds of Section 2.2 would, for an SSP, give the trivial value c, so branch-and-bound algorithms could degenerate into complete enumeration. For the same reason, no effect can be expected from the reduction procedures of Section 2.3. In the next sections we examine specific optimal and heuristic algorithms for SSP.

3.2. Exact algorithms

SSP has been optimally solved through dynamic programming by Faaland [18] and Ahrens and Finke [2], and, for large-size problems, through an enumerative technique by Balas and Zemel [5]. The Faaland algorithm is a pure dynamic programming recursion and is dominated by the Ahrens-Finke method, which is also based on a tournament-sort technique. Because of the large core memory requirements (the Ahrens-Finke algorithm needs about $2^{n/4 + 4}$ words), the dynamic programming approaches can be used only for small values of n. The Balas-Zemel algorithm is a particularization of their method for large 0-1 Knapsack Problems (Section 2.5).

Martello and Toth [59] have presented a mixture of dynamic programming and branch-and-bound, which proved to be more efficient, on the average, than the other methods. The main idea is that of applying the dynamic programming recursions only to a subset of items while, for the remaining items, tree-search is performed. When n is large, this mixed approach is applied only to a core problem, which is theoretically defined similarly to that of the 0-1 knapsack problem, but can be determined much more efficiently.

Algorithm MTS

0. *[initialization]*
    Define the values of parameters M1, M2 and $\bar{c}$;
    if $n \leq 100$ **then** (no core problem is needed, so) set $J1 = \emptyset$, sort the items so that $w_j \geq w_{j+1}$ for $j = 1, \ldots, n - 1$ and go to Step 2
        else define the values of parameters t and Δt, set $c_0 = c$ and find the smallest s such that $\Sigma_{j=1}^{s} w_j > c$.

1. [*core problem*]

    Set $J1 = \{j : 1 \leq j \leq s - t - 1\}$ (the items in $J1$ are supposed to be in the optimal solution); the core problem to be solved is defined by $n = 2t + 1$, $c = c_0 - \Sigma_{j \in J1} w_j$ and by items $s - t, \ldots, s + t$ (in steps 2 and 3 we denote these items with $1, 2, \ldots, n$); sort these items so that $w_j \geq w_{j+1}$ for $j = 1, \ldots, n - 1$.

2. [*dynamic programming phase*]

    A *state* (V, X) is defined by a feasible solution X, represented by the set of items having $x_j = 1$, and by the corresponding weight $V = \Sigma_{j \in X} w_j$.
    Note that X can be stored as a bit string. Two states are considered distinct iff they have different weights. We use two lists of states.
    Given a prefixed value $M1 < n$, list $(V1_i, X1_i)$, $i = 1, \ldots, L1$, contains all the $L1$ distinct states which correspond to the last $M1$ items and whose weight is $V1_i \leq c$. The states are ordered so that $V1_i < V1_{i+1}$ for all i.
    Given two prefixed values $M2$ ($M1 < M2 < n$) and $\bar{c} < c$, list $(V2_i, X2_i)$, $i = 1, \ldots, L2$, contains all the $L2$ distinct states which correspond to the last $M2$ items and whose weight is $V2_i \leq c$. The states are ordered so that $V2_i < V2_{i+1}$ for all i.
    Start by determining list $(V1_i, X1_i)$ as follows. Initially place only states $(V1_1, X1_1) = (0, \emptyset)$ and $(V1_2, X1_2) = (w_n, \{n\})$ in the list. Then, at iteration k ($k = 1, \ldots, M1 - 1$), form a new one from each state $(V1_i, X1_i)$, $(V1_i + w_{n-k}, X1_i \cup \{n - k\})$, if $V1_i + w_{n-k} \leq c$ and $V1_i + w_{n-k} \neq V1_p$ for all p, and place it in the list so that the states are ordered according to decreasing weights. These operations can be implemented efficiently by taking the sorting of the states into account.
    Once list $(V1_i, X1_i)$ has been determined, list $(V2_i, X2_i)$ is obtained as follows. Initially, place all the states $(V1_i, X1_i)$ such that $V1_i \leq \bar{c}$ in list $(V2_i, X2_i)$ by preserving the sorting of the states. Then, at iteration k ($k = 1, \ldots, M2 - M1$), update the list in the same way as was done for $(V1_i, X1_i)$ (replace $(V1_i, X1_i)$ with $(V2_i, X2_i)$, c with $\bar{c}$, and $n - k$ with $n - M1 + 1 - k$).
    Set $N1 = n - M1$ and $N2 = n - M2$ (Figure 3.1 shows the states covered by the two lists: the thick lines approximate the step functions giving, for each item, the maximum state weight obtained at the corresponding iteration).

3. [*tree-search phase*]

    Determine $f_j = \Sigma_{i=j}^{n} w_i$ for $j = 1, \ldots, n$.
    Generate a binary decision-tree by setting x_j to 1 or 0 for $j = 1, \ldots, N1$. Only the first $N1$ items are considered, since all the feasible combinations of the items $N1 + 1, \ldots, n$ are in list $(V1_i, X1_i)$. A forward move starting from an item k consists in finding the first item $\bar{k} \geq k$ which can be added to the current solution and then in adding the largest sequence $\bar{k}, \bar{k} + 1, \ldots$ of consecutive items to the current solution such that the solution remains feasible. A backtracking step consists in removing from the current solution

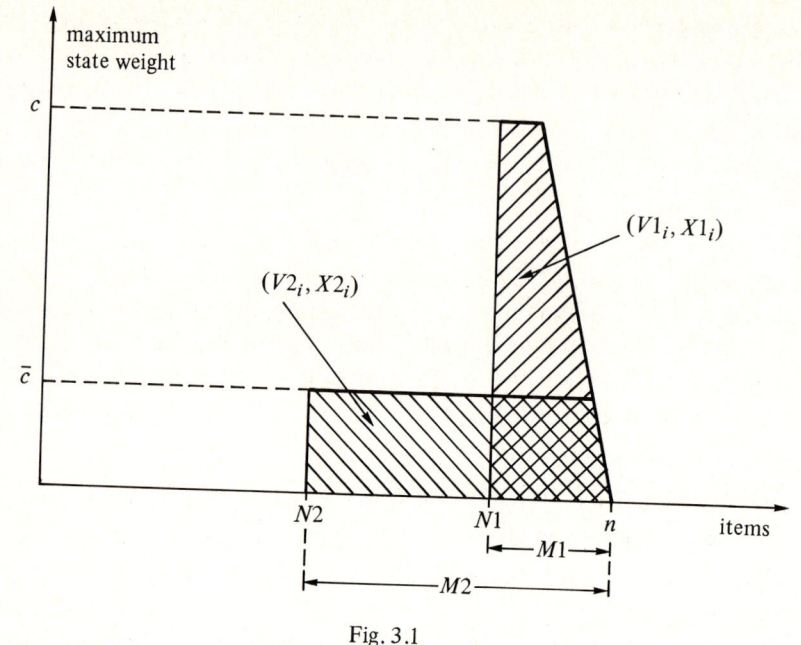

Fig. 3.1

that item $\bar{\bar{k}}$ which was inserted last and in performing a forward move starting from $\bar{\bar{k}} + 1$.

At any node of the decision-tree, let

X = set of the items which are in the current solution;
r = residual capacity = $c - \Sigma_{j \in X} w_j$;
z^* = value of the best solution so far.

Before any forward move (say starting from k), perform the following step:

if $r \geq f_k$ **then** set $z^* = \max\{z^*, \Sigma_{j \in X} w_j + f_k\}$ and perform a backtracking step;
    else determine, through binary search, $q = \min\{j : f_j \leq r\}$ and set $\bar{s} = n - q + 1$;
    if $\Sigma_{j \in X} w_j + (f_k - f_{k+\bar{s}}) \leq z^*$ **then** perform a backtracking step.

(Note that $\bar{s}$ is the maximum number of items which could be added to the current solution by subsequent forward moves, so $f_k - f_{k+\bar{s}} \ (= \Sigma_{j=k}^{k+\bar{s}-1} w_j$ = value of the $\bar{s}$ largest available items) is an upper bound on the value we can add to the current solution).

At the end of a forward move, let m be the next item to be considered (i.e. the forward move inserted items $\bar{k}, \ldots, m-1$) and define:

$$l(r) = \max_{1 \leq i \leq L1} \{i : V1_i \leq r\};$$

$$h(r, X) = \max_{1 \leq i \leq L2} \{i : V2_i \leq r \text{ and } X \cap X2_i = \emptyset\}.$$

($V1_{l(r)}$ is the maximum weight we can add to the current solution by using only the last $M1$ items; $V2_{h(r,X)}$ is the maximum weight not greater than $\bar{c}$ that can be added to the current solution by using only the last $M2$ items. Both $l(r)$ and $h(r, X)$ can be determined through binary search). Then perform the following step:

if $r \leq \bar{c}$ then determine $h(r, X)$ and set $\Delta = V2_{h(r,X)}$;
    else determine $l(r)$ and set $\Delta = V1_{l(r)}$;
        if $\Delta < \bar{c}$ and $z^* < c - r + \bar{c}$ then determine $h(r, X)$ and set $\Delta = V2_{h(r,X)}$.

Now Δ gives the maximum additional weight we can obtain from the lists, so we can set $z^* = \max\{z^*, c - r + \Delta\}$ (if now $z^* = c$, execution stops). Then, if $r \leq \bar{c}$ and $m > N2$, we perform a backtracking step (since no better solution can be found); otherwise, we perform a forward move starting from m.

4. [*halting tests*]

Let X^* be the set of items which are in the optimal solution at the end of Step 3; if $J1 = \emptyset$ then the optimal solution is z^*, X^*; **stop**;
    else if $z^* = c$ then the optimal solution is $z^* + \Sigma_{j \in J1} w_j, X^* \cup J1$; **stop**;
        else set $t = t + \Delta t$ and go to Step 1.

The parameters for the dynamic programming lists have been experimentally determined as the following functions of n and $M = \max\{w_j\}$: $M1 = \min(2\log_{10}M, 0.7n)$, $M2 = \min(2.5\log_{10}M, 0.8n)$, $\bar{c} = 1.3 w_{N2}$. The parameters for the definition of the core problem have been experimentally fixed as $t = 15, \Delta t = 5$.

3.3. Computational experiments

In Table 3.1 algorithm MTS is compared with algorithm AF of Ahrens and Finke [2] on problems with $n \leq 40$, i.e. when no core problem is generated. The following sets of test problems have been considered:

1) w_i uniformly random in range $1 - 10^3$, $c = 10^3 n/4$;
2) w_i uniformly random in range $1 - 10^6$, $c = 10^6 n/4$;
3) w_i uniformly random in range $1 - 10^{12}$, $c = 10^{12} n/4$;
4) w_i even uniformly random in range $2 - 10^3$, $c = 10^3 n/4 + 1$ (odd).

In all data sets the value of c was set to the expected value of $\Sigma_{j=1}^n w_j/2$, so that the solutions would contain about half of the items. For each data set and each value of n, 10 problems were generated and solved, through Fortran IV programs, on a CDC-Cyber 730. Each of the two algorithms was assigned 16000 words of core memory and a time limit of 450 seconds to solve the 90 problems generated for each data set. The entries in Table 3.1 give the average and maximum running times.

The problems of data sets 1) and 2) are comparatively «easy», since they generally admit a large number of feasible solutions of value c. The problems of data sets 3) and 4) are comparatively «hard» for the opposite reason.

Table 3.1
CDC-Cyber 730 seconds. Average (maximum) times for 10 problems.

Data set	n	AF	MTS
w_i uniformly random in $1 - 10^3$ $c = 10^3 n/4$	8	0.012 (0.018)	0.004 (0.013)
	12	0.023 (0.039)	0.010 (0.018)
	16	0.040 (0.067)	0.011 (0.026)
	20	0.069 (0.112)	0.007 (0.021)
	24	0.137 (0.203)	0.010 (0.029)
	28	0.349 (0.531)	0.010 (0.019)
	32	0.940 (1.934)	0.009 (0.024)
	36	2.341 (3.672)	0.009 (0.018)
	40	5.590 (9.306)	0.011 (0.026)
w_i uniformly random in $1 - 10^6$ $c = 10^6 n/4$	8	0.012 (0.021)	0.004 (0.012)
	12	0.029 (0.043)	0.013 (0.025)
	16	0.091 (0.134)	0.049 (0.074)
	20	0.322 (0.543)	0.185 (0.263)
	24	0.640 (1.282)	0.513 (0.992)
	28	1.341 (1.844)	0.647 (1.054)
	32	2.284 (5.923)	0.661 (1.593)
	36	4.268 (12.917)	0.605 (1.732)
	40	9.712 (25.346)	0.663 (2.306)
w_i uniformly random in $1 - 10^{12}$ $c = 10^{12} n/4$	8	0.013 (0.029)	0.004 (0.014)
	12	0.029 (0.051)	0.013 (0.029)
	16	0.092 (0.143)	0.050 (0.093)
	20	0.422 (0.536)	0.232 (0.341)
	24	2.070 (2.296)	1.098 (1.164)
	28	9.442 (10.383)	6.306 (12.663)
	32	time limit	time limit
w_i even and uniformly random in $2 - 10^3$ $c = 10^3 n/4 + 1$	8	0.013 (0.027)	0.005 (0.013)
	12	0.028 (0.060)	0.021 (0.031)
	16	0.090 (0.138)	0.053 (0.087)
	20	0.392 (0.490)	0.190 (0.351)
	24	1.804 (3.312)	0.525 (1.081)
	28	7.091 (9.254)	0.969 (1.744)
	32	21.916 (30.604)	1.496 (1.930)
	36	time limit	2.184 (4.615)
	40	–	2.941 (6.208)

Table 3.2 gives computational results for problems with high values of n. These problems are generally easy to solve if the range of weights is not too large since, in this case, many solutions of value c exist. The Ahrens-Finke algorithm cannot in general be applied to such problems because of the excessive memory requirement. Balas and Zemel [5] have experimented a specialized version (BZS) of their algorithm for large $0-1$ KP's on sets of SSP's obtained by

Table 3.2
w_i uniformly random in $10-M$; $c = 0.5 \sum_{j=1}^{n} w_j$. CDC-Cyber 730 seconds. Average (maximum) times over 20 problems.

Data set	n	BZS	MTS
$M = 10^2$	1000	0.012 (0.048)	0.009 (0.016)
	2500	0.018 (0.036)	0.016 (0.037)
	5000	0.036 (0.164)	0.025 (0.041)
	10000	0.056 (0.074)	0.044 (0.048)
$M = 10^3$	1000	0.054 (0.150)	0.025 (0.041)
	2500	0.060 (0.162)	0.032 (0.047)
	5000	0.060 (0.116)	0.034 (0.062)
	10000	0.084 (0.136)	0.049 (0.082)
$M = 10^4$	1000	0.490 (1.764)	0.083 (0.125)
	2500	0.386 (0.796)	0.092 (0.130)
	5000	0.474 (2.142)	0.100 (0.142)
	10000	0.604 (2.378)	0.122 (0.137)

randomly generating the weights in different ranges $10-M$ (with $M = 10^2$, 10^3 and 10^4) and by setting $c = 0.5 \sum_{j=1}^{n} w_j$. For each pair (n, M) they ran the Fortran IV code of their algorithm on a CDC-6600 over 20 problems. We randomly generated problems according to the same distribution and ran algorithm MTS on a CDC-Cyber 730, which, on problems of this kind, is about two times slower than the CDC-6600. The times given in [5] for BZS were consequently multiplied by 2.

In all cases the initial core problem (of size $n = 31$) was enough to solve the problem.

3.4. Heuristic algorithms

The most important heuristic algorithms for SSP are approximation schemes, which allow one to obtain any *worst-case performance ratio* (see Section 1) by fixing the value of a parameter. As for the 0-1 knapsack problem, we subdivide these algorithms into two classes (see Section 2.8): *polynomial approximation schemes* and *fully polynomial approximation schemes*. We denote by $\hat{X}$ a heuristic solution and by $\hat{z}$ the corresponding value of the objective function.

The first polynomial approximation scheme for SSP was proposed by Johnson [39]. For any prefixed positive integer k, the scheme gives an algorithm J^k of time complexity $O(n^k)$ (plus $O(n \log n)$ for sorting the items), space complexity $O(n)$ and worst-case performance ratio $r(J^k) = \frac{k}{k+1}$.

These results have been improved on by Martello and Toth [60] through a polynomial approximation scheme based on a modified greedy algorithm. The standard greedy algorithm for SSP examines the items according to decreasing weights and adds each w_j to the current solution value g if and only if $w_j \leq c - g$. The modified algorithm assumes that a partial feasible solution M is known, and, for a given $i \leq n$, examines only those items w_j such that $j \notin M$ and $j \geq i$. The corresponding procedure follows.

Procedure $g(M, i)$

comment it is assumed that $w_j \geq w_{j+1}$ for $j = 1, \ldots, n - 1$;
set $g = \Sigma_{j \in M} w_j$, $X = M$;
for $j = i$ to n do
  if $j \notin M$ and $w_j \leq c - g$ then set $g = g + w_j$, $X = X \cup \{j\}$.

It is self-evident that $g(M, i)$ runs is $O(n)$ time. For any prefixed positive integer k, the Martello-Toth approximation scheme is the following:

Algorithm MT^k

Sort the items so that $w_j \geq w_{j+1}$ for $j = 1, \ldots, n - 1$;
if $k = 1$ then set $\hat{z} = g(\emptyset, 1)$, $\hat{X} = X$ and **stop**;
set $\hat{z} = 0$;
for each $M \subset N = \{1, \ldots, n\}$ such that $|M| \leq k - 2$ and $\Sigma_{j \in M} w_j \leq c$ do
  for each $i \in N - M$ do
    if $g(M, i) > \hat{z}$ then set $\hat{z} = g(M, i)$, $\hat{X} = M \cup X$.

For $k = 2$ the algorithm considers only set $M = \emptyset$, so it applies the modified greedy algorithm n times (with $i = 1, 2, \ldots, n$, respectively) with no partial solution. For $k = 3$ it considers, in addition, the n singletons $\{1\}, \{2\}, \ldots, \{n\}$ as sets M, so each item is imposed in turn as a partial solution and, for each partial solution, the modified greedy algorithm is applied $n - 1$ times. For $k = 4$ all feasible pairs of items are imposed as a partial solution, and so on.

The time and space complexities of MT^k are the same as those of Johnsons' J^k scheme, i.e. $O(n^k)$ (plus $O(n \log n)$ for the sorting) and $O(n)$, respectively. As to the worst-case performance, it has been proved in [60] that:

- $r(MT^1) = r(J^1) = 1/2$;
- $r(MT^2) = 3/4$ (while $r(J^2) = 2/3$);
- $r(MT^3) = 6/7$ (while $r(J^3) = 3/4$);
- for $k > 3$, $r(MT^k) \geq (k + 3)/(k + 4)$ (while $r(J^k) = k/(k + 1)$). The exact worst-case performance ratio of MT^k for $k > 3$ is still an open problem, although it is conjectured that $r(MT^k) = \frac{k(k+1)}{k(k+1)+2}$ (note that this is true for $k \leq 3$).

MT^k is related to the polynomial approximation scheme S^k proposed by Sahni [68] for the 0-1 knapsack problem. However, the Sahni scheme, when applied to SSP, is dominated by both MT^k and J^k, in the sense that it has the same time and space complexities but $r(S^k) = (k - 1)/k$. An efficient implementation of the Martello-Toth scheme has been presented in [62].

Let us now consider fully polynomial approximation schemes for SSP. As for the 0-1 knapsack problem, the complexities will be given as functions of n and of $\epsilon = 1 - r$ (*worst-case relative error*).

The first fully polynomial approximation scheme for SSP was derived by Ibarra and Kim [35] from their scheme for the 0-1 knapsack problem (see Section 2.8). They proved that by removing the preliminary sorting of the items from this scheme one has a scheme for SSP which has time complexity $O(n + \log(1/\epsilon)/\epsilon^4)$ and space complexity $O(n + 1/\epsilon^3)$. Lawler [44] modified the Ibarra-Kim scheme to obtain time complexity $O(n + 1/\epsilon^3)$ and space complexity $O(n + 1/\epsilon^2)$.

A different fully-polynomial approximation scheme is the following, proposed by Gens and Levner [24].

Algorithm GL^ϵ

find the largest s such that $P = \sum_{j=1}^{s} w_j \leq c$. Set $P_0 = \max(P, \max_{1 \leq j \leq n}\{w_j\})$; set $V^1 = \{0, w_1\}$;

for $j = 2$ to n do

    set $T = V^{j-1} \cup \{(w_j + w_k) : w_k \in V^{j-1},\ w_j + w_k \leq c,\ (w_j + w_k) \notin V^{j-1}\}$, with the elements $t_i \in T$ ordered so that $t_i < t_{i+1}$ (this can easily be obtained through ordinary dynamic programming techniques);

    set $h = 1, i = 1, v_h = t_i$;

    repeat

        if $t_{i+1} > v_h + \epsilon P_0$ then set $i = i + 1$;

            else set $i = \max\{q : t_q \leq v_h + \epsilon P_0\}$;

        set $h = h + 1, v_h = t_i$

    until $i = |T|$;

    set $V^j = \{v_1, \ldots, v_h\}$;

set $\hat{z} = v_h$ and determine the items in the approximate solution $\hat{X}$ by backtracing from V^n to V^1.

GL^ϵ has time and space complexity $O(n/\epsilon)$, so the scheme can be either more or less efficient than the Lawler scheme, depending on the values of n and ϵ.

The practical behaviour of the approximation schemes of Johnson [39], Martello-Toth [60], Lawler [44] and Gens-Levner [24] has been experimentally analyzed in [62]. The algorithms have been coded in Fortran IV and compared on seven sets of randomly generated problems, as well as an two sets of deterministic hard problems proposed in Chvàtal [13], with n ranging from 10 to 1000 and prefixed worst-case performance ratios $r_1 = 1/2, r_2 = 3/4, r_3 = 6/7$. The experiments showed that

(i) All the approximation schemes have a much better experimental performance than their worst-case performance. So, in practical applications, we can obtain good results with short computing times, i.e. by prefixing small values of the worst-case performance ratio.

(ii) Although polynomial approximation schemes have a worse bound on computing time, their average performance appears superior to that of the fully polynomial approximation schemes, in the sense that they generally give better results with shorter computing times and fewer core memory requirements.

(iii) MT^k turned out to be the most efficient scheme. For $n \geq 50$, the largest average error of MT^2 was 0.0075 percent, that of MT^3 0.0005 percent.

4. 0-1 multiple knapsack problem

4.1. The problem

The *0-1 Multiple Knapsack Problem (MKP)* is a generalization of the 0-1 knapsack problem of Section 2 to the case where m containers with capacities $c_1, \ldots, c_m$ are available. If p_j and w_j are, respectively, the profit and weight of the j-th item ($j = 1, \ldots, n$), and we introduce a boolean variable $x_{i,j}$ which is set to 1 if item j is inserted in container i, or to 0 otherwise, the problem can be formally stated as

$$\text{maximize} \quad z = \sum_{i=1}^{m} \sum_{j=1}^{n} p_j x_{i,j} \tag{4.1}$$

$$\text{subject to} \quad \sum_{j=1}^{n} w_j x_{i,j} \leq c_i \quad (i = 1, \ldots, m) \tag{4.2}$$

$$\sum_{i=1}^{m} x_{i,j} \leq 1 \quad (j = 1, \ldots, n) \tag{4.3}$$

$$x_{i,j} = 0 \text{ or } 1 \quad (i = 1, \ldots, m; j = 1, \ldots, n) \tag{4.4}$$

The problem is clearly NP-hard, since it reduces to the 0-1 knapsack problem when $m = 1$.

In what follows we will assume, without loss of generality

$$\min_{j} \{w_j\} \leq \min_{i} \{c_i\};$$

$$\max_{j} \{w_j\} \leq \max_{i} \{c_i\};$$

$$\sum_{j=1}^{n} w_j > \max_{i} \{c_i\}.$$

Furthermore, we will assume that the items are sorted so that $p_j/w_j \geq p_{j+1}/w_{j+1}$

($j = 1, \ldots, n-1$).

4.2. Relaxations and upper bounds

Two relaxation methods are generally employed to obtain upper bounds for MKP: the Lagrangean relaxation and the surrogate relaxation.

The *Lagrangean Relaxation*, relative to a nonnegative vector (λ), is

$$\text{maximize} \quad z_\lambda = \sum_{i=1}^{m}\sum_{j=1}^{n} p_j x_{i,j} - \sum_{j=1}^{n} \lambda_j \left(\sum_{i=1}^{m} x_{i,j} - 1 \right)$$

$$\text{subject to} \quad \sum_{j=1}^{n} w_j x_{i,j} \leq c_i \quad (i = 1, \ldots, m)$$

$$x_{i,j} = 0 \text{ or } 1 \quad (i = 1, \ldots, m; j = 1, \ldots, n).$$

Since the objective function can be written as

$$z_\lambda = \sum_{j=1}^{n} \lambda_j + \sum_{i=1}^{m}\sum_{j=1}^{n} (p_j - \lambda_j) x_{i,j},$$

the relaxed problem can be decomposed into a series of m single knapsack problems. For $i = 1, \ldots, m$ we can solve the single knapsack problem

$$\text{maximize} \quad z_i = \sum_{j=1}^{n} (p_j - \lambda_j) x_{i,j}$$

$$\text{subject to} \quad \sum_{j=1}^{n} w_j x_{i,j} \leq c_i$$

$$x_{i,j} = 0 \text{ or } 1 \quad (j = 1, \ldots, n)$$

and then set $z_\lambda = \sum_{j=1}^{n} \lambda_j + \sum_{i=1}^{m} z_i$. In order to find the tightest possible bound, vector (λ) should be determined so that z_λ is minimized. An approximation of the optimum (λ) can be obtained through subgradient techniques, which are, however, time consuming. Hung and Fisk [34], who first used this relaxation to obtain upper bounds for MKP, set the λ_j's equal to the optimal dual multipliers of constraints (4.3) in the continuous relaxation given by (4.1), (4.2), (4.3) and

$$0 \leq x_{i,j} \leq 1 \quad (i = 1, \ldots, m; j = 1, \ldots, n),$$

that is, $\lambda_j = p_j - w_j (p_t/w_t)$ if $j < t$, or $\lambda_j = 0$ otherwise, where t is the smallest index such that

$$\sum_{j=1}^{t} w_j > \sum_{i=1}^{m} c_i.$$

The *Surrogate Relaxation* of MKP, relative to a non-negative vector (π), is defined as

$$\text{maximize} \quad z_\pi = \sum_{i=1}^{m} \sum_{j=1}^{n} p_j x_{i,j}$$

$$\text{subject to} \quad \sum_{i=1}^{m} \pi_i \sum_{j=1}^{n} w_j x_{i,j} \leq \sum_{i=1}^{m} \pi_i c_i$$

$$\sum_{i=1}^{m} x_{i,j} \leq 1 \quad (j = 1, \ldots, n)$$

$$x_{i,j} = 0 \text{ or } 1 \quad (i = 1, \ldots, m; j = 1, \ldots, n).$$

In this case too, Hung and Fisk [34] have set the π_i's equal to the optimal dual multipliers of constraints (4.2) in the continuous relaxation, that is, $\pi_i = p_t/w_t$ for all i. Martello and Toth [57] have proved that setting $\pi_i = k$ (k any positive constant) for all i's leads to the minimum value of the optimal solution of the surrogate relaxation, that is, to the tightest upper bound this relaxation can give for MKP. If we set $y_j = \sum_{i=1}^{m} x_{i,j}$ ($j = 1, \ldots, n$), the surrogate relaxation of MKP can thus be expressed as the 0-1 single knapsack problem

$$\text{maximize} \quad z_\pi = \sum_{j=1}^{n} p_j y_j \tag{4.5}$$

$$\text{subject to} \quad \sum_{j=1}^{n} w_j y_j \leq \sum_{i=1}^{m} c_i \tag{4.6}$$

$$y_j = 0 \text{ or } 1 \quad (j = 1, \ldots, n). \tag{4.7}$$

Both the Lagrangean and the surrogate relaxation are NP-hard problems, since they require the solution of 0-1 single knapsack problems. However, we can compute upper bounds on the value of z_λ or z_π (as shown in Section 2.2) and hence upper bounds for MKP in polynomial time.

Neither relaxation dominates the other. In general, it can be expected that z_π gives tighter upper bounds when m is small or the ratio n/m is large, since the surrogate problem allows one to split items between two knapsacks and the number of split items is in this case comparatively small.

4.3. Reduction

The size of a 0-1 multiple knapsack problem can be reduced, in a way similar to that presented for KP in Section 2.3, so as to determine two sets:

$$J1 = \left\{ j : \sum_{i=1}^{m} x_{i,j} = 1 \text{ in an optimal solution to MKP} \right\};$$

$$J0 = \left\{ j : \sum_{i=1}^{m} x_{i,j} = 0 \text{ in an optimal solution to MKP} \right\}.$$

In this case, however, only $J0$ allows one to reduce the size of the problem by eliminating the corresponding items, while $J1$ gives only information useful in reducing the number of nodes of a branch-decision tree, since it cannot specify in which knapsack the items must be inserted.

Ingargiola and Korsh [37] have presented a specific reduction procedure, based on an extension of the dominance between items. Let jDk indicate that item j dominates item k. Set $D_j = \{k : jDk\}$ can be determined, for $j = 1, \ldots, n$, as follows. Let $N = \{1, 2, \ldots, n\}$.

Procedure DOMINANCE

set $D_j = \{k : w_k \geq w_j$ and $p_k \leq p_j$ with both equalities not true, or $w_k = w_j$, $p_k = p_j$ and $k > j\}$;

repeat

    set $d = |D_j|$;

    for each $k \in N - D_j$ **do**

        if there exists a subset $E \subseteq D_j$ such that $w_j + \Sigma_{e \in E} w_e \leq w_k$ and $p_j + \Sigma_{e \in E} p_e \geq p_k$

        then set $D_j = D_j \cup \{k\}$;

until $|D_j| = d$.

Given any two items j and k such that jDk and a feasible solution of value $\bar{z}$ that includes k but excludes j, then there is a feasible solution of value $\bar{\bar{z}} \geq \bar{z}$ that includes j and excludes k. Consequently, if an item j is excluded from the optimal solution, then also the items in D_j must be excluded. It follows that

Proposition 4.1. Given any feasible solution of value $\bar{z}$, all items in $J1 = \{j : \Sigma_{k \in N-(\{j\} \cup D_j)} p_k \leq \bar{z}\}$ must be included in an optimal solution.

In fact, the exclusion of any item $j \in J1$, and hence of all the items in D_j, would give solutions of value no greater than $\bar{z}$. A reduction procedure can thus start by determining sets D_j (note that this computation requires a time which is exponential in n), finding a feasible solution through any heuristic algorithm and then determining set $J1$. Observing now that if an item k is imposed in the optimal solution, then all items j such that jDk must be imposed, set $J0$ can be determined through the following

Proposition 4.2. Given any feasible solution and the corresponding set $J1$ (obtained from Proposition 4.1), let $E_k = \{j : jDk, j \notin J1\}$. Then all the items in $J0 = \{k : w_k + \Sigma_{j \in E_k} w_j > \Sigma_{i=1}^m c_i - \Sigma_{j \in J1} w_j\}$ must be excluded from an optimal solution.

In fact, the imposition of any item $k \in J0$, and hence of all the items in E_k, would produce infeasible solutions.

4.4. Exact algorithms

Algorithms for MKP are generally oriented either to the case of low values of the ratio n/m or to the case of high values of this ratio. Algorithms for the first class, which has applications, for example, when m liquids, that cannot be mixed, have to be loaded into n tanks, have been presented by Christofides, Carpaneto, Mingozzi and Toth [12] and by Neebe and Dannenbring [66]. In the following we will review algorithms for the second class, which has been more intensively studied.

Hung and Fisk [34] have proposed a depth-first branch-and-bound algorithm, where successively higher levels of the decision-tree are constructed by selecting an item and assigning it to knapsacks in decreasing order of their capacities; when all the knapsacks have been considered, the item is assigned to a dummy knapsack, implying its exclusion from the solution: so, each node of the decision-tree generates $m + 1$ descendent nodes. The upper bound associated with each node can be computed either as the solution of the Lagrangean relaxation, or the surrogate relaxation, of the current problem, or as the smaller of the two. As was pointed out in Section 4.2, vectors (λ) and (π) have been set respectively equal to the optimal dual multipliers of constraints (4.3) and (4.2) in the continuous relaxation of MKP.

Martello and Toth [54] have obtained better computational results through a different branching scheme which computes at each node a Lagrangean relaxation with $\lambda_j = 0$ for all j (it can easily be verified that this choice is not dominated and does not dominate the Hung-Fisk choice). In the resulting problem (given by equations (4.1), (4.2) and (4.4)), each knapsack can be solved independently of the others. If no item appears in two or more knapsacks, a feasible solution to the current problem has been found and a backtracking can be performed. Otherwise, an item which appears in m' ($2 \leq m' \leq m$) knapsacks is selected and m' nodes are generated ($m' - 1$ by assigning the item to the first $m' - 1$ knapsacks where it appears, the m'-th one by excluding it from them). The m' upper bounds are computed by solving only m' single knapsacks and utilizing part of the solutions previously found for the ascendent nodes. Each bound can be improved by assuming the smaller between it on the one hand and the solution of the corresponding surrogate relaxation on the other.

The experimental results given in [54] indicate the clear superiority of the Martello-Toth algorithm over the Hung-Fisk one, both with and without previous application of the reduction algorithm of Section 4.3.

A further improvement has been obtained by Martello and Toth [57] through a so-called «bound-and-bound» algorithm. This modification of a branch-and--bound approach, based on the computation at each decision-node of both an upper bound and a lower bound of the current problem, can be used to solve any 0-1 linear programming problem. A complete description of the general approach can be found in [57].

For MKP, the resulting algorithm consists of an enumerative scheme where each node of the decision-tree generates two branches either by assigning an item j to a knapsack i ($x_{i,j} = 1$) or by excluding j from i ($x_{i,j} = 0$). Stack S_i ($i = 1, \ldots, m$) contains those items that are currently assigned to knapsack i or excluded from it. The *current problem* corresponding to (S) is the original problem with the additional constraints given by fixing the items in S_i ($i = 1, \ldots, m$). Let *upper* (S) and *lower* (S) be respectively an upper and a lower bound to this problem.

upper (S) is computed by procedure SIGMA which solves a surrogate relaxation of the current problem.

lower (S) is computed by a heuristic procedure, PI, which finds an optimal solution for the first knapsack, then excludes the items inserted in it and finds an optimal solution for the second knapasck, and so on. The heuristic solution found is stored in $\hat{x}_{i,j}$ ($i = 1, \ldots, m; j = 1, \ldots, n$).

At each iteration, the algorithm selects as branching variable the next $x_{i,j}$ such that $\hat{x}_{i,j} = 1$, according to increasing values of i. It follows that, given the current value of i, knapsacks $1, \ldots, i-1$ are «completely loaded» (i.e. no further item can be inserted in them), knapsack i is «partially loaded», and knapsacks $i+1, \ldots, m$ are «empty».

The main conceptual difference between this approach and a standard depth--first branch-and-bound one is that the branching phase is here performed by updating the partial solution through the solution obtained from the computation of a lower bound. This gives two important advantages:

(a) For all S for which *lower* $(S) =$ *upper* (S), $(\hat{x})$ is obviously an optimal solution to the corresponding current problem, so it is possible to avoid exploration of part of the tree.

(b) For all S for which *lower* $(S) <$ *upper* (S), S is updated through the heuristic solution previously found by procedure PI, so the resulting partial solution can generally be expected to be better than that which would be obtained by a series of forward steps, each fixing a variable independently of the following ones.

A general description of the algorithm follows. It is assumed that items and knapsacks are sorted no that $p_j/w_j \geq p_{j+1}/w_{j+1}$ for $j = 1, \ldots, n-1$ and $c_i \leq c_{i+1}$ for $i = 1, \ldots, m-1$.

Algorithm MKP

1. [*initialization*]
      for $i = 1$ to m do set $S_i = \emptyset$;

set $z^* = 0$, $x_{i,j} = 0$ for all i and j, $i = 1$;
apply procedure SIGMA (i) and let $u = upper\,(S)$;

2. [*heuristic*]
apply procedure PI (i) and let $l = lower\,(S)$, $(\hat{x}) = $ heuristic solution found;
if $l > z^*$ then set $z^* = l$, $(x^*) = (\hat{x})$;
  if $l = u$ then go to step 4;

3. [*updating*]
let $j = \min\{v : v \notin S_i \text{ and } \hat{x}_{i,v} = 1\}$ $(j = 0$ if no such v exists);
if $j = 0$ then if $i < m - 1$ then set $i = i + 1$ and repeat step 3;
      else go to step 4;
 else set $S_i = S_i \cup \{j\}$, $x_{i,j} = 1$;
   apply procedure SIGMA (i) and let $u = upper\,(S)$;
   if $u > z^*$ then repeat step 3;

4. [*backtracking*]
let j be the last item inserted in S_i such that $x_{i,j} = 1$ $(j = 0$ if no such j exists);
if $j = 0$ then if $i = 1$ then **stop**;
      else set $S_i = \emptyset$, $i = i - 1$ and repeat step 4;
 else set $x_{i,j} = 0$, $S_i = S_i - \{v \in S_i : v$ was inserted in S_i after $j\}$;
   apply procedure SIGMA (i) and let $u = upper\,(S)$;
   if $u \leqslant z^*$ then repeat step 4;
    else go to step 2.

Procedure SIGMA (i)

let $c = (c_i - \Sigma_{j \in S_i} w_j x_{i,j}) + \Sigma_{r=i+1}^{m} c_r$, $Q = \{j : x_{r,j} = 0 \text{ for all } r\}$;
solve the single knapsack problem defined by the items in Q and by capacity c and let $\bar{z}$ be the solution value;
set $upper\,(S) = \bar{z} + \Sigma_{r=1}^{i} \Sigma_{j \in S_r} p_j x_{r,j}$.

Procedure PI (i)

let $lower\,(S) = \Sigma_{r=1}^{i} \Sigma_{j \in S_r} p_j x_{r,j}$, $Q = \{j : x_{r,j} = 0 \text{ for all } r\}$;
set $c = c_i - \Sigma_{j \in S_i} w_j x_{i,j}$, $\bar{Q} = Q - S_i$, $r = i$;
repeat
 solve the single knapsack problem defined by the items in $\bar{Q}$ and by capacity c; let $\hat{z}$ be the solution value and store the solution vector in the r-th row of $(\hat{x})$;
 set $lower\,(S) = lower\,(S) + \hat{z}$;
 set $Q = Q - \{j : \hat{x}_{r,j} = 1\}$, $\bar{Q} = Q$, $r = r + 1$;
 if $r \leqslant m$ then set $c = c_r$;
until $r > m$;
for $r = 1$ **to** $i - 1$ **do for** $j = 1$ **to** n **do** set $\hat{x}_{r,j} = x_{r,j}$;
for each $j \in S_i$ **do if** $x_{i,j} = 1$ **then** set $\hat{x}_{i,j} = 1$.

A Fortran implementation of MKP has been presented in [63]. The experimental performance of this code is analyzed in Section 4.6.

4.5. Heuristic algorithms

To our knowledge, no polynomial approximation scheme for MKP has been presented. Nor is any heuristic with guaranteed worst-case performance known.

Fisk and Hung [21] have presented a heuristic approach which is based on the exact solution of the surrogate relaxation (4.5) - (4.7) of MKP and hence requires, in the worst case, a non-polynomial running time.

Martello and Toth [58] have presented heuristic procedures, all polynomial in the problem size $m + n$, which can be combined in different ways according to the size and the difficulty of the problems to be solved, to produce various approximate algorithms for MKP. The basic approach can be outlined as follows (assume that items and knapsacks are arranged so that $p_j/w_j \geqslant p_{j+1}/w_{j+1}$ for $j = 1, \ldots, n - 1$ and $c_i \leqslant c_{i+1}$ for $i = 1, \ldots, m - 1$).

1. Determination of an *initial feasible solution*. The simplest way to do this is to apply the greedy algorithm (see Section 2.8) m times: to the first knapsack, then to the second one by using only the remaining items, and so on. Other approaches are also proposed.

2. *Rearrangement* of the feasible solution found. The purpose of this step is to exchange items between the knapsacks so that each contains items of dissimilar profit per unit weight. In fact, it has been experimentally verified that in this way the subsequent improving algorithms tend to have better performance.

3. *Improvement* of the rearranged feasible solution. This step is performed by applying two procedures. The first considers all pairs of items in the current solution and, if possible, interchanges them between knapsacks should the insertion of a new item into the solution be allowed. The second procedure tries to exclude in turn each item currently in the solution and to replace it with one or more items not in the solution so that the total profit is increased.

The overall complexity of this approach is $O(mn + n^3)$. A Fortran implementation of the method has been given in [58].

4.6. Computational results

In this section we analyze the experimental performance of the algorithms of Section 4.4 and 4.5. We will call MKP the Fortran code [63] of the exact algorithm of Section 4.4, and HMKP the Fortran code [58] of the heuristic algorithm of Section 4.5. MKP has been tested also as a heuristic by halting execution after a prefixed number of iterations.

A series of test problems has been obtained by independently generating the values p_j, w_j and c_i from a uniform distribution according to the conditions:

$$10 \leqslant p_j \leqslant 100 \quad (j = 1, \ldots, n);$$
$$10 \leqslant w_j \leqslant 100 \quad (j = 1, \ldots, n);$$

$$0 \leqslant c_i \leqslant 0.5 \sum_{j=1}^{n} w_j - \sum_{u=1}^{i-1} c_u \quad (i = 1, \ldots, m-1);$$

$$c_m = 0.5 \sum_{j=1}^{n} w_j - \sum_{u=1}^{m-1} c_u.$$

We solved 10 problems for each value of n ($n = 50, 100, 200, 500, 1000$) and of m ($m = 2, 5, 10$). Table 4.1 gives the average times, expressed in seconds, of a CDC-Cyber 730 and comprehensive of the sorting times, for the following cases:

1) exact solution with MKP;

2) heuristic solution with MKP halted after 10 iterations;

3) heuristic solution with MKP halted after 50 iterations;

4) heuristic solution with HMKP.

Table 4.1
CDC-Cyber 730 seconds. Average times (average percentage errors) over 10 problems.

m	n	Exact solution	Heuristic solution		
		MKP	MKP (10 iterations)	MKP (50 iterations)	HMKP
2	50	0.082	0.049 (0.028)	0.070 (0.004)	0.013 (0.170)
	100	0.129	0.089 (0.018)	0.127 (0.000)	0.031 (0.147)
	200	0.153	0.143 (0.000)	0.152 (0.000)	0.057 (0.049)
	500	0.243	0.242 (0.000)	0.242 (0.000)	0.132 (0.020)
	1000	0.503	0.502 (0.000)	0.502 (0.000)	0.266 (0.003)
5	50	1.190	0.157 (0.344)	0.434 (0.312)	0.018 (0.506)
	100	1.014	0.268 (0.076)	0.601 (0.027)	0.040 (0.303)
	200	1.178	0.327 (0.018)	0.687 (0.012)	0.074 (0.148)
	500	0.862	0.659 (0.001)	0.705 (0.001)	0.186 (0.031)
	1000	1.576	1.231 (0.001)	1.576 (0.000)	0.391 (0.016)
10	50	3.852	0.162 (0.287)	0.477 (0.211)	0.035 (0.832)
	100	7.610	0.324 (0.174)	0.950 (0.092)	0.057 (0.437)
	200	32.439	0.659 (0.060)	1.385 (0.039)	0.106 (0.219)
	500	5.198	1.760 (0.009)	3.836 (0.003)	0.535 (0.078)
	1000	9.729	3.846 (0.003)	7.623 (0.001)	0.870 (0.031)

For the heuristic solutions, the times are followed, in brackets, by the average percentage errors.

Table 4.1 shows that the time required to find the exact solution increases much more steeply with m than with n and tends to become impractical for $m > 10$. When used as a heuristic, MKP gives solutions very close to the optimum; the running times are reasonable and increase slowly with n and m. HMKP is faster than MKP, but its solutions are clearly worse.

The test problems of Table 4.1 are relatively «easy», since profits and weights are uncorrelated (see Section 2.7). For correlated profits and weights, exact solutions obviously tend to become impossible when $n \geqslant 100$, while HMKP can solve, with reasonable percentage errors, problems with $m = 100$ and $n = 1000$ (see [63]).

5. Other single knapsack problems

In this section we review some problems strictly connected with the 0-1 knapsack problem, in the sense that they can be either transformed into a 0-1 knapsack problem or solved through techniques similar to those described in Section 2. We will not examine other knapsack-type problems which conceptually differ from those considered in this survey, such as, for example, the Multiple Choice Knapsack Problem, the Fractional Knapsack Problem, the Quadratic Knapsack Problem, and so on.

5.1. Unbounded knapsack problem

When, for each j ($j = 1, \ldots, n$), an infinite number of items of profit p_j and weight w_j is available, we have the so-called *Unbounded Knapsack Problem*:

$$\text{maximize} \quad z = \sum_{j=1}^{n} p_j x_j$$

$$\text{subject to} \quad \sum_{j=1}^{n} w_j x_j \leqslant c, \tag{5.1}$$

$$x_j \geqslant 0 \text{ and integer} \quad (j = 1, \ldots, n).$$

The most efficient dynamic programming algorithm for exact solution of the problem is that of Gilmore and Gomory [28] as improved by Hu [32]. Enumeration methods have been proposed by Gilmore and Gomory [26], Cabot [8] and Martello and Toth [51]. This last method experimentally turned out to be the fastest (see Martello and Toth [49, 51]). In fact, it can exactly solve problems with up to 10000 variables (with p_j, w_j uniformly random in $1 - 1000$ and

$c = 0.5 \, \Sigma_{j=1}^{n} w_j$) in average time of 3.1 seconds, sorting time included, on a CDC--Cyber 730.

The unbounded knapsack problem can also be transformed into an equivalent 0-1 problem (and hence solved with one of the algorithms of Section 2) as follows:

Set $k = 0$;
for $j = 1$ to n do
    set $d = [c/w_j]$, $e = 1$;
    while $e \leq d$ **do** set $k = k + 1, \bar{p}_k = e p_j, \bar{w}_k = e w_j, e = 2e$.

The 0-1 knapsack problem to be solved is

$$\text{maximize} \quad z = \sum_{j=1}^{k} \bar{p}_j \bar{x}_j \tag{5.2}$$

$$\text{subject to} \quad \sum_{j=1}^{k} \bar{w}_j \bar{x}_j \leq c, \tag{5.3}$$

$$\bar{x}_j = 0 \text{ or } 1 \quad (j = 1, \ldots, k), \tag{5.4}$$

where $k = \Sigma_{j=1}^{n} \lceil \log_2 [c/w_j] \rceil$ [1] (the solution vector (x) can easily be obtained from $(\bar{x})$).

5.2. Bounded knapsack problem

When, for each j ($j = 1, \ldots, n$), b_j items of profit p_j and weight w_j are available, we have so-called *Bounded Knapsack Problem*:

$$\text{maximize} \quad z = \sum_{j=1}^{n} p_j x_j$$

$$\text{subject to} \quad \sum_{j=1}^{n} w_j x_j \leq c,$$

$$0 \leq x_j \leq b_j \text{ and integer} \quad (j = 1, \ldots, n),$$

where it is usually assumed, without loss of generality, that $\Sigma_{j=1}^{n} b_j w_j > c$ and $b_j w_j \leq c$ for $j = 1, \ldots, n$.

The most efficient method for the exact solution of this problem is Martello and Toth's branch-and-bound algorithm [51] (see also a recent note by Aittoniemi and Oehlandt [3]). The method can solve (see [49]) randomly generated

---

[1] $\lceil a \rceil$ = smallest integer not less than a.

problems with up to 1000 variables (with p_j, w_j in the range $1 - 1000$, b_j in $1 - 10$ and $c = 0.5 \sum_{j=1}^{n} w_j$) in an average time of 8.5 seconds, sorting time included, on a CDC-Cyber 730. A different branch-and-bound approach has been proposed by Ingargiola and Korsh [38] and corrected by Martello and Toth [56]. A dynamic programming algorithm has been presented by Nemhauser and Ullmann [67].

The bounded knapsack problem too can be transformed into 0-1 form as follows:

Set $k = 0$;
for $j = 1$ to n do
    set $d = b_j$;
    while $d > 0$ do set $e = \lceil d/2 \rceil$, $k = k + 1$, $\bar{p}_k = ep_j$, $\bar{w}_k = ew_j$, $d = [d/2]$.

The equivalent 0-1 knapsack problem is defined by (5.2) - (5.4) with $k = \sum_{j=1}^{n} \lceil \log_2 b_j \rceil$.

5.3. Change-making problems

Consider an unbounded knapsack problem where $p_j = -1$ for $j = 1, \ldots, n$ and where, in condition (5.1), the equality constraint is imposed. The resulting problem can be expressed as

$$\text{minimize} \quad z = \sum_{j=1}^{n} x_j \tag{5.5}$$

$$\text{subject to} \quad \sum_{j=1}^{n} w_j x_j = c, \tag{5.6}$$

$$x_j \geq 0 \text{ and integer} \quad (j = 1, \ldots, n)$$

and is generally called the *Unbounded Change-Making Problem*. It can be viewed, in fact, as the problem of assembling a given change, c, using the least number of coins of specified values w_j ($j = 1, \ldots, n$) in the case where, for each value, an infinite number of coins is available.

A recursive algorithm for the exact solution of the problem in the case where one of the w_j's has value 1 (i.e. a feasible solution always exists) was presented by Chang and Gill [9]. An Algol implementation of this method was presented by the authors [10] and corrected by Johnson and Kernighan [40]. The exact solution of the general case has been obtained by Wright [80] through dynamic programming and by Martello and Toth [50, 55] through branch-and-bound. The Martello-Toth algorithm [55] experimentally proved to be very much faster than all the other methods, solving problems up to 1000 variables (with w_j uniformly random in the range $1 - 2000$ and $c = \sum_{j=1}^{n} w_j$) in average time of 0.3 seconds, sorting time included, on a CDC-Cyber 730.

The greedy algorithm for the unbounded change-making problem consists in sorting the items according to decreasing weights and then, for $j = 1, \ldots, n$, in inserting in the solution as many items of the j-th type as possible. Chang and Korsh [11], starting from the results obtained by Magazine, Nemhauser and Trotter [46] on the greedy solutions of knapsack problems, gave necessary and sufficient conditions for deciding whether a given instance of the problem is exactly solved by the greedy algorithm. The problem of maximum percentage error when the greedy algorithm does not work has been studied by Tien and Hu [73]. The experimental performance of the greedy algorithm has been analyzed by Martello and Toth [55].

The *Bounded Change-Making Problem* is defined by (5.5), (5.6) and by

$$0 \leq x_j \leq b_j \text{ and integer} \quad (j = 1, \ldots, n).$$

The only exact algorithm for this problem is the Martello and Toth's branch-and-bound approach [50], which solves randomly generated problems up to 1000 variables (with w_j in the range $1 - 2000$, b_j in $1 - 5$ and $c = 0.5 \, \Sigma_{j=1}^{n} b_j w_j$) in average time of 0.3 seconds, sorting time included, on a CDC-Cyber 730.

Aknowledgement

This work was supported by Ministero della Pubblica Istruzione, Italy.

References

[1] A.V. Aho, J.E. Hopcroft and J.D. Ullman, *Data Structures and Algorithms*, Addison-Wesley, 1983.
[2] J.H. Ahrens and G. Finke, «Merging and Sorting Applied to the Zero-One Knapsack Problem». *Operations Research* 23, 1099 - 1109, 1975.
[3] L. Aittoniemi and K. Oehlandt, «A Note on the Martello-Toth Algorithm for One-Dimensional Knapsack Problems», *European Journal of Operational Research* 20, 117, 1985.
[4] L.G. Babat, «Linear Functions on the N-dimensional Unit Cube», *Doklady Akademiia Nauk SSSR* 222, 761 - 762, 1975.
[5] E. Balas and E. Zemel, «An Algorithm for Large Zero-One Knapsack Problems», *Operations Research* 28, 1130 - 1154, 1980.
[6] R.S. Barr and G.T. Ross, «A Linked List Data Structure for a Binary Knapsack Algorithm», *Research Report CCS* 232, Centre for Cybernetic Studies, University of Texas, 1975.
[7] R. Bellman and S.E. Dreyfus, *Applied Dynamic Programming*, Princeton University Press, Princeton, N.J., 1962.
[8] A.V. Cabot, «An Enumeration Algorithm for Knapsack Problems», *Operations Research* 18, 306 - 311, 1970.
[9] S.K. Chang and A. Gill, «Algorithmic Solution of the Change-Making Problem», *Journal of ACM* 17, 113 - 122, 1970.
[10] S.K. Chang and A. Gill, «Algorithm 397. An Integer Programming Problem», *Communications of ACM* 13, 620 - 621, 1970.
[11] L. Chang and J.F. Korsh, «Canonical Coin-Changing and Greedy Solutions»,*Journal of ACM* 23, 418 - 422, 1976.
[12] N. Christofides, G. Carpaneto, A. Mingozzi and P. Toth, «The Loading of Liquids into Tanks», *Imperial College Research Report*, London, 1976.
[13] V. Chvàtal, «Hard Knapsack Problems», *Operations Research* 28, 402 - 411, 1980.
[14] J. Cord, «A Method for Allocating Funds to Investment Projects when Returns are Subject to Uncertainty», *Management Science* 10, 335 - 341, 1964.

[15] G.B. Dantzig, «Discrete Variable Extremum Problems», *Operations Research* 5, 266 - 277, 1957.
[16] R.S. Dembo and P.L. Hammer, «A Reduction Algorithm for Knapsack Problems», *Methods of Operations Research* 36, 49 - 60, 1980.
[17] K. Dudzinski and S. Walukiewicz, «Upper Bounds for the 0-1 Knapsack Problem», *Report MPD-10--49/84, Systems Research Institute, Warsaw*, 1984.
[18] B. Faaland, «Solution of the Value-Independent Knapsack Problem by Partitioning», *Operations Research* 21, 332 - 337, 1973.
[19] D. Fayard and G. Plateau, «Resolution of the 0-1 Knapsack Problem: Comparison of Methods», *Mathematical Programming* 8, 272 - 307, 1975.
[20] D. Fayard and G. Plateau, «An Algorithm for the Solution of the 0-1 Knapsack Problem», *Computing* 28, 269 - 287, 1982.
[21] J.C. Fisk and M.S. Hung, «A Heuristic Routine for Solving Large Loading Problems», *Naval Research Logistics Quarterly* 26, 643 - 650, 1979.
[22] M.R. Garey and D.S. Johnson, *Computers and Intractability: a Guide to the Theory of NP-Completeness*, Freeman, San Francisco, 1979.
[23] R.S. Garfinkel and G.L. Nemhauser, *Integer Programming*, John Wiley and Sons, 1972.
[24] G.V. Gens and E.V. Levner, «Fast Approximation Algorithms for Knapsack Type Problems», in K. Iracki, K. Malinowski and S. Walukiewicz, eds., *Optimization Techniques, Part 2*, Lecture Notes in Control and Information Sciences 23, 185 - 194, Springer, Berlin, 1980.
[25] P.C. Gilmore and R.E. Gomory, «A Linear Programming Approach to the Cutting Stock Problem I», *Operations Research* 9, 849 - 858, 1961.
[26] P.C. Gilmore and R.E. Gomory, «A Linear Programming Approach to the Cutting Stock Problem II», *Operations Research* 11, 863 - 888, 1963.
[27] P.C. Gilmore and R.E. Gomory, «Multi-Stage Cutting Stock Problems of Two and More Dimensions», *Operations Research* 13, 94 - 120, 1965.
[28] P.C. Gilmore and R.E. Gomory, «The Theory and Computation of Knapsack Functions», *Operations Research* 14, 1045 - 1074, 1966.
[29] H. Greenberg and R.L. Hegerich, «A Branch Search Algorithm for the Knapsack Problem», *Management Science* 16, 327 - 332, 1970.
[30] D.S. Hirshberg and C.K. Wong, «A Polynomial-Time Algorithm for the Knapsack Problem with Two Variables», *Journal of ACM* 23, 147 - 154, 1976.
[31] E. Horowitz and S. Sahni, «Computing Partitions with Applications to the Knapsack Problem», *Journal of ACM* 21, 277 - 292, 1974.
[32] T.C. Hu, *Integer Programming and Network Flows*, Addison-Wesley, New York, 1969.
[33] P.D. Hudson, «Improving the Branch and Bound Algorithms for the Knapsack Problem», *Queen's University Research Report, Belfast*, 1977.
[34] M.S. Hung and J.C. Fisk, «An Algorithm for 0-1 Multiple Knapsack Problems», *Naval Research Logistics Quarterly* 24, 571 - 579, 1978.
[35] O.H. Ibarra and C.E. Kim, «Fast Approximation Algorithms for the Knapsack and Sum of Subset Problems», *Journal of ACM* 22, 463 - 468, 1975.
[36] G.P. Ingargiola and J.F. Korsh, «A Reduction Algorithm for Zero-One Single Knapsack Problems», *Management Science* 20, 460 - 463, 1973.
[37] G.P. Ingargiola and J.F. Korsh, «An Algorithm for the Solution of 0-1 Loading Problems», *Operations Research* 23, 1110 - 1119, 1975.
[38] G.P. Ingargiola and J.F. Korsh, «A General Algorithm for One-Dimensional Knapsack Problems», *Operations Research* 25, 752 - 759, 1977.
[39] D.S. Johnson, «Approximation Schemes for Combinational Problems», *Journal of Computer and System Science* 9, 256 - 278, 1974.
[40] S.C. Johnson and B.W. Kernighan, «Remarks on Algorithm 397», *Communications of ACM* 15, 469, 1972.
[41] S. Kaplan, «Solution of the Lorie-Savage and Similar Integer Programming Problems by the Generalized Lagrange Multiplier Method», *Operations Research* 14, 1130 - 1136, 1966.
[42] R.M. Karp, «Reducibility among Combinatorial Problems», in R.E. Miller and J.W. Thatcher, eds., *Complexity of Computer Computations*, Plenum Press, 1972.
[43] P.J. Kolesar, «A Branch and Bound Algorithm for the Knapsack Problem», *Management Science* 13, 723 - 735, 1967.
[44] E.L. Lawler, «Fast Approximation Algorithms for Knapsack Problems», *Mathematics of Operations Research* 4, 339 - 356, 1979.

[45] N. Maculan, «Relaxation Lagrangienne: Le Probleme du Knapsack 0-1», *Canadian Journal of Operational Research and Information Processing* 21, 315 - 327, 1984.
[46] M.J. Magazine, G.L. Nemhauser and L.E. Trotter, Jr «When the Greedy Solution Solves a Class of Knapsack Problems», *Operations Research* 23, 207 - 217, 1975.
[47] M.J. Magazine and O. Oguz, «A Fully Polynomial Approximate Algorithm for the 0-1 Knapsack Problem», *European Journal of Operational Research* 8, 270 - 273, 1981.
[48] S. Martello and P. Toth, «An Upper Bound for the Zero-One Knapsack Problem and a Branch and Bound Algorithm», *European Journal of Operational Research* 1, 169 - 175, 1977.
[49] S. Martello and P. Toth, «Computational Experiences with Large-Size Unidimensional Knapsack Problems», *Presented at the TIMS/ORSA Joint National Meeting, San Francisco*, 1977.
[50] S. Martello and P. Toth, «Solution of the Bounded and Unbounded Change-Making Problem», *Presented at the TIMS/ORSA Joint National Meeting, San Francisco*, 1977.
[51] S. Martello and P. Toth, «Branch and Bound Algorithms for the Solution of the General Unidimensional Knapsack Problem», in M. Roubens, ed., *Advances in Operations Research*, North-Holland, Amsterdam, 1977.
[52] S. Martello and P. Toth, «Algorithm for the Solution of the 0-1 Single Knapsack Problem», *Computing* 21, 81 - 86, 1978.
[53] S. Martello and P. Toth, «The 0-1 Knapsack Problem», in N. Christofides, A. Mingozzi, P. Toth and C. Sandi, eds., *Combinatorial Optimization*, John Wiley and Sons, 1979.
[54] S. Martello and P. Toth, «Solution of the Zero-One Multiple Knapsack Problem», *European Journal of Operational Research* 4, 276 - 283, 1980.
[55] S. Martello and P. Toth, «Optimal and Canonical Solutions of the Change-Making Problem», *European Journal of Operational Research* 4, 322 - 329, 1980.
[56] S. Martello and P. Toth, «A Note on the Ingargiola-Korsh Algorithm for One-Dimensional Knapsack Problems», *Operations Research* 28, 1226 - 1227, 1980.
[57] S. Martello and P. Toth, «A Bound and Bound Algorithm for the Zero-One Multiple Knapsack Problem», *Discrete Applied Mathematics* 3, 275 - 288, 1981.
[58] S. Martello and P. Toth, «Heuristic Algorithms for the Multiple Knapsack Problem», *Computing* 27, 93 - 112, 1981.
[59] S. Martello and P. Toth, «A Mixture of Dynamic Programming and Branch-and-Bound for the Subset-Sum Problem», *Management Science* 30, 765 - 771, 1984.
[60] S. Martello and P. Toth, «Worst-Case Analysis of Greedy Algorithms for the Subset-Sum Problem», *Mathematical Programming* 28, 198 - 205, 1984.
[61] S. Martello and P. Toth, «A New Algorithm for the 0-1 Knapsack Problem», *Report OR/85/1, DEIS--University of Bologna*, 1985.
[62] S. Martello and P. Toth, «Approximation Schemes for the Subset-Sum Problem: Survey and Experimental Analysis», *European Journal of Operational Research* (to appear).
[63] S. Martello and P. Toth, «A Program for the the 0-1 Multiple Knapsack Problem», *ACM Transactions on Mathematical Programming* (to appear).
[64] H. Müller-Merback, «An Improved Upper Bound for the Zero-One Knapsack Problem: a Note on the Paper by Martello and Toth», *European Journal of Operational Research* 2, 212 - 213, 1978.
[65] R.M. Nauss, «An Efficient Algorithm for the 0-1 Knapsack Problem», *Management Science* 23, 27 - 31, 1976.
[66] A. Neebe and D. Dannenbring, «Algorithms for a Specialized Segregated Storage Problem», *Technical Report* No. 77 - 5, University of North-Carolina, 1977.
[67] G.L. Nemhauser and Z. Ullmann, «Discrete Dynamic Programming and Capital Allocation», *Management Science* 15, 494 - 505, 1969.
[68] S. Sahni, «Approximate Algorithms for the 0/1 Knapsack Problem», *Journal of ACM* 22, 115 - 124, 1975.
[69] H.M. Salkin, *Integer Programming*, Addison-Wesley, New York, 1975.
[70] H.M. Salkin and C.A. de Kluyver, «The Knapsack Problem: a Survey», *Naval Research Logistics Quartely* 22, 127 - 144, 1975.
[71] U. Suhl, «An Algorithm and Efficient Data Structures for the Binary Knapsack Problem», *European Journal of Operational Research* 2, 420 - 428, 1978.
[72] M.M. Syslo, N. Deo and J.S. Kowalik, *Discrete Optimization Algorithms-with Pascal Programs*, Prentice Hall, 1983.
[73] B.N. Tien and T.C. Hu, «Error Bounds and the Applicability of the Greedy Solution to the Coin--Changing Problem», *Operations Research* 25, 404 - 418, 1977.

[74] P. Toth, «A New Reduction Algorithm for 0-1 Knapsack Problems», *Presented at the ORSA/TIMS Joint National Meeting, Miami*, 1976.
[75] P. Toth, «Dynamic Programming Algorithms for the Zero-One Knapsack Problem», *Computing* 25, 29 - 45, 1980.
[76] P.R.C. Villela and C.T. Bornstein, «An Improved Bound for the 0-1 Knapsack Problem», *Report ES31-83, COPPE-Federal University of Rio de Janeiro*, 1983.
[77] H.M. Weingartner, *Mathematical Programming and the Analysis of Capital Budgeting Problems*, Prentice Hall, Englewood Cliffs, N.J., 1963.
[78] H.M. Weingartner, «Capital Budgeting and Interrelated Projects: Survey and Synthesis», *Management Science* 12, 485 - 516, 1968.
[79] H.M. Weingartner and D.N. Ness, «Methods for the Solution of the Multi-Dimensional 0-1 Knapsack Problem», *Operations Research* 15, 83 - 103, 1967.
[80] J.W. Wright, «The Change-Making Problem», *Journal of ACM* 22, 125 - 128, 1975.
[81] A.A. Zoltners, «A Direct Descent Binary Knapsack Algorithm», *Journal of ACM* 25, 304 - 311, 1978.

Silvano Martello
Paolo Toth
DEIS-University of Bologna
Viale Risorgimento 2
40136 Bologna
Italy

LINEAR ASSIGNMENT PROBLEMS

Silvano MARTELLO and Paolo TOTH

1. Introduction

Suppose n jobs have to be assigned to m machines. Let $c_{i,j}$ be the cost of assigning job j to machine i, $r_{i,j}$ the quantity of resource correspondingly required, and b_i the total amount of resource available for machine i. The *Generalized Assignment Problem (GAP)* consists of minimizing the total cost required to assign each job to exactly one machine without exceeding the machine's available resources. If we introduce a binary variable $x_{i,j}$ taking the value 1 if and only if job j is assigned to machine i, and define $I = \{1, \ldots, m\}$, $J = \{1, \ldots, n\}$, GAP can be mathematically described as:

$$(\text{GAP1}) \quad \text{minimize} \quad z = \sum_{i \in I} \sum_{j \in J} c_{i,j} x_{i,j} \qquad (1.1)$$

$$\text{subject to} \quad \sum_{j \in J} r_{i,j} x_{i,j} \leq b_i \quad (i \in I), \qquad (1.2)$$

$$\sum_{i \in I} x_{i,j} = 1 \quad (j \in J), \qquad (1.3)$$

$$x_{i,j} = 0 \text{ or } 1 \quad (i \in I, j \in J). \qquad (1.4)$$

We will assume, without loss of generality, that $c_{i,j}, r_{i,j}$ and b_i are non-negative integers. GAP is NP-hard, as we can see by transformation from the subset-sum problem, which is known to be NP-hard. Given a set of n positive integers $w_1, \ldots, w_n$, and another positive integer d, the subset-sum problem is to find that subset whose sum is closest to, without exceeding, d. Given any instance of subset-sum, we can define the following instance of GAP: $m = 2, c_{1,j} = r_{1,j} = r_{2,j} = w_j$ and $c_{2,j} = M w_j$ (M a sufficiently large number) for all $j \in J$; $b_1 = d$, $b_2 = \infty$. Now let $J_1 = \{j : x_{1,j} = 1$ in the optimal solution of GAP$\}$. Since $z = \sum_{j \in J_1} w_j + M \sum_{j \notin J_1} w_j$, J_1 gives the optimal solution of the instance of subset-sum.

The most famous particular case of GAP is the classical *Min-Sum Assignment*

Problem (*AP*), which is given by GAP when $n = m$, $b_i = 1$ and $r_{i,j} = 1$ for all $i \in I$ and $j \in J$ (note that, in the resulting formulation, we can replace $\leq$ with $=$ in (1.2)). If the objective function of AP is replaced with

$$\text{minimize} \quad z = \max_{i \in I, j \in J} \{c_{i,j} x_{i,j}\}$$

we have the *Min-Max* (or *Bottleneck*) *Assignment Problem* (*MAP*). If it is replaced with

$$\text{minimize} \quad z = \max_{i \in I, j \in J} \{c_{i,j} x_{i,j}\} - \min_{i \in I, j \in J} \{c_{i,j} x_{i,j}\}$$

we have the *Balanced Assignment Problem* (*BAP*).

All these problems can be solved in polynomial time, as we will see in Sections 2 (AP), 3 (MAP) and 4 (BAP). GAP will be considered in Section 5.

All the problems analyzed in the present paper are *linear*. We will not consider nonlinear assignment problems, such as, for example, the well-known *Quadratic Assignment Problem*, which is reviewed in Finke, Burkard and Rendl [15].

2. Min-sum assignment problem

2.1. The problem

Given a cost matrix (c) of order n, the *Min-Sum Assignment Problem* (*AP*) consists of finding a permutation (f) of the elements of set $N = \{1, 2, \ldots, n\}$ that minimizes

$$z = \sum_{i \in N} c_{i, f_i}.$$

It will be assumed, without loss of generality, that the elements of the cost matrix are non-negative integers.

If $c_{i,j}$ represents the cost of assigning job (column) j to machine (row) i (with $j \in N$, $i \in N$), AP can be considered as the problem of minimizing the total cost required to assign each job to exactly one machine and each machine to exactly one job. By introducing binary variables $x_{i,j}$ taking the value 1 iff column j is assigned to row i, AP can be mathematically formulated as the *integer linear program*

$$(AP) \quad \text{minimize} \quad z = \sum_{i \in N} \sum_{j \in N} c_{i,j} x_{i,j} \tag{2.1}$$

$$\text{subject to} \quad \sum_{j \in N} x_{i,j} = 1 \quad (i \in N), \tag{2.2}$$

$$\sum_{i \in N} x_{i,j} = 1 \quad (j \in N), \tag{2.3}$$

$$x_{i,j} = 0 \text{ or } 1 \quad (i \in N, j \in N).$$

Since the coefficient matrix associated with constraints (2.2) and (2.3) is *totally unimodular*, AP is equivalent to the *continuous linear program* (P) given by (2.1), (2.2), (2.3) and

$$x_{i,j} \geqslant 0 \quad (i \in N, j \in N).$$

The *dual problem* of P is

$$(D) \quad \text{maximize} \quad \bar{z} = \sum_{i \in N} u_i + \sum_{j \in N} v_j$$

$$\text{subject to} \quad u_i + v_j \leqslant c_{i,j} \quad (i \in N, j \in N),$$

where u_i and v_j are the *dual variables* associated, respectively, with row i and column j. It is well known that, since both P and D are continuous problems, solving P is equivalent to solving D.

A third formulation of AP can be obtained by considering the following graph theory problem. Given a *bipartite directed graph* $G = (S \cup T, A)$, where $S = \{s_1, \ldots, s_n\}$, $T = \{t_1, \ldots, t_n\}$ and $c_{i,j}$ is the cost of arc $(s_i, t_j) \in A$, the solution of the *minimum cost perfect matching problem* associated with graph G coincides with the solution of AP. Vertices $s_i \in S$ and $t_j \in T$ correspond, respectively, to row i ($i \in N$) and column j ($j \in N$) of (c).

AP can be solved by any linear programming algorithm, but more efficient polynomial methods exploiting the special structure of the problem can be used. These procedures can be subdivided into three main classes: *primal algorithms* derived from the simplex method (Barr, Glover and Klingman [1], Mc Ginnis [25]), *primal-dual algorithms based on the Hungarian method* (Kuhn [19, 20], Silver [27], Bourgeois and Lassalle [3], Lawler [21], Carpaneto and Toth [6, 8, 9], Bertsekas [2], Mc Ginnis [25]) and *primal-dual algorithms based on the shortest augmenting path methods* (Ford and Fulkerson [17], Tomizawa [28], Edmonds and Karp [14], Dorhout [13], Burkard and Derigs [4]). Efficient Fortran implementations and extensive experimental comparisons of the above procedures are given in Carpaneto, Martello and Toth [5]. In the following, only primal-dual algorithms, which have been shown to be more efficient than the primal ones (see [25] and [5]), will be considered.

2.2. Primal-dual algorithms

Most primal-dual algorithms can be viewed as specializations of a general algorithm which makes use of the bipartite directed graph $G = (S \cup T, A)$ previously described and of a non-negative *reduced cost matrix* (c^*) defined as $c_{i,j}^* = c_{i,j} - u_i - v_j$ ($s_i \in S, t_j \in T$), where (u) and (v) are the dual variables considered in Section 2.1. It can be proved that any optimal solution of the AP associated with the reduced cost matrix (c^*) is optimal for the AP associated with the original cost matrix (c) as well.

Algorithm AP

1. $\overline{S} \leftarrow S$, $\overline{T} \leftarrow T$, $\overline{A} \leftarrow \phi$ ($\overline{S}$ and $\overline{T}$ are the unassigned vertex sets, arcs $(t_j, s_i) \in \overline{A}$ define the current partial assignment);

2. **while** $|\overline{A}| < n$ **do**
   begin

3.   with respect to the reduced cost matrix (c^*), find a minimum cost *augmenting path* $P = \{(p_1, p_2), (p_2, p_3), \ldots, (p_{2h-1}, p_{2h})\}$ (i.e. a directed path between two unassigned vertices p_1 and p_{2h} whose arcs are alternately in sets A and $\overline{A}$ and such that $\Sigma_{k=1}^{h} c^*_{p_{2k-1}, p_{2k}}$ is a minimum);

4.   **for** $k = 1$ to $h - 1$ **do** $\overline{A} \leftarrow \overline{A} - \{(p_{2k}, p_{2k+1})\}$;

5.   **for** $k = 1$ to h **do** $\overline{A} \leftarrow \overline{A} \cup \{(p_{2k-1}, p_{2k})\}$;

6. $\overline{S} \leftarrow \overline{S} - \{p_1\}$, $\overline{T} \leftarrow \overline{T} - \{p_{2h}\}$
   end

7. **for each** $(t_j, s_i) \in \overline{A}$ **do** $f_i \leftarrow j$.

At each of the n iterations, solution (u), (v) of the dual problem is feasible since constraint $c_{i,j}^* \geq 0$ (and hence $u_i + v_j \leq c_{i,j}$) holds for each $s_i \in S$ and $t_j \in T$. The corresponding solution (x) of the primal problem is infeasible since only a subset of constraints (2.2) and (2.3) is satisfied. Hence the algorithm is primal for dual problem D and dual for primal problem P.

The performance of algorithm AP can be improved through the following *initialization* steps. The dual variables can be initialized as

$$v_j = \min\{c_{i,j} : s_i \in S\} \qquad (t_j \in T),$$
$$u_i = \min\{c_{i,j} - v_j : t_j \in T\} \qquad (s_i \in S).$$

An initial partial primal solution can easily be obtained by assigning vertex s_i to vertex t_j if $c_{i,j}^*$ is zero and both vertices s_i and t_j are unassigned (note that, because of the definition of (u) and (v), at least one zero element exists in each row and column of the reduced cost matrix). The number of initial assignments can be increased (see [6]) by performing, for each unassigned vertex s_i, an additional step which tries to reassign a previously assigned vertex in S to a currently unassigned vertex in T so as to allow the assignment of vertex s_i by considering only zero elements of the reduced cost matrix.

All primal-dual algorithms can be derived from AP by specializing the techniques used to find a minimum cost augmenting path and to update the dual variables.

2.3. Hungarian algorithms

The Hungarian algorithms look for an augmenting path by considering only zero elements of the reduced cost matrix (c^*). When no such path exists, the dual variables are updated so as to increase the number of zero elements in (c^*) and the augmenting path search continues.

Algorithm HAP

1. *initialize* $\overline{S}, \overline{T}$ and $\overline{A}$;
2. **while** $|\overline{A}| < n$ **do**
   begin
3.   let $s_r \in \overline{S}$ be any unassigned vertex;
4.   **repeat**
5.   find an augmenting path $P = \{(p_1, p_2), (p_2, p_3), \ldots, (p_{2h-1}, p_{2h})\}$, with $p_1 = s_r$, by considering only zero elements of (c^*) and applying a *labelling procedure*;
6.   **if** such a path P exists **then**
           begin
7.   **for** $k = 1$ **to** $h - 1$ **do** $\overline{A} \leftarrow \overline{A} - \{(p_{2k}, p_{2k+1})\}$;
8.   **for** $k = 1$ **to** h **do** $\overline{A} \leftarrow \overline{A} \cup \{(p_{2k-1}, p_{2k})\}$;
9. $\overline{S} \leftarrow \overline{S} - \{s_r\}, \overline{T} \leftarrow \overline{T} - \{p_{2h}\}$
           end
           else
           begin
10. $d \leftarrow \min\{c_{i,j}^* : s_i \in S \text{ and labelled}, t_j \in T \text{ and unlabelled}\}$;
11.   **for each** labelled vertex $s_i \in S$ **do** $u_i \leftarrow u_i + d$;
12.   **for each** labelled vertex $t_j \in T$ **do** $v_j \leftarrow v_j - d$
            end
13.   **until** $s_r \notin \overline{S}$
    end

The time complexity of HAP depends on the technique used for the computation of d at Step 10, since the remaining steps require $O(n^3)$ time. In fact Step 1 can be performed in $O(n^2)$ time. The while-loop is executed $O(n)$ times. At each iteration of the while-loop, Steps 7 and 8 are performed once, Steps 10, 11 and 12 at most $O(n)$ times (the number of labelled vertices increases by at least two units at each execution), Step 5 globally requires $O(n^2)$ time (each element of the cost matrix is considered at most once). If the computation of d is performed as specified at step 10, each execution requires $O(n^2)$ time, so the overall complexity of HAP is $O(n^4)$. It is possible to achieve better complexity if, at Step 5, for each unlabelled vertex $t_j \in T$, one computes the minimum (say π_j) of the reduced costs of the labelled vertices in S. The complexity of Step 5 is not affected by this modification, while Step 11 requires now $O(n)$ time ($d = \min\{\pi_j : \text{vertex } t_j \in T \text{ is unlabelled}\}$), so the overall complexity is $O(n^3)$.

2.4. Shortest augmenting path algorithms

The bipartite graph formulation can be used to transform AP into a *minimum-cost flow problem* (*MCFP*) on network $\bar{G} = (S \cup T \cup V, A \cup W)$ defined as: $V = \{s, t\}$, $W = \{(s, s_i) : s_i \in S\} \cup \{(t_j, t) : t_j \in T\}$, all arcs have unit capacity, the cost of arcs $(i, j) \in A$ is $a_{i,j}$ and that of arcs in W is zero. It can easily be seen that the minimum cost flow of value n from s to t gives the solution of AP. Because of the special structure of the network, it is possible to solve MCFP by applying a standard *shortest path routine* (for instance the Dijkstra labelling method) to find, in $O(n^2)$ time, a minimum cost augmenting path with respect to the non-negative reduced cost matrix (c^*). The updating of the dual variables is postponed until an augmenting path is found.

Algorithm SPAP

1. *initialize* $\bar{S}, \bar{T}$ and $\bar{A}$;

2. **while** $|\bar{A}| < n$ **do**

    begin

3.   let $s_r \in S$ be any unassigned vertex;

4.   find an augmenting path $P = \{(p_1, p_2), (p_2, p_3), \ldots, (p_{2h-1}, p_{2h})\}$, with $p_1 = s_r$, by applying a shortest path labelling procedure on (c^*) (let π_j be the cost of the path from vertex s_r to the labelled vertex $t_j \in T$ and t_q the last vertex of P);

5.   **for each** labelled vertex $t_j \in T - \{t_q\}$ **do**

        begin

6.   let s_i be the vertex currently assigned to t_j;

7. $v_j \leftarrow v_j + \pi_j - \pi_q, u_i \leftarrow u_i - \pi_j + \pi_q$

    end

8.   for $k = 1$ to $h - 1$ do $\overline{A} \leftarrow \overline{A} - \{(p_{2k}, p_{2k+1})\}$;
9.   for $k = 1$ to h do $\overline{A} \leftarrow \overline{A} \cup \{(p_{2k-1}, p_{2k})\}$;
10. $\overline{S} \leftarrow \overline{S} - \{s_r\}, \overline{T} \leftarrow \overline{T} - \{t_q\}$
end

The time complexity of SPAP is clearly $O(n^3)$.

A combination of the Hungarian and shortest augmenting path methods has been proposed in [9].

2.5. Sparse cost matrices

In many practical cases the cost matrix (c) can be represented as a sparse matrix. All the primal-dual algorithms described in the previous sections can be modified so as to consider this case (see [8, 5]). Since the elements of the cost matrix are always scanned only along the rows and never along the columns, the entries of row i $(i \in N)$ of the sparse matrix can efficiently be represented by two sets J_i and C_i containing, respectively, the corresponding column indices and costs.

The computational results show (see Section 2.6) that the computing times for sparse problems are much smaller than those for the corresponding complete problems. Moreover, the entries of the cost matrix associated with the optimal assignment, that is, entries c_{i,f_i} $(i \in N)$, generally have very small values with respect to the other entries. On the basis of these considerations Carpaneto and Toth [9] have proposed an algorithm (APM) for the complete problem which transforms the complete cost matrix (c) into a sparse one (c') (by removing all the entries of the initial reduced cost matrix greater than a given threshold) and computes, by applying a sparse cost matrix procedure, the optimal solutions (x') and (u'), (v') corresponding to the primal and dual problems associated with (c'). If the dual solution (u'), (v') is feasible for the original dual problem D, that is, if $c_{i,j} - u'_i - v'_j \geq 0$ holds for all $i \in N$ and $j \in N$, then (u'), (v') is optimal for D and the primal solution (x') is feasible and optimal for the original primal problem P.

2.6. Computational results

Two implementations of the Hungarian method described in Section 2.3 have been considered-algorithms HAP1, requiring $O(n^3)$ time, and HAP2, requiring $O(n^4)$ time. The second algorithm, derived from the procedure proposed in [6], utilizes a pointer technique to locate the unexplored zero elements of the reduced cost matrix and can be faster for particular instances.

The Fortran codes corresponding to algorithms HAP1, HAP2, SPAP (Section 2.4), APM (Section 2.5) for complete cost matrices, and to the sparse matrix implementation of HAP1, were run on a CDC-Cyber 730 on randomly-generated test problems. Two classes of problems were considered by generating the entries of the cost matrix as uniformly random integers in the ranges

$(1-10^2)$ and $(1-10^6)$ respectively. For the sparse cost matrices, two different values (5% and 20%) of the density of the number of the sparse matrix elements with respect to the number of elements in the corresponding complete matrix, were considered. Six values of n (40, 80, 120, 160, 200, 240) for complete matrices and eight values (120, 160, 200, 240, 280, 320, 360, 400) for sparse matrices were utilized. For each cost range, density and value of n, 5 problems were solved. Tables 2.1 to 2.4 give the corresponding average computing time (expressed in seconds and comprehensive of the initalization phase) for each algorithm and the average time and number of assignments of the initialization procedure.

Tables 2.1 and 2.2 clearly show that, for complete cost matrices, algorithm APM has the best performance, especially for large values of n. Algorithms HAP1 and SPAP are almost equivalent, while algorithm HAP2 has good computing times only for small values of the cost range.

For sparse cost matrices, Tables 2.3 and 2.4 show that the total computing times increase more steeply with the cost range than with density.

Table 2.1.
Complete matrices. $1 \leqslant c_{i,j} \leqslant 10^2$. CDC-Cyber 730 seconds over 5 problems.

n	Total computing time				Initialization phase	
	HAP1	HAP2	SPAP	APM	Time	Number of assignments
40	0.086	0.214	0.086	0.070	0.026	34.0
80	0.403	0.722	0.446	0.272	0.099	67.6
120	1.041	1.367	1.225	0.625	0.213	101.2
160	1.912	2.054	2.366	1.022	0.362	140.4
200	3.294	2.821	4.134	1.593	0.569	175.0
240	5.709	3.724	7.503	3.078	0.766	212.0

Table 2.2.
Complete matrices. $1 \leqslant c_{i,j} \leqslant 10^6$. CDC-Cyber 730 seconds over 5 problems.

n	Total computing time				Initialization phase	
	HAP1	HAP2	SPAP	APM	Time	Number of assignments
40	0.099	0.448	0.079	0.071	0.025	34.2
80	0.488	3.215	0.403	0.282	0.096	66.2
120	1.356	10.703	1.183	0.785	0.218	96.8
160	2.673	23.453	2.345	1.409	0.373	131.4
200	4.843	46.618	4.363	2.320	0.601	160.0
240	7.317	82.424	6.721	3.233	0.834	197.2

Table 2.3.
Algorithm HAP1 for sparse matrices. $1 \leq c_{i,j} \leq 10^2$. CDC-Cyber 730 seconds over 5 problems.

	Density 5%			Density 20%		
	Total time	Initialization phase		Total time	Initialization phase	
n		Time	Number of assignments		Time	Number of assignments
120	0.518	0.021	98.6	0.602	0.059	97.6
160	0.942	0.035	130.8	0.965	0.101	132.2
200	1.576	0.050	166.4	1.666	0.154	168.2
240	2.384	0.073	197.6	2.216	0.203	201.2
280	3.371	0.092	225.8	3.363	0.282	230.0
320	4.329	0.118	258.4	4.576	0.369	267.2
360	5.874	0.138	292.6	5.931	0.456	301.4
400	6.680	0.168	328.4	7.229	0.586	338.0

Table 2.4.
Algorithm HAP1 for sparse matrices. $1 \leq c_{i,j} \leq 10^6$. CDC-Cyber 730 seconds over 5 problems.

	Density 5%			Density 20%		
	Total time	Initialization phase		Total time	Initialization phase	
n		Time	Number of assignments		Time	Number of assignments
120	0.646	0.020	101.0	0.875	0.062	99.0
160	1.277	0.035	128.2	1.639	0.103	128.2
200	2.188	0.052	162.6	2.659	0.149	162.6
240	3.284	0.069	195.0	4.185	0.223	195.0
280	4.593	0.088	228.6	5.857	0.285	228.6
320	6.442	0.116	257.4	7.878	0.379	257.4
360	7.817	0.132	290.0	10.448	0.463	290.0
400	10.281	0.170	322.6	13.450	0.596	322.6

More extensive computational results and a statistical analysis of the performance of the algorithms consider in this Section are given in [5].

3. Min-max assignment problem

3.1. The problem

Given a cost matrix (c) of order n, the *Min-Max* (or *Bottleneck*) *Assignment Problem* (*MAP*) consists of finding a permutation (f) of the elements of set

$N = \{1, 2, \ldots, n\}$ that minimizes:

$$z = \max\{c_{i,f_i} : i \in N\}.$$

By introducing boolean variables $x_{i,j}$ taking the value 1 iff column j is assigned to row i (i.e. if $f_i = j$), MAP can be mathematically formulated as the integer linear program:

$$\text{(MAP)} \quad \text{minimize} \quad z \tag{3.1}$$

$$\text{subject to} \quad c_{i,j} x_{i,j} \leq z \quad (i \in N, j \in N), \tag{3.2}$$

$$\sum_{j \in N} x_{i,j} = 1 \quad (i \in N), \tag{3.3}$$

$$\sum_{i \in N} x_{i,j} = 1 \quad (j \in N), \tag{3.4}$$

$$x_{i,j} = 0 \text{ or } 1 \quad (i \in N, j \in N).$$

Since, contrary to what occurs for the min-sum problem AP (see Section 2.1), the coefficient matrix associated with constraints (3.2), (3.3) and (3.4) is not totally unimodular, the optimal solution value of the continuous relaxation of MAP gives only a *lower bound* on the value of z.

A third formulation of MAP can be obtained by considering the bipartite directed graph $G = (S \cup T, A)$ defined in Section 2.1 for AP. It can easily be shown that the solution of the *min-max cost perfect matching problem* associated with graph G coincides with the solution of MAP.

Several polynomial methods have been proposed for the exact solution of the problem. As for AP, these algorithms can be subdivided into primal approaches (Gross [18]) and into primal-dual procedures based on the Hungarian method (Carpaneto and Toth [7]) and on the shortest augmenting path method (Derigs and Zimmermann [12], Burkard and Derigs [4], Derigs [11]). In this case too, the primal approach has been shown to be much less efficient than the primal-dual ones. These last procedures differ from those proposed for the min-sum case (see Sections 2.2, 2.3 and 2.4) mainly because the original cost matrix (c) is considered in the search for min-max cost augmenting paths, instead of the reduced cost matrix (c^*). The computational complexities remain the same.

3.2. Hungarian method

In the Hungarian method (see [7]), the zero elements of the reduced cost matrix correspond to the elements of set $C_{\bar{z}} = \{c_{i,j} : c_{i,j} \leq \bar{z}, s_i \in S, t_j \in T\}$, where $\bar{z}$ is a lower bound (*threshold value*) on the value of z. So the search for the initial partial assignment (initialization phase) and the labelling procedure to find an augmenting path consider only elements of $C_{\bar{z}}$.

The threshold value $\bar{z}$ can be inizialized as

$$\bar{z} = \max\{\bar{z}', \bar{z}''\}$$

where

$$\bar{z}' = \max\{\min\{c_{i,j} : s_i \in S\} : t_j \in T\},$$
$$\bar{z}'' = \max\{\min\{c_{i,j} : t_j \in T\} : s_i \in S\}.$$

When no augmenting path with respect to the elements of $C_{\bar{z}}$ has been found by the labelling procedure, the threshold value $\bar{z}$ can be updated, i.e. increased, as follows (see Step 10 of algorithm HAP):

$$\bar{z} = \min\{c_{i,j} : s_i \in S \text{ and labelled, } t_j \in T \text{ and unlabelled}\}.$$

3.3. Shortest augmenting path method

Given an augmenting path P starting from the unassigned vertex $s_i \in S$ and a lower bound $\bar{z}$ on z, let

$$b_{\bar{z}}(P) = \max\{\bar{z}, \max\{c_{i,j} : (s_i, t_j) \in P\}\}$$

be the *bottleneck cost* of path P.

Path P is a *shortest bottleneck augmenting path* with respect to $\bar{z}$ if condition

$$b_{\bar{z}}(P) \leq b_{\bar{z}}(P')$$

holds for any augmenting path P' starting from s_i.

In the shortest augmenting path method (see [12], [4], [11]), at each iteration a search for a shortest bottleneck augmenting path P starting from an unassigned vertex $s_i \in S$ is performed, and the value of lower bound $\bar{z}$ is then updated by setting $\bar{z} = b_{\bar{z}}(P)$.

3.4. Computational results

The Fortran codes BASS [7] (derived from the Hungarian method) and BAP1 [12], LBAP [4], LBAP2 [11] (derived from the shortest augmenting path method) were run on a CDC-Cyber 730 by solving randomly generated test problems. As for AP, two cost ranges of the entries of matrix (c) ($(1-10^2)$ and $(1-10^6)$) and six values of n (40, 80, ..., 240) were considered. For each cost range and value of n, 5 problems were solved. Table 3.1 gives the corresponding average computing times (expressed in seconds) for each algorithm.

The table shows that LBAP2 is slightly faster than BASS and that BAP1 is clearly worse than the other codes. Moreover, the computing times are practically independent of the cost range.

Table 3.1.
CDC-Cyber 730 seconds over 5 problems.

	$1 \leq c_{i,j} \leq 10^2$				$1 \leq c_{i,j} \leq 10^6$			
n	BASS	BAP1	LBAP	LBAP2	BASS	BAP1	LBAP	LBAP2
40	0.069	0.101	0.080	0.053	0.087	0.099	0.083	0.057
80	0.182	0.383	0.234	0.154	0.185	0.431	0.281	0.160
120	0.271	0.812	0.422	0.230	0.293	0.940	0.474	0.259
160	0.365	1.448	0.616	0.318	0.374	1.763	0.647	0.332
200	0.518	2.431	0.728	0.445	0.530	2.854	0.806	0.494
240	0.673	3.776	0.983	0.606	0.707	4.320	1.073	0.651

4. Balanced assignment problem

4.1. Balanced optimization problems

Suppose we are given a finite set E, a cost c_e associated with each $e \in E$ and a family $\mathscr{F}$ of feasible subsets of E. The general *Balanced Optimization Problem* is to find $S^* \in \mathscr{F}$ which minimizes $z = \max\{c_e : e \in S\} - \min\{c_e : e \in S\}$ over all $S \in \mathscr{F}$. If E is the set of cells of an $n \times n$ assignment matrix, c_e the value contained in cell e and $\mathscr{F}$, the family of all subsets of cells which constitute assignments, then we obtain the *Balanced Assignment Problem*: find an assignment which minimizes the cost difference of the most expensive and least expensive cell used.

Assume that a *feasibility procedure* accepts as input a set $E' \subseteq E$ and either produces some $S \in \mathscr{F}$ such that $S \subseteq E'$ or else states that no such S exists. Martello, Pulleyblank, Toth and de Werra [24] have proved that if we know a polynomially bounded feasibility procedure then we are able to solve the associated balanced optimization problem in polynomial time as follows. Let $v_1 < v_2 < \ldots < v_k$ be the (sorted) list of different values appearing as element costs, and for any l, u satisfying $1 \leq l \leq u \leq k$, let $E(l, u) = \{e \in E : v_l \leq c_e \leq v_u\}$.

Algorithm BOP

1. $l \leftarrow 1, u \leftarrow 1, z \leftarrow \infty$;
2. **while** $u \leq k$ **do**
   begin
3.   apply the feasibility procedure to $E(l, u)$;
4.   **if** $E(l, u)$ contains no member of $\mathscr{F}$ **then** $u \leftarrow u + 1$ **else**
       begin
5.   **if** $v_u - v_l < z$ **then** $l^* \leftarrow l, u^* \leftarrow u, z = v_u - v_l$;
6. $l \leftarrow l + 1$
       end
   end

If the feasibility procedure has time complexity $O(f(|E|))$ then the overall complexity of the algorithm is $O(k \cdot f(|E|))$.

4.2. An algorithm for the balanced assignement problem

Given an $n \times n$ cost matrix (c), we could solve the balanced assignment problem through the algorithm of the previous section by using, as feasibility procedure, any algorithm for the assignment problem. Using, for example, algorithm HAP of Section 2.3, would give $O(n^5)$ as overall time complexity. We can, however, solve the problem in $O(n^4)$ through the following specialized algorithm:

Algorithm BAP

1. $z \leftarrow \infty$;

2. solve a min-max assignment problem on (c) and let v_l, v_u be respectively the minimum and maximum cost in the solution ($v_u = \infty$ if no assignment exists);

3. **while** $v_u < \infty$ **do**
   begin
4.   **if** $v_u - v_l < z$ **then**
       begin
5. $u^* \leftarrow u; l^* \leftarrow l; z \leftarrow v_u - v_l$;
6.   **if** $z = 0$ **then** stop
       end
7.   set to ∞ all cells (i, j) of (c) such that $c_{i,j} \leq v_l$;
8.   solve a min-max assignment problem on (c) starting with the current (partial) assignment and define v_l, v_u as in step 2
   end

The correctness of the algorithm can easily be proved (see [24]). The time complexity can be determined as follows. Step 2 can be performed in $O(n^3)$ time (see Section 3.2). Step 8 requires the search of an augmenting path for each row which, after execution of step 7, results unassigned. Since this can occur at most $O(n^2)$ times, and the computational effort to find an augmenting path is $O(n^2)$ (see Section 2.4), the overall complexity of the step and, hence, of the algorithm, is $O(n^4)$.

5. Generalized Assignment Problem

5.1. The problem

The *Generalized Assignment Problem (GAP)* has been defined in Section 1.1 (relations (1.1)-(1.4)) as a minimization problem (GAP1). GAP can also be defined as a maximization problem by manipulating the objective function as follows (Martello and Toth [23]). Let t be any integer constant such that $t > \max_{i \in I, j \in J} \{c_{i,j}\}$, and define, for all $i \in I$ and $j \in J$

$$p_{i,j} = t - c_{i,j}.$$

Then

$$\text{(GAP2)} \quad \text{maximize} \quad w = \sum_{i \in I} \sum_{j \in J} p_{i,j} x_{i,j} \tag{5.1}$$

$$\text{subject to} \quad \sum_{j \in J} r_{i,j} x_{i,j} \leq b_i \quad (i \in I), \tag{5.2}$$

$$\sum_{i \in I} x_{i,j} = 1 \quad (j \in J), \tag{5.3}$$

$$x_{i,j} = 0 \text{ or } 1 \quad (i \in I, j \in J) \tag{5.4}$$

is equivalent to (GAP1). The objective function of (GAP1) is given by $z = nt - w$. Note that $p_{i,j}$ can be interpreted as the profit obtained by assigning job j to machine i. In the following sections we will always consider the (GAP2) version of the problem.

Chalmet and Gelders [10] have studied a different formulation of GAP, where the = sign is replaced with the $\leq$ sign in (5.3). Since it is no longer necessary for all jobs to be assigned, a feasible solution to such a problem always exists (this formulation is more frequent in real-world problems). However, it has been shown in [23] that any problem of this kind can easily be transformed into a (GAP2).

5.2. Upper bounds

GAP is generally solved through branch-and-bound algorithms (remember that the problem is NP-hard). In this section we review the most effective bounds proposed in the literature.

Ross and Soland [26] have proposed the following upper bound. First the relaxed problem (5.1), (5.3), (5.4) is exactly solved by determining, for each $j \in J$, an index $i(j)$ such that

$$p_{i(j),j} = \max_{i \in I} \{p_{i,j} : r_{i,j} \leq b_i\}$$

and setting $x_{i(j),j} = 1$ and $x_{i,j} = 0$ for all $i \in I - \{i(j)\}$. This initial upper bound, of value $\sum_{j \in J} p_{i(j),j}$ is then refined as follows. Let

$$J_i = \{j \in J : x_{i,j} = 1\} \qquad (i \in I),$$

$$d_i = \sum_{j \in J_i} r_{i,j} x_{i,j} - b_i \qquad (i \in I),$$

$$I' = \{i \in I : d_i > 0\}.$$

Since I' is the set of those machines for which condition (5.2) has been violated,

$$q_j = \min_{k \in I - \{i(j)\}} \{p_{i(j),j} - p_{k,j} : r_{k,j} \leq b_k\}$$

is the minimum penalty that will be incurred if job j is reassigned. Hence, for each $i \in I'$, a lower bound on the loss of profit to be paid in order to satisfy condition (5.2) can be obtained by solving the 0-1 knapsack problem

$$(K_i^1) \quad \text{minimize} \quad z_i = \sum_{j \in J_i} q_j y_{i,j}$$

$$\text{subject to} \quad \sum_{j \in J_i} r_{i,j} y_{i,j} \geq d_i,$$

$$y_{i,j} = 0 \text{ or } 1 \qquad (j \in J_i),$$

and the refined upper bound is

$$u_1 = \sum_{j \in J} p_{i(j),j} - \sum_{i \in I'} z_i.$$

Martello and Toth [23] have obtained an upper bound for (GAP2) by solving the relaxed problem (5.1), (5.2), (5.4). This relaxation would have no meaning for (GAP1), since the removal of condition (1.3) would give a trivial problem (solution $x_{i,j} = 0$ for all $i \in J$ and $j \in J$) with a useless lower bound of value 0. For (GAP2) we have a non-trivial problem, whose solution can be obtained by solving, for each machine $i \in I$, the 0-1 knapsack problem

$$(K_i^2) \quad \text{maximize} \quad w_i = \sum_{j \in J} p_{i,j} x_{i,j}$$

$$\text{subject to} \quad \sum_{j \in J} r_{i,j} x_{i,j} \leq b_i,$$

$$x_{i,j} = 0 \text{ or } 1 \qquad (j \in J).$$

In this case, too, the initial upper bound, of value $\Sigma_{i \in I} w_i$, can be refined by computing a lower bound on the penalty to be paid in order to satisfy the violated constraints. Let

$$J^0 = \left\{ j \in J : \sum_{i \in I} x_{i,j} = 0 \right\};$$

$$J^1 = \left\{ j \in J : \sum_{i \in I} x_{i,j} > 1 \right\};$$

$$I_j = \{i \in I : x_{i,j} = 1\} \qquad (j \in J);$$

$w_{i,j}^0 =$ upper bound on (K_i^2) if $x_{i,j} = 0 \quad (i \in I, j \in J)$;

$w_{i,j}^1 =$ upper bound on (K_i^2) if $x_{i,j} = 1 \quad (i \in I, j \in J)$,

where $w_{i,j}^0$ and $w_{i,j}^1$ can easily be obtained, for example, from the continuous solution to (K_i^2). We can now compute

$$v_{i,j}^0 = \min\{w_i, w_{i,j}^1\} \qquad (j \in J^0, i \in I),$$

$$v_{i,j}^1 = \min\{w_i, w_{i,j}^0\} \qquad (j \in J^1, i \in I_j)$$

and determine a lower bound l_j on the penalty to be paid for job j in order to satisfy condition (5.3) as

$$l_j = \begin{cases} \min_{i \in I}\{w_i - [v_{i,j}^0]\} & \text{if } j \in J^0, \\ \sum_{i \in I_j}(w_i - [v_{i,j}^1]) - \max_{i \in I_j}\{w_i - [v_{i,j}^1]\} & \text{if } j \in J^1. \end{cases}$$

The refined upper bound is thus

$$u_2 = \sum_{i \in I} w_i - \max_{j \in J^0 \cup J^1}\{l_j\}.$$

Fisher, Jaikumar and Van Wassenhove [16] have recently proposed a new upper bound for (GAP2), obtained by dualizing constraint (5.3) with a vector (s) of multipliers. The resulting Lagrangean problem separates into m 0-1 single knapsack problems (one for each $i \in I$) of the form

$$(K_i^3) \quad \text{maximize} \quad w_i(s) = \sum_{j \in J}(p_{i,j} - s_j)x_{i,j}$$

$$\text{subject to} \quad \sum_{j \in J} r_{i,j} x_{i,j} \leq b_i,$$

$$x_{i,j} = 0 \text{ or } 1 \qquad (j \in J)$$

and it can be easily verified that

$$u_3(s) = \sum_{j \in J} s_j + \sum_{i \in I} w_i(s)$$

is a valid upper bound on w for any s. Note that the initial Martello-Toth upper bound is given by this Lagrangean relaxation when $s_j = 0$ for all $j \in J$. The Fisher-Jaikumar-Van Wassenhove method for determining the multipliers begins by initializing each s_j to the second largest $p_{i,j}$. With this choice, $p_{i,j} - s_j > 0$ for at most one $i \in I$, so there is a Lagrangean solution satisfying $\sum_{i \in I} x_{i,j} \leq 1$ for all $j \in J$. If $\sum_{i \in I} x_{i,j} = 1$ for all $j \in J$, then this solution is feasible and hence optimal. Otherwise, under certain conditions, it is possible to select a $\bar{j}$ for which $\sum_{i \in I} x_{i,\bar{j}} = 0$ and decrease $s_{\bar{j}}$ by an amount large enough to ensure $\sum_{i \in I} x_{i,\bar{j}} = 1$ in the new Lagrangean solution, but small enough for $\sum_{i \in I} x_{i,j} \leq 1$ to continue to hold for all other j in the new Lagrangean solution. The method is iterated until either the Lagrangean solution becomes feasible or the required conditions fail.

It has been proved in [16] that u_3 dominates the Ross and Solaud bound in the sense that the value of u_3 obtained in the first iteration of this method equals u_1.

5.3. Exact algorithms

The most important algorithms in the literature for the exact solution of GAP are depth-first branch-and-bound algorithms.

In the Ross and Soland scheme [26], upper bound u_1 is computed at each node of the decision-tree. If branching is needed, the variable chosen to separate on, x_{i^*,j^*}, is the one among those with $y_{i,j} = 0$ in the optimal solution to the (K_i^1) for which the quantity

$$q_j / \left(r_{i,j} / \left(b_i - \sum_{j \in J_i} r_{i,j} x_{i,j} \right) \right)$$

is a maximum. Two branches are then generated by imposing $x_{i^*,j^*} = 1$ and $x_{i^*,j^*} = 0$.

In the Martello and Toth scheme [23], at each node of the decision-tree, both upper bounds u_1 and u_2 are computed. If branching is needed, the separation is performed on the job j^* for which

$$l_{j*} = \max_{j \in J^0 \cup J^1} \{l_j\}.$$

Branching is then determined by the type of infeasibility given by job $j*$ in the solution of the relaxed problem solved to obtain u_2. If $j* \in J^0$, m branches are generated by assigning $j*$ to each machine in turn, as shown in figure 5.1. If $j* \in J^1$, $|J^1|$ branches are generated as follows. Let $I_{j*} = \{i_1, i_2, \ldots, i_{\overline{m}}\}$; $\overline{m} - 1$ branches are generated by assigning $j*$ to $i_1, i_2, \ldots, i_{\overline{m}-1}$, and another branch by excluding $j*$ from $i_1, i_2, \ldots, i_{\overline{m}-1}$ (see figure 5.2). It can be proved that, with this branching strategy, it is necessary to solve m (K_i^2) problems to compute the upper bound associated with the root node of the decision-tree, but only one new (K_i^2) problem for each other node.

In the Fisher, Jaikumar and Van Wassenhove scheme [16], at each node of the decision-tree, upper bound u_3 is computed. If branching is needed, the variable chosen to separate on, $x_{i*,j*}$, is the one for which $r_{i*,j*}$ is maximum. Two branches are then created by imposing $x_{i*,j*} = 1$ and $x_{i*,j*} = 0$.

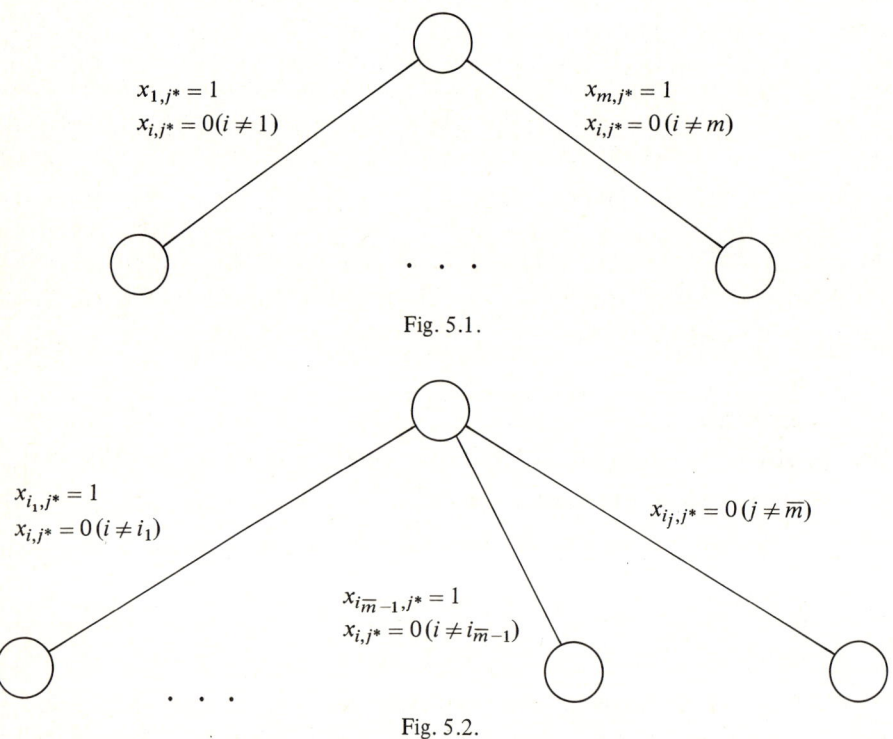

Fig. 5.1.

Fig. 5.2.

5.4. Heuristic and reduction algorithms

Martello and Toth [23] have proposed the following algorithm to find a heuristic solution ($\hat{x}$) to (GAP2). Let $f(i, j)$ be a *weight function* relative to the assignment of job j to machine i. The algorithm iteratively considers all the unassigned jobs and determines that job j^*, which has the maximum difference between the largest and the second largest $f(i,j):j^*$ is then assigned to the machine for which $f(i,j^*)$ is a maximum. The second part of the algorithm (steps 11-14) attempts to improve on this solution through local exchanges.

Algorithm HGAP

1. $F \leftarrow J$;

2. **for each** $i \in I$ **do** $a_i \leftarrow b_i$;

3. **while** $F \neq \emptyset$ **do**

    begin

4. $g \leftarrow -\infty$;

5.   **for each** $j \in F$ **do**

    begin

6. $f(\bar{i},j) \leftarrow \max_{i \in I}\{f(i,j):r_{i,j} \leq a_i\}$ (if this set is empty **then** stop: no feasible solution has been found);

7. $d \leftarrow f(\bar{i},j) - \max_{i \in I-\{\bar{i}\}}\{f(i,j):r_{i,j} \leq a_i\}$ (if this set is empty **then** $d \leftarrow +\infty$);

8.   **if** $d > g$ **then** $g \leftarrow d, j^* \leftarrow j, i^* \leftarrow \bar{i}$

    end

9. $\hat{x}_{i^*,j^*} \leftarrow 1, F \leftarrow F - \{j^*\}, a_{i^*} \leftarrow a_{i^*} - r_{i^*,j^*}$;

10.   **for each** $i \in I - \{i^*\}$ **do** $\hat{x}_{i,j^*} \leftarrow 0$

    end

11. **for each** $j \in J$ **do**

    begin

12.   let i^* be the machine for which $\hat{x}_{i^*,j} = 1$;

13. $p_{\bar{i},j} \leftarrow \max_{i \in I-\{i^*\}}\{p_{i,j}:r_{i,j} \leq a_i\}$;

14.   **if** $p_{\bar{i},j} > p_{i^*,j}$ **then** $\hat{x}_{i^*,j} \leftarrow 0, a_{i^*} \leftarrow a_{i^*} + r_{i^*,j}, \hat{x}_{\bar{i},j} \leftarrow 1, a_{\bar{i}} \leftarrow a_{\bar{i}} - r_{\bar{i},j}$

    end

HGAP can be implemented to run in $O(nm \log m + n^2)$ time. Computational experiments have shown that good results are given by the following choices for $f(i,j): f(i,j) = p_{i,j}, f(i,j) = p_{i,j}/r_{i,j}, f(i,j) = -r_{i,j}, f(i,j) = -r_{i,j}/b_i$.

The following two algorithms, also presented in [23], can be used to reduce the size of a GAP. Let $(\hat{x})$ be any feasible solution (for example that found by HGAP) and $\hat{w} = \Sigma_{i \in I} \Sigma_{j \in J} p_{i,j} \hat{x}_{i,j}$ the corresponding value. The first algorithm determines the initial upper bound of Ross and Soland and fixes to 0 those variables $x_{i,j}$ which, if set to 1, would give an upper bound not better than $\hat{w}$.

Algorithm R1GAP

1. **for each** $j \in J$ **do** find $p_{i(j),j} = \max_{i \in I}\{p_{i,j} : r_{i,j} \leq b_i\}$;

2. $u \leftarrow \Sigma_{j \in J}\, p_{i(j),j}$;

3. **for each** $i \in I$ **do**

4.   **for each** $j \in J$ **do**

5.   **if** $\hat{w} \geq u - (p_{i(j),j} - p_{i,j})$ or $r_{i,j} > b_i$ **then** $x_{i,j} \leftarrow 0$

R1GAP has time complexity $O(nm)$.

The second algorithm applies the same reduction criterion to the initial Martello and Toth upper bound $\bar{u} = \Sigma_{i \in I}\, w_i$ (where w_i is the solution of problem (K_i^2) in the relaxation of (GAP2)).

Algorithm R2GAP

1. **for each** $i \in I$ **do**

2.   **for each** $j \in J$ **do**
        begin

3. $\delta \leftarrow \min\{w_i, \text{(upper bound on } (K_i^2) \text{ if } x_{i,j} \text{ is set to } 1 - \hat{x}_{i,j})\}$;

4.   **if** $\hat{w} \geq \bar{u} - (w_i - [\delta])$ **then** $x_{i,j} \leftarrow \hat{x}_{i,j}$
        end

The value δ has the same meaning as $v_{i,j}^0$ and $v_{i,j}^1$ in Section 5.2. R2GAP requires, in the worst case, a non-polynomial running time since finding each w_i requires the solution of a 0-1 knapsack problem. However, such w_i's must be determined in the first iteration of the Martello and Toth branch-and-bound algorithm of Section 5.3. Hence, if R2GAP is applied as preprocessing of this algorithm, the extra computational effort involved is $O(mng(n))$, where $g(n)$ is the time complexity for finding each upper bound on (K_i^2).

5.5. Computational results

We coded the algorithms of Ross and Soland and of Martello and Toth (including HGAP, R1GAP and R2GAP) in Fortran IV and ran them on a CDC-6600. All the 0-1 knapsack subproblems were solved trhough the algorithm of Martello and Toth [22]. The Fisher, Jaikumar and Van Wassenhove algorithm was coded by the authors in Fortran IV and run on a DEC 10. Since the CDC-6600 is considered to be about 5 times faster than the DEC 10, the Fisher-Jaikumar-Van Wassenhove times in Table 5.1 are DEC 10 seconds divided by 5.

Computational experiments were executed on the following problem generations:

A. Ross-Soland generation [26]: $r_{i,j}$ and $p_{i,j}$ uniformly random between 5 and 25 and between 1 and 40 respectively and $b_i = 0.6\,(n/m)\,15 + 0.4\,\max_{i \in I}\{\Sigma_{j \in J_i}\,r_{i,j}\}$.

B. Same as A, but with b_i set to 70 percent of the value given in A.

C. Same as A for $r_{i,j}$ and $p_{i,j}$ but $b_i = 0.8\,\Sigma_{i \in J}\,r_{i,j}/m$.

D. $r_{i,j}$ uniformly random between 1 and 100; $p_{i,j} = 10 + r_{i,j} - e$, with e uniformly random in $(-10, 10)$; $b_i = 0.8\,\Sigma_{j \in J}\,r_{i,j}/m$.

Problems of type A generally admit many feasible solutions, while in problems of types B and C constraints (5.2) are tight; in problems of type D a correlation between profits and resource requirements (often found in real-world applications) is introduced.

For each data generation we considered two values of m ($m = 3, 5$) and two values of n ($n = 10, 20$). We solved 10 problems for each data generation and

Table 5.1.
CDC-6600 seconds; 10 problems for each entry.

Data set	m	n	Ross-Soland		Martello-Toth		Fisher-Jaikumar-Van Wassenhove	
			Average running time	Average number of nodes	Average running time	Average number of nodes	Average running time	Average number of nodes
A	3	10	0.012	5	0.030	2	0.034	0
A	3	20	0.031	8	0.054	3	0.053	0
A	5	10	0.029	13	0.093	7	0.050	1
A	5	20	0.077	17	0.464	24	0.129	2
B	3	10	0.227	127	0.288	42	0.307	4
B	3	20	32.852(6)	11801	5.977	555	1.028	12
B	5	10	0.230	110	0.312	28	0.393	5
B	5	20	Time limit	–	18.203	1098	2.986	28
C	3	10	0.321	191	0.165	19	0.337	2
C	3	20	61.185(4)	25175	7.961	718	1.116	13
C	5	10	0.370	162	0.215	22	0.508	3
C	5	20	Time limit	–	15.084	1027	2.750	25
D	3	10	0.793	510	0.409	50	0.361	4
D	3	20	Time limit	–	16.289	1235	3.872	31
D	5	10	1.353	669	0.767	90	0.788	8
D	5	20	Time limit	–	100.932(2)	5889	13.521	126

for each pair (m, n). The entries in Table (5.1) give average times and average number of nodes solved in the branch-decision tree. 250 seconds were assigned to each algorithm to solve the 20 problems generated for each data type and value of m. For the cases with a time limit, we give the number of problems solved in brackets.

The table shows that the Ross and Soland algorithm is faster for data set A, but much slower than the other methods for data sets B, C and D. For these (harder) data sets, the Martello and Toth algorithm appears slightly faster when $n = 10$, while the Fisher, Jaikumar and Van Wassenhove algorithm is clearly superior when $n = 20$. The analysis of the average number of nodes solved in the branch-decision tree indicates that the good performance of the Fisher-Jaikumar-Van Wassenhove algorithm depends on the quality of the upper bound (their branch-and-bound being a very simple dichotomic scheme). Hence good results can be expected if this bound is imbedded in the more sophisticated Martello-Toth algorithm.

Table 5.2 shows the performance of heuristic algorithm HGAP on large-size problems of type A, the only problems which can be solved exactly, allowing an analysis of the quality fo the heuristic solutions found. The table shows that HGAP gives very good solutions with acceptable running times. Further experiments indicated that the running times of HGAP are practically independent of the data set, while the quality of the solutions decreases for data sets B, C and D.

Reduction procedures R1GAP and R2GAP showed good experimental performance for data set A (more than 90% of the variables were fixed) and a rather poor performance for the other data sets (about 10% of the variables were fixed).

Table 5.2.
Algorithm HGAP; CDC-6600 seconds; 10 problems for each entry.

Data set	m	n	Average running time	Average error	Number of optimal solutions found
A	5	50	0.096	0.07%	7
A	5	100	0.240	0.05%	5
A	5	200	0.569	0.02%	7
A	10	50	0.134	0.06%	9
A	10	100	0.334	0.06%	6
A	10	200	0.692	0.03%	5
A	20	50	0.243	0.13%	5
A	20	100	0.540	0.03%	6
A	20	200	1.239	0.04%	5

Aknowledgement

This work was supported by Consiglio Nazionale delle Ricerche (C.N.R.), Italy.

References

[1] R.S. Barr, F. Glover and D. Klingman, «The Alternating Basis Algorithm for Assignment Problems», *Mathematical Programming* 13, 1 - 13, 1977.
[2] D.P. Bertsekas, «A New Algorithm for the Assignment Problem», *Mathematical Programming* 21, 152 - 171, 1981.
[3] F. Bourgeois and J.C. Lassalle, «An Extension of the Munkres Algorithm for the Assignment Problem to Rectangular Matrices», *Communications of ACM* 14, 802 - 804, 1971.
[4] R.E. Burkard and U. Derigs, *Assignment and Matching Problems Solutions Methods with FORTRAN Programs*, Springer, Berlin, 1980.
[5] G. Carpaneto, S. Martello and P. Toth, *Il Problema dell'Assegnamento: Metodi ed Algoritmi*, Progetto Finalizzato Informatica, SOFMAT, C.N.R., Roma, 1984.
[6] G. Carpaneto and P. Toth, «Algorithm 548, Solution of the Assignment Problem», *ACM Transactions on Mathematical Software* 6, 104 - 111, 1980.
[7] G. Carpaneto and P. Toth, «Algorithm for the Solution of the Bottleneck Assignment Problem», *Computing* 27, 179 - 187, 1981.
[8] G. Carpaneto and P. Toth, «Algorithm for the Solution of the Assignment Problem for Sparse Matrices», *Computing* 31, 83 - 94, 1983.
[9] G. Carpaneto and P. Toth, «Primal-Dual Algorithms for the Assignment Problem», Report OR/84/2, DEIS, University of Bologna, 1984.
[10] L. Chalmet and L. Gelders, «Lagrange Relaxations for a Generalized Assignment-Type Problem», in M. Roubens, ed., *Advances in Operations Research*, North-Holland, Amsterdam, 1977.
[11] U. Derigs, «Alternate Strategies for Solving Bottleneck Assignment Problems - Analysis and Computational Results», *Computing* 33, 95 - 106, 1984.
[12] U. Derigs and U. Zimmermann, «An Augmenting Path Method for Solving Linear Bottleneck Assignment Problems», *Computing* 19, 285 - 295, 1978.
[13] B. Dorhout, «Het Lineaire Toewijzingsproblem, Vergelijking van Algoritmen», Report BN 21, Stichting Mathematisches Centrum, Amsterdam, 1973.
[14] J. Edmonds and R.M. Karp, «Theoretical Improvements in Algorithmic Efficiency of Network Flow Problems», *Journal of ACM* 19, 248 - 261, 1972.
[15] G. Finke, R.E. Burkard and F. Rendl, «Quadratic Assignment Problems», this volume.
[16] M.L. Fisher, R. Jaikumar and L.N. Van Wassenhove, «A Multiplier Adjustment Method for the Generalized Assignment Problem», *Management Science* (to appear).
[17] L.R. Ford and D.R. Fulkerson, *Flow in Networks*, Princeton University Press, Princeton, New York, 1962.
[18] O. Gross, «The Bottleneck Assignment Problem», P-1630, The Rand Corporation, Santa Monica, California, 1959.
[19] N.W. Kuhn, «The Hungarian Method for the Assignment Problem», *Naval Research Logistics Quarterly* 2, 83 - 97, 1955.
[20] N.W. Kuhn, «Variants of the Hungarian Method for the Assignment Problem», *Naval Research Logistics Quarterly* 3, 253 - 258, 1956.
[21] E.L. Lawler, *Combinatorial Optimization: Networks and Matroids*, Holt, Rinehart and Winston, New York, 1976.
[22] S. Martello and P. Toth, «Algorithm for the Solution of the 0-1 Knapsack Problem», *Computing* 21, 81 - 86, 1978.
[23] S. Martello and P. Toth, «An Algorithm for the Generalized Assignment Problem», in J.P. Brans, ed., *Operational Research '81*, North-Holland, Amsterdam, 1981.
[24] S. Martello, W.R. Pulleyblank, P. Toth and D. de Werra, «Balanced Optimization Problems», *Operations Research Letters* 3, 275 - 278, 1984.
[25] L.F. McGinnis, «Implementation and Testing of a Primal-Dual Algorithm for the Assignment Problem», *Operations Research* 31, 277 - 291, 1983.

[26] G.T. Ross and R.M. Soland, «A Branch and Bound Algorithm for the Generalized Assignment Problem», *Mathematical Programming* 8, 91 - 103, 1975.
[27] R. Silver, «An Algorithm for the Assignment Problem», *Communications of ACM* 3, 605, 1960.
[28] N. Tomizawa, «On Some Techniques Useful for Solution of Transportation Network Problems», *Networks* 1, 173 - 194, 1971.

Silvano Martello
Paolo Toth
DEIS, University of Bologna
Viale Risorgimento 2
40136 Bologna
Italy

NETWORK SYNTHESIS AND DYNAMIC NETWORK OPTIMIZATION

Michel MINOUX

1. Introduction

Determining a minimum cost network designed in such a way as to meet a given set of specifications (flowing prescribed traffic requirements, achieving a quality of service, etc. ...), is a fundamental class of problems which arise in a wide variety of contexts of applications such as Transportation Science, Telecommunication network engineering, distribution systems, energy networks, water distribution networks etc. .

This paper is intended as a survey of past work concerning two important and quite general problems arising in the area of distributed telecommunication (telephone and/or data processing) networks: (a) network synthesis under non-simultaneous single-commodity or multicommodity flow requirements (section 2); (b) determining an optimal investment policy for meeting increasing multi-commodity flow requirements over a given time period. (sections 3 to 9). The importance of the latter problem stems from the fact that most optimum network design problems are essentially *dynamic* in nature, in the sense that the time factor should be explicitly taken into account. Since most research effort in the network optimization area has been directed so far towards static rather than dynamic models, we have felt it preferable here to insist on dynamic models and methods for their solution.

Optimum network design and synthesis problems have attracted much interest in the OR Community since the early 60's, and this is mainly due to the importance and variety of practical applications involved: Transportation systems, Telecommunications systems, teleprocessing and Computer Communication networks, distribution systems, energy distribution systems, water supply networks and so on. Surveys covering applied and/or computational issues related to networks synthesis and optimum network design problems will be found for instance in Frank and Frisch [24], Zadeh [79, 80], Steenbrink [73], Minoux [57], Boorstyn and Frank [8], Schwartz [71], Dionne and Florian [16], Gavish [28], Kobayashi [48], Magnanti and Wong [52], Minoux [65].

Even if we restrict to a specific context of application such as Telecommunications (i.e. telephone and/or data transmission networks) the diversity of models (and solution methods), which have been studied in connection with optimum design issues for systems and networks is striking and, without the pretention of being exhaustive, one can mention:

— minimum concave cost multicommodity network flow models (with applications to the determination of an optimum network structure and/or network planning): see Yaged [77, 78]; Zadeh [79, 80], Minoux [57, 58]. In the special case of linear with fixed costs: see Billheimer and Gray [6]; Minoux [58]; Boffey and Hinxman [7];

— minimum concave cost single commodity network flow models (with application to optimum design of centralized networks): Zangwill [81], Gallo and Sodini [25], Gallo, Sandi and Sodini [26];

— minimum linear cost multicommodity flows with a budget constraint (the so-called «optimum network problem»): see Scott [72]; Boyce and al [9]; Hoang [41]; Dionne and Florian [16];

— optimum network synthesis with nonsimultaneous *single-commodity* flow requirements: see Gomory and Hu [34, 35];

— optimum network synthesis with nonsimultaneous *multicommodity* flow requirements (with application to optimum design of networks under reliability constraints or time-varying requirement matrix): see Minoux [61, 64], Minoux and Serreault [67, 68];

— capacitated minimum spanning trees (with application to optimum design of homogeneous centralized data processing networks): see Chandy and Lo [10], Kershenbaum [46], Kershenbaum and Boorstyn [47], Gavish [29, 30];

— concentrator location problems (in connection with optimum design of heterogeneous centralized data processing networks): see Chandy and Russel [11], Elias and Ferguson [19], Tang, Woo and Bahl [75], Gavish [28];

— Steiner tree problems in graphs (related to applications such as optimal location of switching centers or concentrators in networks): see Dreyfus and Wagner [18], Hakimi [39], Aneja [1], Claus and Maculan [15], Beasley [4] and the survey paper by Maculan in the present book.

As can be observed from the above references, most of the work carried out in the area of optimal network design concerns *static models*, in the sense that an optimal network structure is looked for, given the traffic flow requirements *at some specific time instant*, but *without taking into account the evolution of the traffic requirements over a (sufficiently long) time period*. Section 2 below provides a brief survey of basic static network synthesis models.

However, it turns out that, partly due to the fundamentally discontinuous nature of the possible investment decisions (for increasing capacity on links and/or nodes, choice is usually limited to a few distinct physical devices corresponding to a few [capacity, cost] pairs), examples are easily found (see Minoux [56], and section 3) where the sequence of optimum static solutions (obtained by solving the static model at the various successive instants $t = 0, 1, \ldots, T$ of the time period $[0, T]$ under consideration) is not only *far from the optimum*, but *practically unfeasible* (e.g. because investments decided at instant t, should be removed at instant $t + 1$). In view of this, it is readily seen that good investment policies can only be obtained by taking account of the time factor, in other words (even if it is more often implicit than explicit) by taking account of the essentially *dynamic* nature of network optimization problems. However, — and this is likely to be due to the intrinsic complexity of realistic dynamic models — it appears that only a limited amount of research effort has been devoted so far to such models; this is precisely one of the main purposes of this paper to try and provide some state-of-the art in the area of dynamic network design problems describing (Sections 3 to 9 below) various models and approaches for solving them. Moreover, it will be shown how the most practically applicable solution methods are related to the work on static models cited above.

2. Static network synthesis models

As already alluded to in the introduction, a great deal of work has been devoted in the past to network optimization problems viewed as *static problems* i.e. without taking into account the evolution, over several successive time periods, of the networks under consideration. Thus an extensive survey of this area (the interested reader may refer to Minoux [65]) would be beyond the scope of this paper which, as already said, will focus on dynamic rather than static models. However, before turning to a full treatment of this issue, we felt it worth while giving at least a flavor of some of the most basic static network optimization models which are commonly referred to as *network synthesis problems* (this terminology was suggested by Gomory and Hu [35]). This is a class of problems in which a minimum cost network is looked for (assuming that the costs are linear functions of installed capacity) with the requirement that the capacities installed on the various links of the network be large enough to meet a number of given distinct and independent single-commodity (or multicommodity) flow requirements.

2.1. The single-commodity case (Gomory and Hu [34, 35])

Considering first, the single commodity case, the problem may be more specifically stated as follows. The structure of the network is represented by a nondirected (connected) graph $G = [X, U]$ where X is the set of nodes ($|X| = N$) and U the set of edges ($|U| = M$). Moreover, we assume that we are given a set of p (distinct) single-commodity flow requirements on the above graph, the kth of these ($1 \leq k \leq p$) being described by:
 — an origin-destination pair (s_k, t_k);
 — the requested flow value $r^k > 0$ between s_k and t_k.

(Since the problem is basically nondirected, s_k and t_k may be indifferently considered as source or sink of the flow). Also, with each edge $u \in U$, we associate a linear cost γ_u, the cost of one unit capacity on link u. It is required to determine capacities $y_u \geqslant 0$ to be assigned to the edges $u \in U$ of G in such a way that:

(i) the total cost $\Sigma_{u \in U} \gamma_u \cdot y_u$ is minimized;

(ii) for each node pair (s_k, t_k) $(1 \leqslant k \leqslant p)$ there exists a feasible flow of value r^k between s_k and t_k.

In other words, a set of capacities $y = (y_u)_{u \in U}$ on the graph will be feasible if and only if it allows nonsimultaneously meeting each one of the given flow requirements (independently of the others). This is why the problem is often referred to as «*network synthesis with nonsimultaneous single-commodity flow requirements*». A linear programming formulation can easily be set up by using the basic definition of a single-commodity flow in terms of the *node-arc incidence matrix A*: choose any orientation on each edge $u = (i, j)$ of G thus obtaining an *arc*, still denoted by u (arc u is (i, j) if the orientation chosen is from i to j, and (j, i) if it is from j to i). Each column of A corresponds to an arc, $u = (i, j)$ and has exactly two nonzero entries $a_{iu} = +1$ and $a_{ju} = -1$ corresponding to the two endpoints of u. Now a vector $\varphi^k = (\varphi_u^k)_{u \in U}$ represents a flow of value r^k between two nodes s_k and t_k if and only if:

$$A \varphi^k = r^k \cdot b^k$$

where b^k is a N-vector with exactly two nonzero entries $+1$ and -1 respectively corresponding to rows s_k and t_k of A. (note here that φ has negative components on arcs for which the flow happens to be sent in the direction opposite to the orientation on the arc, but constraints (ii) only concern the absolute values $|\varphi_u|$ of the flow on the arcs).

The single-commodity network synthesis problem can then be expressed as the following linear program

$$(LP) \begin{cases} \text{Minimize} \quad \sum_{u \in U} \gamma_u y_u \\ \text{subject to} \\ \forall k = 1, \ldots, p: \\ \qquad A \cdot \varphi^k = r^k \cdot b^k \\ \forall u \in U: |\varphi_u^k| \leqslant y_u \end{cases}$$

The matrix representation of the problem is shown on *Table 1* featuring a block-diagonal structure corresponding to the φ^k variables ($k = 1, \ldots, p$), the p network flow subproblems being coupled through the y variables (coupling variables). Note that the capacity constraints $|\varphi_u^k| \leqslant y_u$ have been rewritten there: $-y_u \leqslant \varphi_u^k \leqslant y_u$.

Gomory and Hu [34] have shown how problem (LP) can be solved through *generalized linear programming* by reformulating it in terms of a large scale linear program with only M variables (the coupling variables y) but a large number of

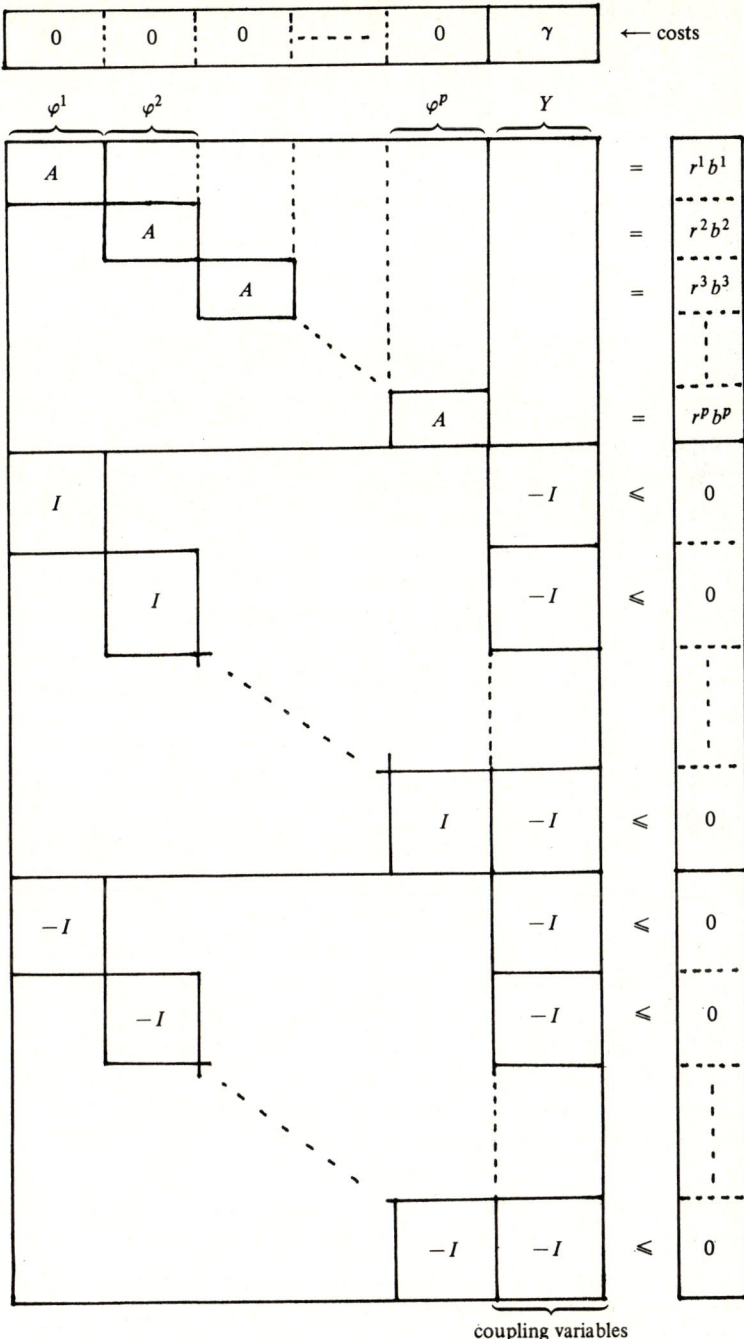

Table 1. Matrix representation of the single-commodity network synthesis problem (A is the node-arc incidence matrix of the graph).

constraints of the form:

$$\sum_{u \in C^k} y_u \geq r^k$$

where C^k is a $s_k - t_k$ cut in the network (thus the above inequalities state that for each pair (s_k, t_k), the capacity of any cut C^k should be greater than or equal to the flow requirement r^k. Note that the equivalence with the original problem stems from the max-flow-min-cut theorem (Ford and Fulkerson [23])). The objective is still to minimize $\Sigma_{u \in U} \gamma_u y_u$. Of course, due to the large number of constraints (of cuts in the network) the problem is *relaxed* by considering only a (small) subset of cut constraints; constraints which are violated by the current solution (optimal solution to the current relaxed problem) are then easily identified by maximum network flow computations, and iteratively appended to the current relaxed problem (this process is often referred to as *constraint-generation*, and is seen to be the dual of a *column generation scheme*). Since there are finitely many cuts in the network and each generated cut is necessarily different from those already belonging to the current relaxed problem, the process necessarily terminates in a finite number of iterations with a solution to the relaxed problem (implicitly) satisfying all the cut constraints: it is then an optimal solution to the single-commodity network synthesis problem. It is interesting here to point out that this constraint generation process can be interpreted as applying the *Benders decomposition procedure* (Benders 1962) to the linear program (LP) (note that the original Benders procedure was designed to handle mixed integer programming problems — problems in which the coupling variables are constrained to be integral — but applies as well and even better from the point-of-view of computational efficiency, to «ordinary» linear programs: it can then be used as a decomposition technique provided that the problem to be solved display the appropriate structure, namely diagonal blocks with coupling constraints).

A special case of the single-commodity network synthesis problem is worth mentioning: the one where all the costs are equal to 1. This is the «*basic Gomory and Hu network synthesis problem*» for which a purely combinatorial (and polynomial) graph theoretic algorithm exists (Gomory and Hu [35]). This algorithm is based on the fact that only a subset of (at most $N-1$) requirements need be considered, the ones belonging to the so-called «*dominant requirement tree*» (considering the graph G_R on X where there exists an edge (i, j) iff there is a strictly positive flow requirement from i to j, the weight on the edge being the value of the requirement, a dominant requirement tree is nothing but a maximum weight spanning tree on G_R).

The dominant requirement tree is then decomposed into a sum of subtrees with *uniform requirements*, and the optimal network is directly obtained («synthetized») by associating with each uniform requirement subtree a cycle running through all the nodes spanned by the subtree. For more details, see the original paper by Gomory and Hu [35] or Lawler [49], Minoux [65].

2.2. The multicommodity case (Minoux [61])

The single-commodity network synthesis problem considered by Gomory and Hu can be generalized to multicommodity network flows in the following way. Here p distinct independent multicommodity flow requirements are given on the graph $G = [X, U]$ representing the network structure. Each of the individual multicommodity flow requirement k, is described by a set of origin-destination pairs (s_k^1, t_k^1) (s_k^2, t_k^2), $(s_k^{q_k} t_k^{q_k})$ together with the corresponding source-sink requirements $r_k^1 r_k^2, \ldots, r_k^{q_k}$.

Here, a set of capacities $y = (y_u)_{u \in U}$ assigned to the links of the network is called *feasible* with respect to multicommodity flow k if the capacities allow *simultaneously* flowing all the single-commodity flow requirements r_k^1 (between s_k^1 and t_k^1), r_k^2 (between s_k^2 and t_k^2) ... $r_k^{q_k}$ (between $s_k^{q_k}$ and $t_k^{q_k}$).

The purpose is to find a minimum cost $(\Sigma_{u \in U} \gamma_u y_u)$ assignment of capacities which is feasible with respect to any one of the p given multicommodity flow requirements *independently of the others*. This is the reason why in Minoux [61] the problem has been referred to as «*network synthesis with nonsimultaneous multicommodity flow requirements*».

Again, using a *node-arc formulation* for each individual feasible multicommodity flow problem, this can be formulated as a (large scale) linear program with block-diagonal structure and M coupling variables y_u ($u \in U$). However, contrasting with the single-commodity case, each block now displays the matrix structure of a feasible multicommodity flow problem in node-arc formulation i.e.: a block diagonal substructure composed of repeated node-arc incidence matrices of the graph with coupling constraints (the capacity constraints). Again Benders decomposition can be applied to such a structure with the more complex feature that now, each subproblem is a feasible multicommodity flow problem (significantly harder to solve than the maximum flow problem in the single-commodity case). Minoux [61] showed how the multicommodity network synthesis problem can be reformulated as a large scale linear program in the coupling variables y_u only, but with very many constraints, by using a proper generalization of the max-flow-min-cut theorem to multicommodity flows, and then proposed to solve this large scale program by applying a *constraint generation scheme*. Minoux and Serreault [67, 68] applied this methodology to a telecommunication problem involving multicommodity network optimization when *security constraints* are imposed (the network to be «synthetized» should be able to flow prescribed traffic requirements, even in case of failure of some of its constitutive elements – links or nodes –). Due to the very high dimensionality of the problems to be solved in real applications, their implementation made use of subgradient algorithms for (approximately) solving both the master problem and the feasible multicommodity flow subproblems at each iteration. The proof of convergence to an approximate solution (to within a controllable accuracy) is provided by Minoux [64] in the more general context of Benders decomposition, when approximations are allowed in the solution of both the master problem and the subproblems.

Another interesting application of network synthesis with nonsimultaneous multicommodity flow requirements will be found in Minoux [65].

3. The dynamic network optimization problem: formulation as a minimum cost dynamic multicommodity flow problem with discontinuous cost functions

We now turn to the study of dynamic network optimization problems, by describing in this section a very general formulation where the minimum PW of AC (Present Worth of Actual Cost) of network expansion over a given time period $[0, T]$ is to be determined. Contrary to many others described in the litterature, the model presented here is intended to stick to reality as much as possible, in particular by deliberately coping with the difficulty, constantly encountered in practice, of discrete discontinuous cost functions (due to the fact that capacity expansion on the links can only be performed by choosing among a finite set of possible capacity values). We think that, due to its great generality and realism, both the model and solution methods described here should have wide applicability and not only in the Telecommunication field. The *structure of the network* will be represented by an (undirected) graph $G = [X, U]$ where X is the set of nodes ($|X| = N$) and U is the set of edges (or links) ($|U| = M$). In case where a transmission link may be created between every pair of nodes (i, j) $i = 1, \ldots, N, j = 1, \ldots, N, i < j$, then $M = N(N-1)/2$. However in practice we have usually $M \ll N(N-1)/2$. The time period $[0, T]$ is supposed to be divided into T intervals of unit duration (in general each time unit will correspond to one year), and it is required that at each time instant t ($0 \leq t \leq T$) the capacity installed on the links of the network be sufficient to allow flowing prescribed traffic requirements at instant t. The traffic requirements at instant t may be characterized as a *multicommodity flow* composed of $K = N(N-1)/2$ distinct (independent) single-commodity flows, one for each possible source-sink pair (i, j) $i = 1, \ldots, N, j = 1, \ldots, N, i < j$. Each individual single commodity flow associated with a source-sink pair (i, j) will be represented at instant t by a M-vector $\varphi^{ij}(t) = (\varphi_u^{ij}(t))_{u \in U}$ (an arbitrary orientation being chosen on each link of the network, it is well-known that the M-vector $\varphi^{ij}(t)$ is a flow if and only if it satisfies the so-called first Kirchhoff's law) and the value of the flow $v(\varphi^{ij}(t))$ is equal to the total flow leaving node i (source) or, equivalently, to the total flow into node j (sink). (Observe that in most applications, the $(i-j)$ flow will represent the total traffic from i to j *and* from j to i, so the distinction between source and sink among the pair (i, j) is purely conventional). When all the K single commodity flows $\varphi^{ij}(t)$

are simultaneously present on the network, then the total flow value on each link $u \in U$ at instant t is

$$\psi_u(t) = \sum_{i=1}^{N} \sum_{\substack{j=1 \\ j>i}}^{N} |\varphi_u^{ij}(t)|$$

In what follows, a minimum cost network expansion over the time period $[0, T]$ is looked for under the constraint that, at each instant t ($t = 0, 1, \ldots, T$) the value of each $(i-j)$ flow $\varphi^{ij}(t)$ be equal to some desired value $r^{ij}(t)$ ($i = 1, \ldots, N$, $j = 1, \ldots, N$, $i<j$). The $N \times N$ matrix $R(t)$ whose entries are the values $r^{ij}(t)$ for all (i,j) $i<j$ will be called the *traffic requirement matrix* at time t.

Thus, from now on, and $\forall t = 0, 1, \ldots, T$, $\varphi^{ij}(t)$ will denote a single-commodity flow *of value* $r^{ij}(t)$ from source i to sink j in the network.

A *solution* to the network expansion problem will be described by a sequence of $T+1$ M-vectors : $Y(0), Y(1), \ldots, Y(T)$ where, for each t ($0 \leq t \leq T$):

$$Y(t) = [Y_u(t)]_{u \in U}$$

and $Y_u(t)$ is the total capacity (of the transmission facilities) installed on link u at instant t.

In many situations, where network expansion requires, on each link, a succession of investments where new transmission systems are progressively added to face increasing requirements, an investment, once decided, is not reconsidered later on, resulting in total installed capacity values $Y_u(t)$ which are *nondecreasing functions of time*.

In that case, which will be the one considered here, the various vectors $Y(t)$ which compose a solution should meet the following *dominance relations*:

$$Y(0) \leq Y(1) \leq \ldots \leq Y(T) \tag{1}$$

where, for any two M-vectors Y^1 and Y^2, the *dominance relation* $\leq$ is defined by:

$$Y^1 \leq Y^2 \text{ if and only if } Y_u^1 \leq Y_u^2 \quad (\forall u \in U).$$

A solution $Y = [Y(0), Y(1), \ldots, Y(T)]$ will be called *feasible* if, for each $t = 0, 1, \ldots, T$, the capacities $Y_u(t)$ ($u \in U$) installed on the links of the network are sufficient to *simultaneously* flow all the required amounts of flows $r^{ij}(t)$, in other words, if there exist M-vectors $\varphi^{ij}(t)$ such that

$$\begin{cases} \forall u \in U : \psi_u(t) = \sum_i \sum_{j>i} |\varphi_u^{ij}(t)| \leq Y_u(t) & (2) \\ \forall (i,j), i<j, \varphi^{ij}(t) \text{ is an } (i-j) \text{ flow of value } v(\varphi^{ij}(t)) = r^{ij}(t). & (3) \end{cases}$$

(2) are called the *capacity constraints* and (3) the *requirement constraints*.

No we define the *cost* of a solution $Y = [Y(0), \ldots, Y(T)]$ satisfying the dominance relations (1) in the following way. Each time extra capacity is needed on some link $u \in U$, we assume that we have choice among a given (finite) family of equipments, each equipment $s = 1, 2, \ldots, S$ in the family being characterized by its capacity $Q(s)$ and total cost $P(s)$ which can be quite accurately approximated by a formula of the type: $P_u(s) = F(s) + l_u V(s)$ where:
- $F(x)$ is the fixed cost of equipments located at the terminal endpoints of the link (independent from the length of the link).
- $V(s)$ is the variable cost per unit distance of transmission system s of capacity $Q(s)$.
- l_u is the total physical length of link u.

(Of course, it would not be conceptually difficult to consider more complex dependence of cost on link length or possibly other parameters associated with the link).

Thus for a given link u, l_u is fixed, and the family of possible transmission equipments which may be installed on link u (to increase capacity) may be represented on a capacity versus cost chart as a set of points, each point $(Q(s), P(s))$ corresponding to a specific equipment of the family. *Figure 1* illustrates this on a family of equipments composed of 5 systems with the following capacity/cost characteristics (cost being given for, say, a 100 km link).

System number	Total capacity (number of channels)	Total cost (for a 100 km link) Millions US $
$s = 1$	$Q(1) = 500$	$P(1) = 4.5$
$s = 2$	$Q(2) = 1000$	$P(2) = 7$
$s = 3$	$Q(3) = 2000$	$P(3) = 10$
$s = 4$	$Q(4) = 3000$	$P(4) = 13$
$s = 5$	$Q(5) - 10000$	$P(5) = 27$

It can be checked that the cost roughly follows a law of variation of the form $0.11 (Q)^{0.6}$. (This example will be considered again in §7).

As it can be observed from these figures (and this is exactly what happens in reality) the above family obeys the so-called *economy of scale phenomenon* namely that the mean cost (cost per line) is decreasing with the total capacity of the systems installed (see figure 2). If, as is the case on figure 1, the cost follows a law of variation of the form $\beta(Q)^{0.6}$ where β is a constant, then mean costs follow a law of variation of the form $\beta(Q)^{-0.4}$.

In what follows, we will denote by $\Gamma_u(x)$ the cost function on link u which is defined only for $x = 0$ and for $x = Q(s)$ ($s = 1, \ldots, S$) and whose value is

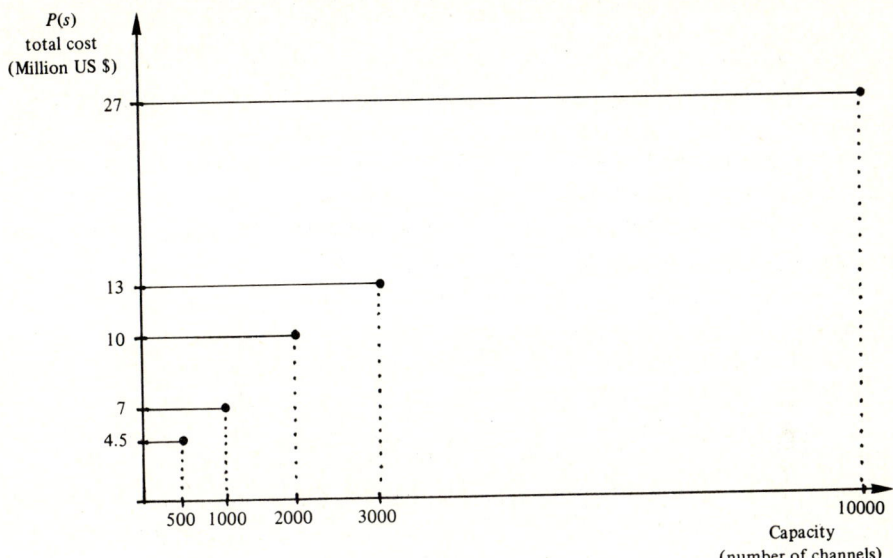

Fig. 1. Example of a capacity versus cost chart associated with a family of 5 transmission systems. It can be checked that the cost as a function of capacity is well approximated by a function of the form: $P \simeq 0.11 \, (Q)^{0.6}$.

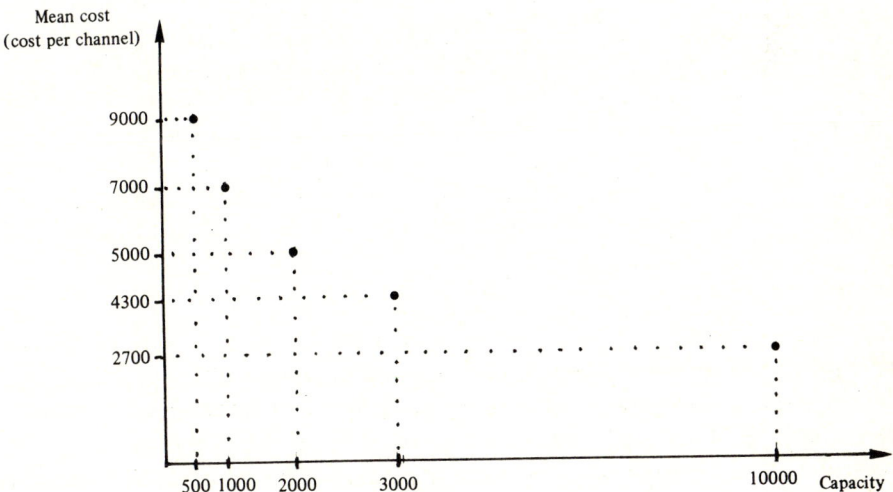

Fig. 2. The «economy-of-scale» phenomenon: among a given family of transmission systems, mean cost (cost per channel) is a decreasing function of the capacity of the system. Since on figure 1, the cost is well approximated by the formula $0.11 \, (Q)^{0.6}$, the mean cost variation is well approximated by the formula $0.11 \, (Q)^{-0.4}$.

$$\Gamma_u(x) \begin{cases} = P_u(s) = F(s) + l_u V(s) \text{ if } x \text{ is equal to } Q(s) \text{ for some } s \ (1 \leq s \leq S); \\ = 0 \text{ if } x = 0; \\ \text{undefined, otherwise.} \end{cases}$$

Now, in order to determine the best network expansion strategy, we shall resort to what economists have been using for years, namely the *Present Worth of Actual Cost* (PWAC) *criterion*, which if we consider any sequence of investments of values $I(1), I(2), \ldots, I(T)$ performed at each instant of the time period $[0, T]$ is defined by:

$$C_{\text{PWAC}} = \frac{I(1)}{(1+\tilde{c})} + \frac{I(2)}{(1+\tilde{c})^2} + \ldots + \frac{I(T)}{(1+\tilde{c})^T}.$$

The parameter $\tilde{c}$ is an exogenous parameter usually chosen in the range [0.05, 0.15] and is called the *actualization rate*; its value depends on a great number of economic factors such as: interest rates, rates of growth of the firm, and so on. It will be considered here as a fixed known external parameter.

Using the above cost criterion, we are now in a position to define the total PW of AC of a feasible solution $Y = [Y(0), Y(1), \ldots, Y(T)]$ to the network expansion problem.

For the sake of simplicity in the presentation, we will assume that, on each link of the network, no more than one system of the family at hand need be installed at a time. Of course, this simplifying assumption could easily be dropped, but it turns out that it is quite realistic since the capacities of the transmission systems are usually chosen sufficiently large so that fill-in takes on several years to occur. Under these conditions the solution Y is such that, for each link $u \in U$, the extra capacity $Y_u(t) - Y_u(t-1)$ added between any two consecutive instants $t-1$ and t is either equal to 0 (no investment on link u at time $t-1$), or corresponds to the capacity $Q(s)$ of one of the systems of the family considered. Thus, using the Γ_u function introduced earlier, the investment cost on link u at time $t-1$ is exactly

$$\Gamma_u(Y_u(t) - Y_u(t-1))$$

In view of this, the total PW of AC of network expansion over the time period $[0, T]$ for the solution Y is the sum of the PW of AC of expansion of the links i.e.:

$$\Gamma(Y) = \sum_{u \in U} \sum_{t=1}^{T} \frac{\Gamma_u(Y_u(t) - Y_u(t-1))}{(1+\tilde{c})^{t-1}}.$$

To summarize, the problem is to find optimal values of $Y_u(t) \, \psi_u(t)$ and $\varphi^{ij}(t)$ solving the following *minimum cost dynamic multicommodity flow problem*:

$$\text{(MCDMF)} \begin{cases} \text{Min } \Gamma(Y) = \sum_{u \in U} \sum_{t=1}^{T} \dfrac{\Gamma_u(Y_u(t) - Y_u(t-1))}{(1+\widetilde{c}\,)^{t-1}} \\[2pt] \text{under the constraints:} \\[2pt] \forall t \in [1, T], \; \forall u \in U : Y_u(t-1) \leqslant Y_u(t) \quad (1) \\ \text{(dominance constraints)} \\[2pt] \forall t, \; \forall u \in U : \psi_u(t) = \sum_{i<j} |\varphi_u^{ij}(t)| \leqslant Y_u(t) \quad (2) \\ \text{(capacity constraints)} \\[2pt] \forall t, \; \forall (i,j) : \varphi^{ij}(t) \text{ is an } (i-j) \text{ flow of value } v(\varphi^{ij}(t)) = r^{ij}(t) \quad (3) \\ \text{(flow requirement constraints)} \\[2pt] \forall t \in [1,T], \; \forall u \in U : Y_u(t) - Y_u(t-1) \in \{0, Q(1), Q(2), \dots, Q(S)\} \quad (4) \\[2pt] \forall u \in U : Y_u(0) \text{ (the capacity existing on link } u \text{ at time 0)} \text{ is supposed} \\ \text{to be given} \end{cases}$$

Constraints (4) have been added here to recall that capacity expansion on the links cannot be performed but by using one of the transmission systems in the family considered.

In order to get a better understanding of the structure of the problem, and in particular of its highly combinatorial nature, we will first reformulate it using a *dynamic programming* approach (Bellman [5]).

4. A dynamic programming formulation of the network expansion problem

We show, in this section, that problem (MCDMF) can be formulated as a Dynamic Programming problem, in other words as a shortest path (minimum cost path) problem in a (large) sequential graph $\overline{G}$ defined in the following way.

The nodes of $\overline{G}$ are divided into T *stages* corresponding to the time instants $t = 1, t = 2, \dots, t = T$. At each stage t, the associated nodes of $\overline{G}$ correspond to the various possible *feasible states* of the network, each feasible state being characterized by a M-vector $Z^j(t)$ with components $Z_u^j(t)$ ($u \in U$) where $Z_u^j(t)$ is the total capacity installed, up to the instant t, on link u. Moreover, the M-vector $Z^j(t)$ associated with a feasible state at time t should be such that the capacities $Z_u^j(t)$ on the links of the network are sufficient to flow all the single-commodity flow requirements $r^{ij}(t)$ *simultaneously*. We note here that the problem of deciding whether a given state vector $Z^j(t)$ is *feasible* with respect to the requirements $r^{ij}(t)$ amounts to looking for a *feasible multicommodity flow* for which various efficient exact or approximate algorithms have been described: see for instance Assad [2] Kennington [45], Gondran and Minoux [36] chapter 6. An efficient approximate algorithm for finding feasible integral multicommodity flows has been devised by Minoux [55].

Since only a finite number, S, of distinct capacity values can be used to augment capacity on each link, and since (according to the assumption made in §3), at each instant t, at most one such capacity value can be added, each component of $Z_u^j(t)$ can take on at most S^t distinct values. Thus there is a *finite number of possible states* (of possible nodes in $\bar{G}$) at each stage t (say, at most $M^{S^t} \leqslant M^{S^T}$).

We also consider in $\bar{G}$ a special node (called the origin) corresponding to the state $Y(0)$ of the network at time 0 (the beginning of the study period). (Remember that the capacity $Y_u(0)$ on each link u in the initial state of the network is supposed to be given).

The arcs of $\bar{G}$ are now defined in a natural way: only arcs between nodes corresponding to two consecutive stages $t-1$ and t are allowed; moreover there will be an arc $[Z^i(t-1), Z^j(t)]$ between two nodes $Z^i(t-1)$ and $Z^j(t)$, if and only if, for each component $u \in U$:

$$Z_u^j(t) - Z_u^i(t-1) \in \{0, Q(1), Q(2), \ldots, Q(S)\}$$

in which case, the cost of the arc will be taken equal to:

$$\sum_{u \in U} \frac{\Gamma_u(Z_u^j(t) - Z_u^i(t-1))}{(1+\mathscr{C})^{t-1}}$$

i.e. the total present worth of actual cost of all the investments performed at time $t-1$ on all the links.

Figure 3 provides a graphical representation of the sequential (dynamic programming) graph $\bar{G}$ (for $1 \leqslant t \leqslant T$, the number of states at stage t has been denoted by v_t). Note that in $\bar{G}$, not every node (state) is reachable by a path from the origin; on the other hand, that there are nodes which can be reached by several distinct paths.

Solving the minimum cost network expansion problem is then equivalent to looking for a shortest (minimum cost) path between the initial state $Y(0)$ (taken as the origin) and the set of nodes corresponding to the final stage T. It is important, for what follows, to realize that, solving the minimum cost path problem, simultaneously provides two important kinds of information: (a) an optimal terminal state which we denote by $Z^*(T)$ (b) a shortest (minimum cost) path between $Y(0)$ (the initial state) and $Z^*(T)$ (the optimal terminal state).

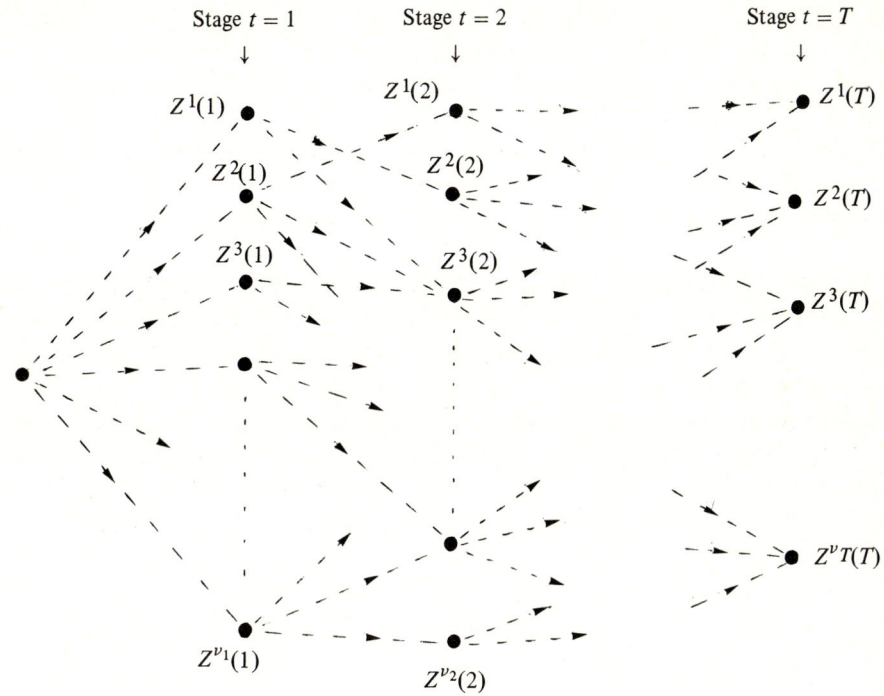

Fig. 3. The minimum cost network expansion problem viewed as a shortest path (minimum cost path) problem on a sequential graph $\overline{G}$.

5. The short-term-long-term decomposition approach

If the Dynamic Programming formulation has the interest of providing a new way of looking at the problem (on which we shall rely later on), it can by no means provide a practically applicable solution method for network optimization problems of realistic size. To be convinced of this, it is enough to say that in practical applications:

- M (the number of possible links) is usually well above 100;
- S (the number of possible distinct transmission systems) is commonly around 10 to 20;
- T (the number of time intervals in the time period considered) is in the range 5 - 10.

Taking only the minimum values of the above figures, and considering that M^{S^T} is a good estimate of the number of states in the dynamic programming formulation (number of nodes in $\overline{G}$) (¹), we obtain a number of the order of $100^{100\,000}$ which is very very far beyond the possibilities of resolution of Dynamic Programming techniques (and even of the most sophisticated ones, e.g.: combining Branch and Bound and Dynamic Programming as suggested in Weingartner and Ness [76], Morin and Marsten [70]; or the so-called state space relaxation techniques as in Christofides, Mingozzi and Toth [13, 14]).

In view of the highly combinatorial nature, and intrinsic difficulty of the problem, pointed out above, the search for exact solutions is likely to remain hopeless for many years, hence the need for devising approximate algorithms both computationally efficient, and providing solutions as close to optimality as possible.

Along this line, it turns out that the Dynamic Programming formulation of the network expansion problem — and this is one of its main interests in the present exposition — naturally leads to a good approximation scheme (already suggested in Minoux [55, 56]) based on a two-stage decomposition of the problem where the two following subproblems (P1) and (P2) are successively solved:

(P1) Find a good approximate final state $Z^f(T)$ $(1 \leqslant f \leqslant \nu_T)$ in the sense that the cost of a minimum cost path between $Y(0)$ and $Z^f(T)$ is close to the cost of the optimal path between $Y(0)$ and the (unknown) $Z^*(T)$;

(P2) Find a path between $Y(0)$ and $Z^f(T)$ which closely approximates the (unknown) minimum cost path between $Y(0)$ and $Z^f(T)$;

In economical terms, problem (P1) might be called the *long term planning subproblem* and problem (P2) the *short-to-medium term planning subproblem*. $Z^f(T)$ may be called the «*target network*». The basic ideas underlying the two--stage decomposition approach are that (i) owing to some simplifying (though reasonable) assumptions, problem (P1) can be reduced to a *static* (nonlinear) minimum cost multicommodity flow problem, significantly simpler than the dynamic multicommodity flow problem initially considered (see §6); (ii) once a good target solution has been determined, it becomes much easier to generate good approximate solutions to the dynamic network expansion problem.

To be a little more specific, the implementation of the two-stage decomposition approach, to be described later on, makes essential use of the following remarks:

(a) in solving the long term subproblem it is not very important to take account of all the details of how network expansion has to be performed, year by year. Only *trends* are important and this will lead to considering *rates of growth of the required traffic values* $r^{ij}(t)$ instead of the traffic requirement

---

(¹) One might object that a large proportion among all these M^{S^T} states would not be feasible. However, since there is no simple straightforward characterization of feasible state vectors among the whole set of a priori possible states, a *feasibility test* will have to be systematically applied to all the states examined, without knowing in advance whether they will be feasible or not. Thus a large number of unfeasible states will have to be taken into consideration too. It this sense, the value M^{S^T} can be considered as a good estimate of the state space cardinality of the problem.

values themselves;

(b) Once a good approximate final state $Z^f(T)$ has been determined according to the above principle, many variables of the problem become fixed; in particular we get the structure of the optimal (or suboptimal) expansion subnetwork, i.e., we know *on which links* capacity has to be added. Moreover, since we know the required growth rate of the flow values on each link of the expansion subnetwork, it is possible to *deduce the nature of the best possible transmission system* (among the S possible systems of the family) which should be installed on each link. Also good approximate dates for performing the various necessary capacity expansions on each link can easily be deduced.

(c) Once we know, for each link of the expansion subnetwork, the sequence of capacity expansions to be performed, and approximate dates of installation, problem (P2) is solved by considering each time instant $t \in [1, T]$ and trying to reduce the total investment cost necessary at instant t at a minimum, while maintaining feasibility (this leads to delaying some of the investments which were part of the long term plan). This can be achieved, as suggested in Minoux [55] or in Minoux [57] by means of a heuristic procedure analogous to the so--called *backward dynamic programming scheme*, beginning from the end of the time period ($t = T$) and ending at $t = 0$ (²). At each t, a number of trials are made for eliminating (i.e. delaying) the most expensive investments, feasibility being tested for by means of a procedure solving, (either exactly or approximately) the feasible multicommodity flow problem.

In the following sections, we will focus on the solution of problem (P1) (the long term planning subproblem) which essentially aims at determining an optimal (or suboptimal) stucture for the expansion subnetwork. For details on the solution of problem (P2) (the short-to-medium term planning subproblem) the reader is referred to Minoux [55, 56].

6. Determining a good approximation to the optimal final state (target network): a (static) minimum cost multicommodity flow model with nonlinear (concave) cost functions

We show in this section how, by means of some simplifying (though reasonable) assumptions, the problem of determining a good approximation to the optimal final state $Z^*(T)$ (see §5) can be reduced to a (static) minimum cost multicommodity flow problem with continuous (but nonlinear) cost functions.

The basic idea here is to repalce each function $r^{ij}(t)$ specifying the evolution

---

(²) Heuristic procedures based on backward dynamic programming usually lead to better results than those based on forward dynamic programming. See Minoux [56] for an example.

of the traffic flow requirement between i and j over time, by its *trends* which will be characterized by the *mean speed of growth over the period* $[0, T]$, denoted by σ^{ij} and defined, in a natural way, as:

$$\sigma^{ij} = \frac{r^{ij}(T) - r^{ij}(0)}{T}.$$

Though considering for each source-sink pair the linear function of time

$$\tilde{r}^{ij}(t) = r^{ij}(0) + \sigma^{ij} \cdot t$$

may not always very accurately account for the given values $r^{ij}(t)$, it will be assumed (assumption A1) that this adequately accounts for the *trends* of variation of the traffic flows over time. This assumption is justified by the practical observation that, since the $r^{ij}(t)$ values are usually obtained from forecasting models, which have a well-known *smoothing* effect, they rarely exhibit sharp and irregular variations.

The second assumption (A2) which will be made is that for every source-sink pair $(i - j)$, the routing of the flow φ^{ij} remains the same all over the time period $[0, T]$. In other words, if at time $t = 1$, we decide to send the flow on some $(i-j)$ path in the network, then it will go on following the same route at every other subsequent time instant.

It follows from assumptions (A1) and (A2) that we can replace the (dynamic) multicommodity flow problem of flowing the $r^{ij}(t)$ requirements at the various successive instants $t = 1, 2, \ldots, T$ by the *static* multicommodity flow problem of flowing the σ^{ij} values through the network; indeed, proceeding that way, we note that the speed of growth of the total flow through any link $u \in U$, θ_u, is equal to the *sum of the speeds of growth* σ^{ij} of all the $(i-j)$ flows running through link u. ($\Sigma(r^{ij}(0) + \sigma^{ij}t) = \Sigma r^{ij}(0) + (\Sigma \sigma^{ij})t = \Sigma r^{ij}(0) + \theta_u t$, where summation is taken over the subset of source-sink pairs corresponding to the flows using link u).

Now the problem which has to be solved is the following: given some (arbitrary) routing of the σ^{ij} flows on the network, resulting on each link $u \in U$ in a total speed of growth $\theta_u \geq 0$, determine the Present Worth of Actual cost of network expansion corresponding to this routing strategy.

In what follows, in order to simplify the discussion, we will assume that all the initial flow values $r^{ij}(0)$ are zero and that no capacity has been installed on the network yet ($Y(0) = 0$). This is the situation where the network has to be built from scratch. Of course, all that will be said easily generalizes to the (more frequent) situation where we start from an existing network with installed capacities $Y(0) \neq 0$ and nonzero flow requirements $r^{ij}(0)$.

For all $u \in U$, we thus have to determine the (minimum) PW of AC of expansion of link u in order to satisfy total flow requirement of the form:

$$\psi_u(t) = \theta_u \cdot t$$

when capacity expansions are performed using only transmission systems belonging to the family $(P(s) Q(s), s = 1 \ldots S)$ (see §3). This minimum PW of AC of expansion of link u when the speed of growth on link u is θ_u will be denoted by $\Phi_u(\theta_u)$. As we will see in section 7 below, the $\Phi_u(\theta_u)$ are nonlinear functions of θ_u which can be closely approximated by concave differentiable functions.

In view of this, and with the assumptions (A1) (A2), the problem of determining a minimum cost expansion network over the time period $[0, T]$ amounts to solving the following (nonlinear concave) static multicommodity flow problem:

(P1) $\begin{cases} \text{Min } \Phi(\theta) = \displaystyle\sum_{u \in U} \Phi_u(\theta_u) \\[4pt] \text{under the constraints:} \\[4pt] \theta_u = \displaystyle\sum_{i<j} |\varphi_u^{ij}| \\[4pt] \forall (i,j) \; \varphi^{ij} \text{ is an } (i-j) \text{ flow of value } v(\varphi^{ij}) = \sigma^{ij}. \end{cases}$

7. Determining the cost functions on the links in the static multicommodity flow model

The problem which we address here is to determine an optimum sequence of investments on some link where the total flow requirements increase linearly with time, $\theta > 0$ being the speed of increase, and assuming that we can choose only among a finite set of transmission sytems, characterized by capacity-cost pairs $Q(s), P(s)$ $(s = 1, \ldots, S)$.

In order to perform an optimal choice among the several possible available systems, we will introduce a simple rule based on the concept of *equivalent cost* (cf. Minoux and Guerard [66]).

Consider, first, the following simplified situation where we have to choose between two equipments E_1, E_2 of capacity Q_1 and Q_2 respectively, and total cost P_1 and P_2 respectively. Since it is desirable that the choice between E_1 and E_2 be as independent as possible from the actual duration T of the time period under consideration, the two solutions should be examined in comparable situations, namely on a sufficiently long time period T' (possibly longer than T) in order that no residual (installed but unused) capacity be left at the end of the time period (see Figure 4). Thus, since the fill-in times for E_1 and E_2 under a growth rate of θ units per year are

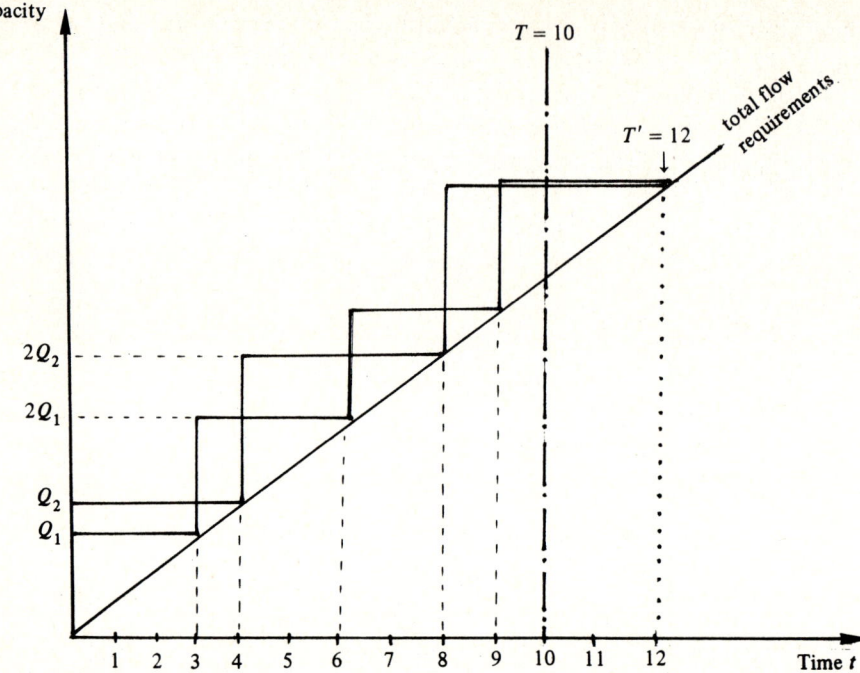

Fig. 4. Comparison of two transmission systems E_1 and E_2 for satisfying a total flow requirement increasing linearly with time (growth rate θ). Taking $T = 10$, neither of the two systems are saturated. $T' = 12$ is the earliest date at which saturation occurs for the two systems.

$$n_1 = \frac{Q_1}{\theta} \qquad n_2 = \frac{Q_2}{\theta}$$

respectively, we'll have to choose T' such that

$$T' = k_1 n_1 = k_2 n_2$$

with integer k_1, k_2.

Then the PW of AC on $[0, T']$ in case E_1 is chosen, is:

$$z_1(\theta, T') = P_1 + \frac{P_1}{(1+\mathscr{C})^{n_1}} + \frac{P_1}{(1+\mathscr{C})^{2n_1}} + \ldots + \frac{P_1}{(1+\mathscr{C})^{(k-1)n_1}}$$

which can be rewritten as:

$$z_1(\theta, T') = P_1 \frac{1 - (1+\mathscr{C})^{-T'}}{1 - (1+\mathscr{C})^{-n_1}}$$

or equivalently:

$$z_1(\theta, T') = \sum_{t=0}^{T'-1} \frac{\theta \gamma_1(\theta)}{(1+\mathscr{C})^t} \tag{5}$$

where

$$\gamma_1(\theta) = \frac{P_1}{\theta} \frac{1-(1+\mathscr{C})^{-1}}{1-(1+\mathscr{C})^{-Q_1/\theta}}.$$

Similarly, in case E_2 is chosen, the PW of AC would be:

$$z_2(\theta, T') = P_2 \frac{1-(1+\mathscr{C})^{-T'}}{1-(1+\mathscr{C})^{-n_2}}$$

which can be rewritten as:

$$z_2(\theta, T') = \sum_{t=0}^{T'-1} \frac{\theta \gamma_2(\theta)}{(1+\mathscr{C})^t} \tag{6}$$

where

$$\gamma_2(\theta) = \frac{P_2}{\theta} \frac{1-(1+\mathscr{C})^{-1}}{1-(1+\mathscr{C})^{-Q_2/\theta}}.$$

It is then seen from (5) and (6) that the comparison between E_1 and E_2 amounts to comparing the two quantities $\gamma_1(\theta)$ and $\gamma_2(\theta)$, which indeed, appear to be *independent from the duration T' on which the comparison is performed.*

Now, the above reasoning can be extended, in a straightforward manner, to the case where choice has to be performed among a family of S equipments of capacities $Q(s)$ ($s = 1, \ldots, S$) and costs $P(s)$ ($s = 1, \ldots, S$). Indeed, by carrying out all possible pairwise comparison, the best choice will correspond to the equipment s such that the quantity:

$$\gamma_s(\theta) = \frac{P(s)}{\theta} \frac{1-(1+\mathscr{C})^{-1}}{1-(1+\mathscr{C})^{-Q(s)/\theta}} \tag{7}$$

is a minimum i.e.:

$$\gamma_s(\theta) = \underset{r=1,2,\ldots,S}{\mathrm{Min}} \{\gamma_r(\theta)\}.$$

We will call $\gamma_s(\theta)$ given in (7) the *equivalent cost* of equipment s under growth rate θ (see Minoux and Guerard [66]). Note that, extending (5) and (6) to any s ($1 \leq s \leq S$) we have the following relation between the equivalent cost and the PW of AC over $[0, T']$:

$$z_s(\theta, T') = \sum_{t=0}^{T'-1} \frac{\theta \gamma_s(\theta)}{(1+\mathscr{C})^t} \tag{8}$$

for any T' which is a multiple of $Q(s)/\theta$ (the fill-in time).

Equation (8) readily suggests another interesting feature of the equivalent cost, which is to allow extending the definition of the PW of AC function $z_s(\theta, T')$ to all values of T', even to those values *which are not multiples of $Q(s)/\theta$* (thus resulting in interpolating and smoothing the discrete set of values $z_s(\theta, Q(s)/\theta) \, z_s(\theta, 2Q(s)/\theta) \ldots z_s(\theta, kQ(s)/\theta) \ldots$). It can be seen that this way of approximating the PW of AC amounts to deduce from the total PW of AC of the equipments *actually installed*, a certain amount accounting for the *residual capacities* existing at the end of the study period. More specifically, if we use the formula

$$z_s(\theta, T) = \sum_{t=0}^{T-1} \frac{\theta \gamma_s(\theta)}{(1+\widetilde{c})^t} \tag{9}$$

for a value of T which is not a multiple of $Q(s)/\theta$, i.e. such that $(k-1)Q(s)/\theta < T$ and $T < kQ(s)/\theta$, this corresponds to implicitly assigning to the residual (unused) capacity $kQ(s) - \theta \cdot T$ at time T a cost equal to:

$$\sum_{t=T}^{T'-1} \frac{\theta \gamma_s(\theta)}{(1+\widetilde{c})^t} \quad \text{where} \quad T' = \frac{kQ(s)}{\theta}.$$

In view of this, we will use (9) for approximating for each value θ of the growth rate, and for fixed T, the PW of AC of a link when system s $(1 \leq s \leq S)$ is being used for performing the capacity expansion of the link.

Since

$$\sum_{t=0}^{T-1} \frac{1}{(1+\widetilde{c})^t} = \frac{1-(1+\widetilde{c})^{-T}}{1-(1+\widetilde{c})^{-1}}$$

(9) can be rewritten as

$$z_s(\theta, T) = P(s) \frac{1-(1+\widetilde{c})^{-T}}{1-(1+\widetilde{c})^{-Q(s)/\theta}} \tag{10}$$

T being a fixed parameter, we study the family of functions $z_s(\theta, T)$ defined by (10). To each system s $(1 \leq s \leq S)$ corresponds one such cost function which, as a function of θ, has the following characteristic:
- for $\theta = 0, z_s(0, T) = P(s)[1-(1+\widetilde{c})^{-T}]$
- for $\theta \to \infty, z_s(\theta, T)$ has an asymptote which is the linear function

$$P(s) \cdot \frac{1-(1+\widetilde{c})^{-T}}{Q(s) \cdot \widetilde{c}} \theta = \frac{z_s(0, T)}{Q(s) \widetilde{c}} \theta.$$

In other words, there is a *fixed cost* at the origin, which is equal to $P(s)[1-(1+\widetilde{c})^{-T}]$ and a *linear asymptotic cost* equal to $1/Q(s) \, \widetilde{c}$ times the fixed cost.

To illustrate this, consider the family of systems ($s = 1, 2, \ldots, 5$) introduced in §3. For $T = 10$ and an actualization rate $\tilde{c} = 0.1$, we obtain the following values of $z_s(0, T)$ and $z_s(0, T)/Q(s)\,\tilde{c}$:

System number	$P(s)$	$Q(s)$	Fixed cost at $\theta = 0$ $z_s(0, T) =$ $P(s)\,[1 - (1 + \tilde{c})^{-T}]$	Asymptotic linear cost $\dfrac{z_s(0, T)}{Q(s)\tilde{c}}$
$s = 1$	4.5	500	2.76	5.5 10^{-2}
$s = 2$	7	1000	4.3	4.3 10^{-2}
$s = 3$	10	2000	6.1	3 10^{-2}
$s = 4$	13	3000	8	2.7 10^{-2}
$s = 5$	27	10000	16.6	1.7 10^{-2}

The family of corresponding curves is displayed on *figure 5* where the bold line indicates the lower envelope of the family, i.e. the graph of the function

$$\Phi(\theta) = \min_{s = 1, \ldots, S} \{z_s(\theta, T)\} \tag{11}$$

which, for any value of θ, represents the minimum PW of AC for meeting the flow requirements growing at rate θ (number of channels per year).

A good approximation to the function $\Phi(\theta)$ above can be obtained as follows. As pointed out in §3, the dependence between the costs $P(s)$ and the capacities $Q(s)$ of the family can be approximated by a function of the form $P \simeq 0.11\,(Q)^{0.6}$.

The idea is to assume that we have precisely a *continuous* (infinite) *family of systems*, with capacity-cost pairs related by a formula of the type

$$P = \mu \cdot (Q)^\alpha \qquad (0 < \alpha < 1) \tag{12}$$

From (10) it follows that for given θ and for the system of capacity Q the PW of AC is equal to:

$$z(\theta, T) = \mu(Q)^\alpha \frac{1 - (1 + \tilde{c})^{-T}}{1 - (1 + \tilde{c})^{-Q/\theta}} \tag{13}$$

Now, for each θ, the best system to use is the one achieving the minimum with respect to Q in (13). The condition $\partial z(\theta, T)/\partial Q = 0$ reads:

$$\frac{\alpha(Q)^{\alpha-1}}{1 - (1 + \tilde{c})^{-Q/\theta}} - \frac{(Q)^\alpha \dfrac{1}{\theta} [\log(1 + \tilde{c})][(1 + \tilde{c})^{-Q/\theta}]}{[1 - (1 + \tilde{c})^{-Q/\theta}]^2} = 0$$

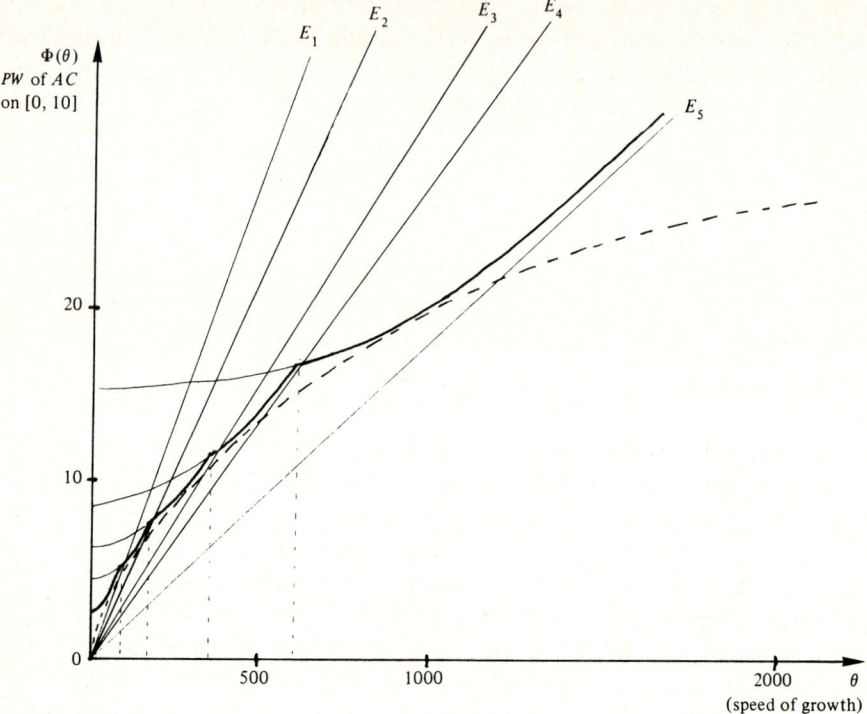

Fig. 5. Building the function $\Phi(\theta)$ giving the minimum PW of AC of expansion corresponding to growth rate θ on a link, as the lower envelope of the family of curves $z_s(\theta, T)$ (here $T = 10$ and $\tilde{\mathscr{C}} = 0.1$).

which yields:

$$(1 + \tilde{\mathscr{C}})^{Q/\theta} = 1 + \frac{Q \log(1 + \tilde{\mathscr{C}})}{\alpha \theta} \qquad (14)$$

Relation (14) shows that, in this case, the optimal fill-in time Q/θ is a constant n^* defined as the solution of equation

$$(1 + \tilde{\mathscr{C}})^n = 1 + n \frac{\log(1 + \tilde{\mathscr{C}})}{\alpha},$$

With the values considered above ($\alpha = 0.6$ and $\tilde{\mathscr{C}} = 0.1$) we obtain: $n^* = 10$.

In view of this, the optimum PW of AC, $z^*(\theta, T)$, on the entire family, is then obtained by substituting n^* to Q/θ and $n^*\theta$ to Q in (13) and we obtain

$$z^*(\theta, T) = \mu(n^*\theta)^\alpha \frac{1 - (1 + \tilde{\mathscr{C}})^{-T}}{1 - (1 + \tilde{\mathscr{C}})^{-n^*}} \qquad (15)$$

This shows that, in this case, the optimum PW of AC function $z^*(\theta, T)$ assumes the same form as the cost versus capacity relation (12). To see that this is indeed a good approximation to $\Phi(\theta)$ we have plotted on figure 5 (in dotted lines) the function corresponding to the values $n^* = T = 10$, $\mu = 0.11$ and $\alpha = 0.6$, i.e.: $z^*(\theta, T) = 0.11 \times 0.614 \times 3.98 \ (\theta)^{0.6} = 0.268 \ (\theta)^{0.6}$; we get for instance

for $\theta = 200$ $z^* = 6.4$

$\theta = 500$ $z^* = 11.15$

$\theta = 1000$ $z^* = 16.9$

$\theta = 2000$ $z^* = 25.6$.

As already pointed out in §3, in most practical applications, the families of available equipments follow relations of type (12), hence it is reasonable to use the approximation (15) of the actual curve $\Phi(\theta)$ defined by (11). Referring back to the network expansion problem, since all the above reasoning can be applied to each link of the network, this leads to the reasonable assumption that, on each link $u \in U$, the cost function $\Phi_u(\theta_u)$ is a *concave* cost function of the parameter θ_u (the growth rate on link u). The following section will now be devoted to solving the minimum cost multicommodity flow problem (P1) where the cost functions $\Phi_u(\theta_u)$ on the links are *concave* and differentiable.

Remark. It should be noted, however, that an interesting feature of the solution methods which will be described below is that they can be applied to the exact cost function $\Phi(\theta)$ defined by (11) (which may not be exactly *concave* nor differentiable as shown on figure 5), thus providing good heuristics for a model which may be closer to reality.

8. Solving the minimum concave cost multicommodity flow problem

Assuming that, on each link $u \in U$ of the network, the PW of AC of link expansion is (or can be closely approximated by) a nonlinear concave differentiable function Φ_u of θ_u (the speed of growth of flow requirements on link u) problem (P1) (the long term subproblem) is a minimum concave cost (static) multicommodity flow problem which, as already indicated in §6, can be stated as:

$$(P1) \begin{cases} \text{Min } \Phi(\theta) = \sum_{u \in U} \Phi_u(\theta_u) \\ \text{subject to:} \\ \theta_u = \sum_{i<j} |\varphi_u^{ij}| \quad (16) \\ \forall (i,j) \; \varphi^{ij} \text{ is an } (i-j) \text{ flow of value } v(\varphi^{ij}) = \sigma^{ij} \quad (17) \end{cases}$$

(remember that $\forall (i,j) \; \sigma^{ij}$ denotes the rate of increase of the flow requirements between i and j, see §6).

In order to illustrate the various procedures which will be described, or to which we will refer, we will consider the example shown on *figure 6*. Figure *6a* indicates the set of all possible links (where extra-capacity can be added); Figure *6b* provides the speeds of growth of the various $(i-j)$ flow requirements (expressed for instance in number of additional channels per year); Figure *6c* shows the speeds of traffic growth on the various links of the network when each flow $(i-j)$ is routed on the link (i,j) (direct route). We will take as cost function on each link of the network a function derived from the approximation $\Phi(\theta) = 0.268 \, (\theta)^{0.6}$ obtained in section 7 and which corresponds to the family of transmission systems considered in §3 for a 100 km link. Assuming, to simplify, that cost is simply proportional to link length, we easily deduce the cost function on each link u of the network of figure 6 by:

$\Phi_u(\theta_u) = 0.268 \, l_u/100 \, (\theta_u)^{0.6}$ (where l_u is the length in kilometers). For instance, computing the total cost of the solution of Figure 6c (where each flow follows the direct route) we get:

cost of link (1,2) : 91.9
cost of link (1,3) : 79.2
cost of link (1,4) : 41.8
cost of link (2,3) : 84.1
cost of link (2,4) : 111.2
cost of link (3,4) : 95.1
and total cost is : 503.3.

Of course, this is not an optimal solution to the minimum (concave) cost multicommodity flow problem, and we now proceed to show how it can be improved while describing various solution methods applicable to the problem. We first begin by mentioning an important property of optimal solutions which is derived from a general well-known result about concave minimization problems over polyhedra, namely that: in case of a unique optimal solution (nondegeneracy), the optimal solution *always lies at an extreme point of the polyhedron*

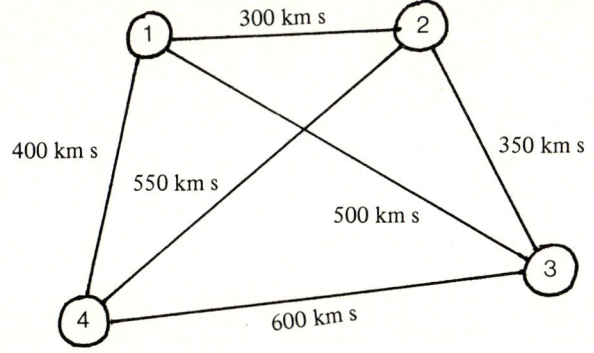

6a. The set of all possible links where capacity can be added and their lengths (kms) (Complete undirected graph on 4 vertices)

	1	2	3	4
1		300	100	50
2			200	150
3				100
4				

6b. Speeds of growth of the various $(i-j)$ flow requirements (number of additional channels per year).

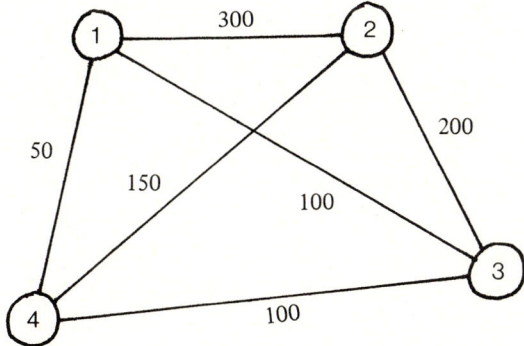

6c. The speeds on the links of the network when each flow $(i-j)$ is routed on the link (i,j) (direct route)

Total cost of this solution: 503.3

Fig. 6. An example of a 4-node network.

(in the degenerate case, there is always an optimal solution which is at an extreme point). For the problem considered here, it turns out that the extreme points of the multicommodity flow polyhedron (defined by conditions (16) and (17) in (P1) above) can be easily characterized as solutions obtained by routing each flow $(i-j)$ on a single route (chain) between i and j in the network (see e.g. Minoux [58]). From this we deduce that (assuming non degeneracy) an optimal solution to the minimum concave cost multicommodity flow problem is such that each flow is routed along a single route (moreover, in case of degeneracy, global optimality is not lost by restricting to such single-routing solutions).

It can also be shown (cf. e.g. Yaged [77]) that, for the minimum concave cost multicommodity network flow problem, the Kühn-Tucker necessary conditions for (local) optimality take on a very simple form. Let $\bar{\theta} = (\bar{\theta}_u)_{u \in U}$ be an extreme point of the multicommodity flow polyhedron (16)-(17) obtained by sending each $(i-j)$ flow along a single chain, say L_{ij}. With each link $u \in U$ we associate the number (marginal cost):

$$\lambda_u = \frac{d\Phi_u}{d\theta_u}(\bar{\theta}_u) = 0.16 \frac{l_u}{100} (\bar{\theta}_u)^{-0.4}.$$

Then, *for $\bar{\theta}$ to be a (local) optimum to the problem, it is necessary that for each pair (i, j), L_{ij} be a shortest $i-j$ chain with respect to the values λ_u on the links.*

As shown by Yaged [77], when a solution $\bar{\theta}$ fails to meet the Kühn-Tucker conditions, then it is possible to generate a new solution $\bar{\theta}'$ improving strictly the cost criterion by carrying out the following operations:

(a) determine the marginal costs λ_u associated with the links $u \in U$;
(b) $\bar{\theta}'$ is the (extreme point) solution obtained by sending each $(i-j)$ flow on the shortest $i-j$ chain with respect to the lengths λ_u on the links (minimum marginal cost chain).

If we denote by T the transformation which maps $\bar{\theta}$ onto $\bar{\theta}'$ ($\bar{\theta}' = T(\bar{\theta})$) then a (local) optimum solution is a *fixed point* of the transformation T above. Thus Yaged's procedure can be viewed as a fixed point computation where, starting from an initial solution θ^0, a sequence of solutions θ^k with strictly decreasing total cost is generated according to $\theta^k = T(\theta^{k-1})$, until a fixed point (a local optimum solution) is reached.

This fixed point computation is illustrated on *figure 7* taking as starting solution the one shown on *figure 6c*. Routing each of the flows on a shortest (minimum marginal cost) chain leads to the new solution shown on *figure 7b*, the total cost of which is: 464.5. This solution satisfies the Kühn-Tucker conditions as is easily verified by one more application of the procedure. The main advantages of the fixed-point approach are: ease of implementation (it reduces to shortest path computations); computational efficiency (even for large scale problems, the number of necessary iterations for reaching a local optimum almost never exceeds 8 - 10).

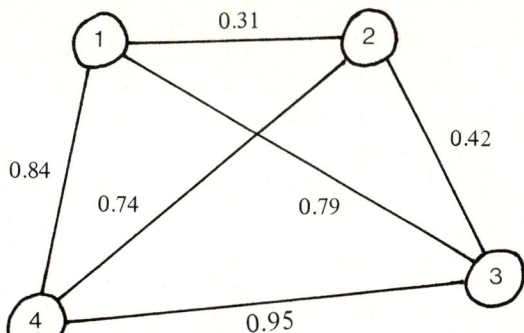

7a. The marginal costs on the links for the solution of Figure 6c

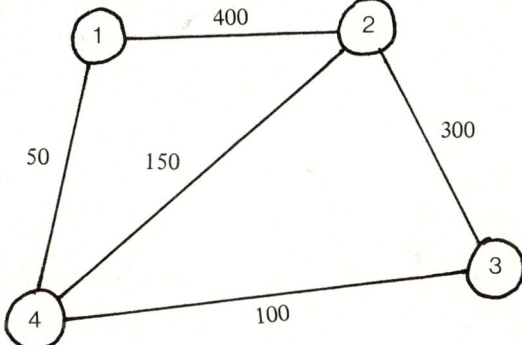

7b. The new solution obtained by routing each $(i-j)$ flow along the shortest chain according to the criterion of marginal cost
Total cost of this solution: 464.5

Fig. 7. Testing for (local) optimality with the Kühn-Tucker conditions and improving a solution. The solution obtained on figure 7b meets the Kühn-Tucker conditions, hence is a local optimum.

However, it suffers important drawbacks due to the fact that (as is more generally the case for concave minimization problems over polyhedra), the number of local optima is usually enormous [3] and in such a context, the Fixed Point Algorithm systematically tends to produce a local minimum which is (just as in our example) quite close to the starting point: in other words, the solution obtained will be good if the starting solution already was, but getting a starting solution already close to (global) optimality is nearly as difficult as solving the problem! This pathology is illustrated on the example of figures 6 and 7, where

---

[3] Indeed it is easy to generate problems for which each extreme point of the polyhedron is a local minimum.

the Fixed Point Approach converges in only 1 iteration but leads to only $38.8/464 \simeq 8\%$ improvement upon the starting solution, whereas $107.1/396 \simeq$ $\simeq 27\%$ improvement would be necessary to get the global optimal solution (of cost 396.1, see figure 8). Other examples of this pathology will be found in Minoux [65].

To get rid of this pathology, Minoux [58] proposed a greedy-type algorithm based on a much more restrictive necessary condition for optimality which can be stated as follows.

Let $\bar{\theta} = (\bar{\theta}_u)$ be an extreme point solution, and for any link $v = (k, l)$ such that $\bar{\theta}_v > 0$, assign to the links of the network the following lengths:

$$\begin{cases} \gamma_u = \Phi_u(\bar{\theta}_u + \bar{\theta}_v) - \Phi_u(\bar{\theta}_u) & u \neq v \\ \gamma_v = +\infty \end{cases}$$

(k, l) being the endpoints of v, let $L(\bar{\theta}, v)$ be the length of the shortest $k-l$ chain with respect to the values γ_u on the links. Then:

Theorem 1. (Minoux [58]). *A necessary condition for $\bar{\theta}$ to be a (global) optimum solution is that, for each $v \in U$ such that $\bar{\theta}_v > 0$:*

$$L(\bar{\theta}, v) - \Phi_v(\bar{\theta}_v) \geq 0. \tag{18}$$

Proof. If $L(\bar{\theta}, v) - \Phi_v(\bar{\theta}_v) < 0$ for some link $v = (k, l)$ such that $\bar{\theta}_v > 0$, then by rerouting all the flow passing through v on the shortest $k-l$ chain (w.r.t. the γ_u) then one obtains a better solution. QED. □

Checking the condition of *theorem 1* is computationally effective since it requires at most $O(M)$ shortest path computations. Moreover, it is seen that the proof given above provides a simple constructive way of improving a solution which does not satisfy the necessary condition for optimality. This lead

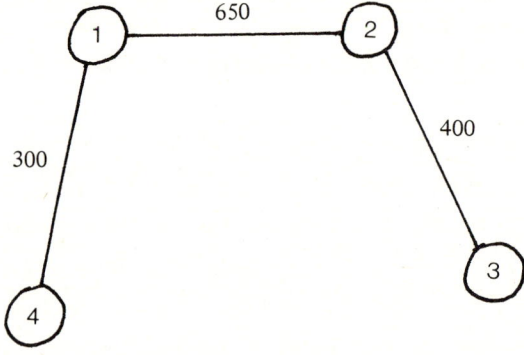

Figure 8. Final solution produced by the greedy algorithm on the example of figure 6 (cost 396.1). This is actually the global optimum to the problem of figure 6.

Minoux [58] to suggest the following greedy-type algorithm:

Greedy algorithm for the minimum concave cost multicommodity flow problem

(a) Let θ^0 be the initial (starting) solution; set $k \leftarrow 0$.

(b) At step k, let θ^k be the current solution. For every $u \in U$ such that $\theta_u^k > 0$ compute:

$$\Delta^k(u) = L(\theta^k, u) - \Phi_u(\theta_u^k)$$

(c) Select $v \in U$ such that

$$\Delta^k(v) = \min_{u/\theta_u^k > 0} \{\Delta^k(u)\}.$$

If $\Delta^k(v) \geq 0$ STOP: the current solution θ^k satisfies the necessary optimality condition of theorem 1;

Otherwise:

(d) Let (k, l) be the endpoints of v, and let $\wedge_{kl}$ be the shortest $k - l$ chain obtained at step (b). Then generate a new solution θ^{k+1} by rerouting all the flow through v on the chain $\wedge_{kl}$ i.e. set:

$$\theta_u^{k+1} \leftarrow \theta_u^k \qquad \forall u \neq v, \ u \notin \wedge_{kl}$$

$$\theta_u^{k+1} \leftarrow \theta_u^k + \theta_v^k \qquad \forall u \neq v, \ u \in \wedge_{kl}$$

$$\theta_v^{k+1} \leftarrow 0.$$

Set $k \leftarrow k + 1$ and return to (b).

Computational experiments show that the above greedy algorithm provides approximate solutions which are almost always very close to optimality if not optimal (out of 60 test problems treated in Minoux [58], exact optimal solutions were obtained on about half of them; moreover, in case of nonoptimal solutions, the differences in cost with the exact optimal solutions were quite small, about 1 or 2%).

As an illustration, the greedy algorithm will be applied to the example of *figure 6* starting with the solution of figure 6c. At the first step, computing the $\Delta^0(u)$ values leads to the following results:

	(1,2)	(1,3)	(1,4)	(2,3)	(2,4)	(3,4)
Δ^0	+ 72.5	− 38.8	+ 6.3	+ 22.9	− 31.6	− 32.2
		↑				

The best improvement corresponds to deleting link (1,3) (marked with an arrow above). The resulting solution θ^1 is obtained by rerouting the total flow (100) through (1,3) on links (1,2) and (2,3) and is the same as the one shown on *Figure 7b* (cost 464.5). Computing the values Δ^1 corresponding to this solution leads to:

	(1,2)	(1,3)	(1,4)	(2,3)	(2,4)	(3,4)
Δ^1	+ 136.5	✕	− 12.9	+ 118.9	− 33.9	− 35.1 ↑

The best improvement now corresponds to deleting link (3,4) and the new solution θ^2 is obtained by rerouting the total flow (100) through (3,4) on links (4,2) and (2,3). The corresponding cost is: 429.4. The values Δ^2 corresponding to this solution are then

	(1,2)	(1,3)	(1,4)	(2,3)	(2,4)	(3,4)
Δ^2	+ 122.3	✕	− 16.3	+ ∞	− 33.2 ↑	✕

Now the best improvement is obtained by deleting link (2,4) which leads to the solution shown on *figure 8* of cost 396.1 and the algorithm stops with a solution satisfying the necessary condition of theorem 1 (no link can now be deleted without disconnecting the network so all the Δ values would be $+\infty$). (For this example, it can be checked — by simply trying all possible solutions — that this is indeed the global optimal solution to the problem of figure 6). In spite of the fact that the greedy algorithm above tends to produce high quality results (i.e. very close to global optimality) it may be quite time-consuming on medium-to-large scale network problems (say $N \geqslant 50$ to fix ideas). Indeed, each step of the algorithm requires $O(M)$ shortest path calculations (as many as there are links carrying strictly positive flow) and as many as $O(M)$ steps may be necessary, hence an $O(M^2 N^2)$ complexity (in case where the starting solution has the complete undirected graph as its *support*, i.e. all links of the complete graph carry strictly positive flow, then $M = N(N-1)/2$ and the complexity is $O(N^6)$). However, looking at things carefully, it can be realized that the transformations performed at each step are *local* in the sense that they only affect the link which is deleted and a few neighbouring nodes and links, and the rest of the network (the major part) remains unaffected. As a consequence, many of the $\Delta^k(v)$ values remain the same (i.e. shouldn't be recomputed) from one step to the next. Minoux [58] suggested a way of taking account of this phenomenon which dramatically improves the efficiency of the basic greedy algorithm, and which has been referred to as the *accelerated greedy algorithm*. Indeed, as shown in Minoux [59, 60, 62] this acceleration technique is quite widely applicable and can be used, more generally, for solving large scale combinatorial problems involving the *minimization of a supermodular set function*. In fact the supermodularity property is equivalent to assuming that (for fixed v) the values $\Delta^k(v)$ are nondecreasing with the iteration number k (monotonicity property) and, in that case, it can be proved that the accelerated greedy algorithm produces the same solution as the basic greedy algorithm (4). However it does the job much more efficiently: extensive computational experiments carried out on the minimum concave cost multicommodity flow problem showed that, *on the average, only 2 or 3 Δ values need be recomputed at each step*, resulting in an average speeding-up factor of about $M/3$ (thus the speeding-up factor itself increases linearly with the number of links in the initial network!). This allowed getting very good approximate solutions to fairly large scale problems (networks with more than 200 nodes and several hundreds of links in the starting solution) within very reduced computing times (2 - 3 minutes) on medium-power computer systems.

9. A new model and solution method for the dynamic network optimization problem

We introduce here a new model and solution method for the dynamic network optimization problem, which extend the whole approach described so far in a number of ways. First, the new model is more general than the one presented in §6 and basically differs from it by the fact that the traffic requirements $r^{ij}(t)$ are no longer approximated by the linear function of time

$$\tilde{r}^{ij}(t) = r^{ij}(0) + \sigma^{ij} \cdot t \quad \left(\text{where } \sigma^{ij} = \frac{r^{ij}(T) - r^{ij}(0)}{T} \right).$$

Contrasting with that, the new model will make use of the very data at hand, namely, for each pair (i,j), the sequence of $T+1$ values $r^{ij}(0), r^{ij}(1), \ldots, r^{ij}(T)$ without any need for assuming regularity in the evolution (for instance, decrease in the traffic requirement may occur from one time instant to the next). Thus instead of routing (static) flows representing speeds of growth of the traffic requirements, on the network, we will consider routing *multidimensional $(T+1)$-component flows* each component t $(0 \leq t \leq T)$ of a given $(i-j)$ flow representing the traffic requirement at time t for that flow. Accordingly, the total flow through

---

(4) It can be proved (cf. Minoux [59, 60]) that the monotonicity property holds for the minimum cost multicommodity flow problem when the cost functions are *linear* with a *fixed cost*; but it doesn't necessarily hold in the case of arbitrary concave cost functions. However, experience on medium-sized problems showed that the solutions obtained by the two algorithms (the basic one and the accelerated one) were systematically the same, thus suggesting that the cost functions should be in a sense *very close to supermodular functions*.

each link u of the network, resulting from some routing strategy, will be represented by a $(T+1)$-dimensional vector denoted (as in §3) by $\psi_u(t)$ ($t = 0, 1, \ldots, T$). Computing the corresponding expansion cost of the network then requires defining, for each link $u \in U$, a cost function in $(T+1)$ variables say:

$$\tilde{\Phi}_u(x_0, x_1, \ldots, x_T)$$

giving the (minimum) cost for link expansion over the considered time period $[0, T]$ when the total flow requirements $x_0, x_1, \ldots, x_T$ ($x_t \geq 0, \forall t$) have to be met at each instant of the period. If the functions $\tilde{\Phi}_u$ are available, then the cost corresponding to any routing strategy where, $\forall u \in U$, the total flows through the links are $\psi_u(t)$ ($t = 0, \ldots, T$) can be easily evaluated by

$$\sum_{u \in U} \tilde{\Phi}_u(\psi_u(0), \psi_u(1), \ldots, \psi_u(T)).$$

It can easily be shown that determining *one function value* $\tilde{\Phi}_u(x_0, \ldots, x_T)$ for any given sequence of requirements $(x_0, x_1, \ldots, x_T)$ on link u can be efficiently done by *dynamic programming*.

Zadeh [80], even showed how to convert the problem into a standard shortest path problem solvable via Dijkstra's algorithm. As an example of how such a computation can be carried out, consider a typical (100 km long) link on which capacity expansion can be obtained by installing any one of the 3 following systems (taken out the 5 described in §3).

	capacity	total cost (for 100 km link)
$s = 1$	$Q(1) = 500$	$P(1) = 4.5$
$s = 2$	$Q(2) = 1000$	$P(2) = 7$
$s = 3$	$Q(3) = 2000$	$P(3) = 10.$

The requirements to be met over the time period $[0,10]$ are given in the following table:

Time instant	0	1	2	3	4	5	6	7	8	9	10
requirement	0	600	1000	1800	1500	2000	2700	3100	3400	3700	3800

and are plotted on *figure 9* (here, $x_1 = 600$ means that the total requirement at the end of the first time interval $[0,1]$ will be equal to 600).

The problem of minimizing the PW of AC of link expansion for satisfying these

Network synthesis and dynamic network optimization

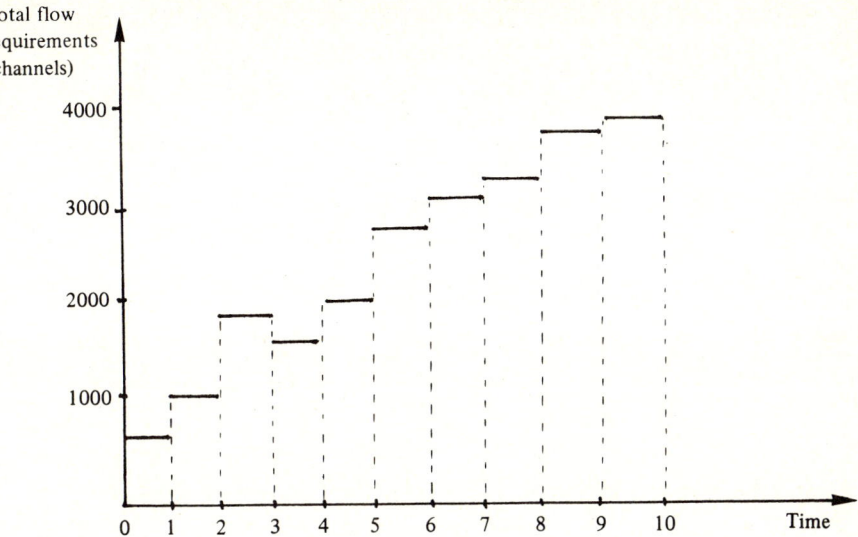

9a. The requirements $x_0, x_1, \ldots, x_{10}$ over $[0,10]$.

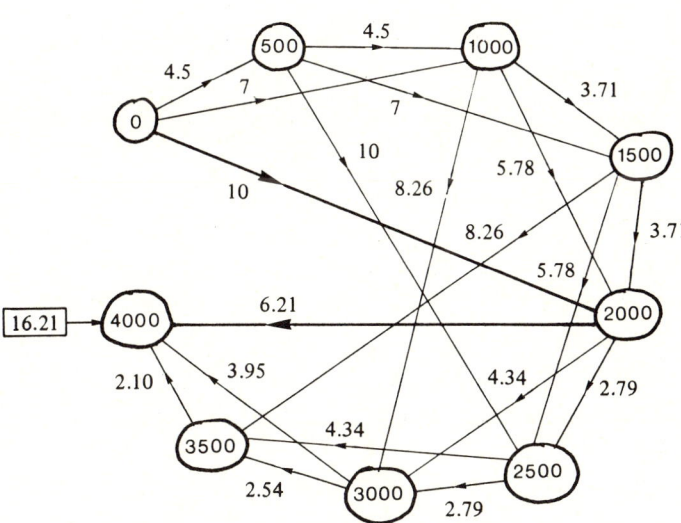

9b. The graph representing all capacity expansion possibilities. The optimum expansion cost corresponds to a shortest path between node 0 and node 4000.

Fig. 9. Determining the optimum PW of expansion cost $\tilde{\Phi}(x_0, \ldots, x_{10})$ and its formulation as a shortest path problem in a circuitless graph.

requirements reduces to a shortest path computation in the (circuitless) graph of *figure 9b* where:
— the nodes represent all the possible values of total capacity obtained by combining systems 1, 2 and 3, up to the maximum (4000);
— there is an arc between two nodes associated with capacity values c_a c_b ($c_b > c_a$) when $c_b - c_a = Q(s)$ for some s, and the cost associated with the arc is $P(s)/(1+\overline{\mathcal{C}})^{t_a}$ where t_a is the first time instant at which extra capacity is needed. For instance there is an arc between (1000) and (2000) of cost $7/(1 + \overline{\mathcal{C}})^2$ because a total capacity 1000 meets the requirements on [0,2] but not on a longer time period.

On the example of *figure 9* the minimum cost path between 0 and 4000 corresponds to a PW of AC equal to 16.21, thus

$$\tilde{\Phi}(x_0, x_1, \ldots, x_{10}) = 16.21.$$

It is seen that, even for larger families of systems, this computation of each function value of the cost functions $\tilde{\Phi}_u$ can be made very efficiently.

Now, in this new context, both the greedy algorithm and accelerated greedy algorithm described in §8 enjoy the interesting property that they only require evaluating the cost function $\Phi_u(\theta_u)$ for some definite values of the parameters θ_u; contrary to other methods (such as Yaged's fixed point algorithm), they do not require the cost functions to be differentiable, nor do they assume that the cost functions take on a simple analytical form allowing computation of derivatives.

In view of that, and referring to §6 above, it is then realized that the only assumption actually needed is assumption (A2) stating that for every source--sink pair $(i-j)$ the routing of the flow φ^{ij} should be kept the same all over the time period $[0, T]$. Owing to it, the basic greedy algorithm of §8 can be readily extended to multidimensional $(T+1)$-component flows, according to the following procedure:

Extended Greedy algorithm for the minimum cost multidimensional multicommodity network flow problem

(a) Let $[\psi^0] = [\psi^0(0), \psi^0(1), \ldots, \psi^0(T)]$ be a starting solution (obtained by routing each individual $(T+1)$-component flow requirement $[r^{ij}(0), r^{ij}(1), \ldots, r^{ij}(T)]$ on a single $(i-j)$ chain in the network). Set $k \leftarrow 0$.

(b) At step k, let $[\psi^k] = [\psi^k(0), \psi^k(1), \ldots, \psi^k(T)]$ be the current solution. For every $v = (i, j) \in U$ such that $\psi_v^k(t) > 0$ for some $t \in [0, T]$ compute:

$$\Delta^k(v) = L([\psi^k], v) - \tilde{\Phi}_v(\psi_v^k(0), \psi_v^k(1), \ldots, \psi_v^k(T)) \tag{19}$$

where $L([\psi^k], v)$ is the length of the shortest $i-j$ chain on the network, where link v is assigned a length $\gamma_v = +\infty$ and links $u \neq v$ are assigned lengths

$$\gamma_u = \tilde{\Phi}_u(\psi_u^k(0) + \psi_v^k(0); \psi_u^k(1) + \psi_v^k(1); \ldots \psi_u^k(T) + \psi_v^k(T)) \\ - \tilde{\Phi}_u(\psi_u^k(0), \psi_u^k(1), \ldots, \psi_u^k(T)). \tag{20}$$

(c) Select $v \in U$ such that:

$$\Delta^k(v) = \underset{u}{\text{Min}}\{\Delta^k(u)\}$$

(where the minimum above is taken over those u such that $\psi_u^k(t) > 0$ for some t).

If $\Delta^k(v) \geq 0$ STOP: the current solution $[\psi^k]$ is a (local) optimum. Otherwise:

(d) Let (i, j) be the endpoints of v (selected at (c)) and let $\wedge_{ij}$ be the shortest $i - j$ chain obtained in (b). Then generate a new solution ψ^{k+1} by rerouting all the flows through v on the chain $\wedge_{ij}$, i.e. set:

$$\psi_u^{k+1}(t) \leftarrow \psi_u^k(t) \quad \forall u \neq v \; u \notin \wedge_{ij}, \; \forall t \in [0, T]$$

$$\psi_u^{k+1}(t) \leftarrow \psi_u^k(t) + \psi_v^k(t) \quad \forall u \neq v, u \in \wedge_{ij}, \; \forall t \in [0, T]$$

$$\psi_v^{k+1}(t) \leftarrow 0 \quad \forall t \in [0, T]$$

Set $k \leftarrow k + 1$ and return to (b).

In the above procedure, each time a cost function $\tilde{\Phi}_u$ needs be evaluated (as in expressions (19) and (20)) a dynamic programming procedure (or the equivalent shortest path procedure, such as the one described earlier in this section) is applied. Thus, only function values are computed at specified points, without having to assume any particular analytical form. It should also be observed that, just as was the case of the basic greedy algorithm in §8, the above procedure considers only routing strategies in which *each flow is sent along a single route* in the network.

Though no formal proof can be given that optimality is not lost by restricting to such a routing strategy in the case of the multidimensional multicommodity flow model, this can be justified by referring to the approximate model presented in sections 5 to 8 (which lead to concave cost minimization over the multicommodity flow polyhedron) for which the single route property of optimal solutions could be proved quite rigorously. Due to the fact that the new approach presented here, *essentially aims at solving a refined version of the same problem*, it seems reasonable to keep on assuming the single route property for optimal solutions.

Concerning computational efficiency, the need for reducing the number of cost function evaluations appears to be crucial here, since each of these requires a shortest path calculation. Thus, solving medium to large scale network problems requires incorporating into the greedy algorithm above the acceleration techniques mentionned in §8. Though, here again, no formal proof can be given that the *monotonicity property* on the Δ^k values (see §8) indeed holds, one can infer from the case of static minimum concave cost multicommodity flows, that the accelerated greedy algorithm should produce the same solutions as the basic greedy algorithm in a high percentage of cases. On the other hand, it should be noted that this problem of computational efficiency, to which the accelerated greedy algorithm appears to bring a very appropriate answer, is certainly the fundamental reason for which so few publications have made concerning dynamic multicommodity network flow problems. Besides B. Yaged's (rather unsuccessful) attempt to generalize the fixed point approach to the dynamic case (see Yaged [78]), Zadeh [80] contains a thorough study of various optimality tests which can be applied to dynamic multicommodity network solutions, and which can be used to generate improved solutions. However these various tests do not seem to have ever been put together into an algorithm, and we think that one reason for this is that a straightforward implementation of these tests would have been prohibitively time consuming.

The algorithm suggested here, and which is based on an extension of the accelerated greedy algorithm devised in Minoux [58] for static minimum cost multicommodity flows, thus appears to be the first computational scheme efficient enough for getting solutions very close to global optimality to dynamic network optimization problems up to the large sizes which are commonly encountered in practice.

References

[1] P. Aneja, «An Integer Linear Programming Approach to the Steiner Problem in Graphs», *Networks* 10, 167 - 178, 1980.
[2] A. Assad, «Multicommodity Network Flows - A Survey», *Networks* 8, 37 - 91, 1978.
[3] M.L. Balinski, «Fixed Cost Transportation Problems», *Naval Research Logistics Quarterly* 8, 41 - 54, 1961.
[4] J.E. Beasley, «An Algorithm for the Steiner Problem in Graphs», *Networks* 14, 147 - 159, 1984.
[5] R. Bellman, *Dynamic Programming*, Princeton University Press, Princeton N.J., 1957.
[6] J. Billheimer and P. Gray, «Network Design with Fixed and Variable Cost Elements», *Transportation Science* 7, 49 - 74, 1973.
[7] T.B. Boffey and A.I. Hinxman, «Solving for Optimal Network Problem», *European Journal of Operational Research* 3, 386 - 393, 1979.
[8] R.R. Boorstyn and H. Frank, «Large Scale Network Topological Optimization», *IEEE Transactions on Communications* COM-25, 29 - 47, 1977.
[9] D.E. Boyce, A. Farhi and R. Weischedel, «Optimal Network Problem: a Branch and Bound Algorithm», *Environment and Planning* 5, 519 - 533, 1973.
[10] K.M. Chandy and T. Lo, «The Capacitated Minimum Spanning Tree», *Networks* 3, 173 - 181, 1973.
[11] K.M. Chandy and R.A. Russel, «The Design of Multipoint Linkages in a Teleprocessing Tree Network», *IEEE Transactions on Computers* C-21, 10, 1062 - 1066, 1972.
[12] N. Christofides and P. Brooker, «Optimal Expansion of an Existing Network», *Mathematical Programming* 6, 197 - 211, 1974.
[13] N. Christofides, A. Mingozzi and P. Toth, «State Space Relaxations for Combinatorial Problems», *Internal Report IC-OR 79-09 Imperial College, London*, 1979.
[14] N. Christofides, A. Mingozzi and P. Toth, «State Space Relaxation Procedure for the Computation of Bounds to Routing Problems», *Networks* 11, 145 - 164, 1981.
[15] A. Claus and N. Maculan, «Une nouvelle Formulation du problème de Steiner sur un graphe», *Prepublication #280, Centre de Recherche sur les Transports, Université de Montreal*, 1983.
[16] R. Dionne and M. Florian, «Exact and Approximate Algorithms for Optimal Network Design», *Networks* 9, 37 - 59, 1979.
[17] P.J. Doulliez and M.R. Rao, «Optimal Network Capacity Planning: A Shortest Path Scheme», *Operations Research* 23, 810 - 818, 1975.
[18] S.E. Dreyfus and R.A. Wagner, «On the Steiner Problem in Graphs», *Networks* 1, 195 - 207, 1972.

[19] D. Elias and M.J. Ferguson, «Topological Design of Multipoint Teleprocessing Networks», *IEEE Transactions on Communications* COM-22, 1753 - 1762, 1974.
[20] L.W. Ellis, «La loi des volumes économiques appliquée aux télécommunications», *Revue des Télécommunications* 1, 50, 4 - 20, 1975.
[21] L.W. Ellis, «Scale Economy Coefficients for Telecommunications», *IEEE Transactions on Systems, Man and Cybernetics* SMC-10, 1, 8 - 15, 1980.
[22] M. Florian and P. Robillard, «An Implicit Enumeration Algorithm for the Concave Cost Network Flow Problem», *Management Science* 18, 184 - 193, 1971.
[23] L.R. Ford and D.R. Fulkerson, Flows in Networks, *Princeton University Press*, 1962.
[24] H. Frank and I.T. Frisch, «The Design of Large Scale Networks», *Proceedings IEEE* 60, 1, 6 - 11, 1972.
[25] G. Gallo and C. Sodini, «Concave Cost Minimization on Networks», *European Journal of Operational Research* 3, 239 - 249, 1979.
[26] G. Gallo, C. Sandi and C. Sodini, «An Algorithm for the Min Concave Cost Flow Problem», *European Journal of Operational Research* 4, 248 - 255, 1980.
[27] M.R. Garey and D.S. Johnson, *Computers and Intractability: a Guide to the Theory of NP-Completeness*, Freeman, San Francisco, 1979.
[28] B. Gavish, «Topological Design of Centralized Computer Networks: Formulations and Algorithms», *Networks* 12, 355 - 377, 1982.
[29] B. Gavish, «Formulations and Algorithms for the Capacitated Minimal Directed Tree Problem», *Journal of ACM* 30, 1, 118 - 132, 1983.
[30] B. Gavish, «Augmented Lagrangean Based Algorithms for Centralized Network Design», *Working Paper QM 8321. The Graduate School of Management, The University of Rochester, NY 14927,* 1984.
[31] A.M. Geoffrion and G. Graves, «Multicommodity Distribution System Design by Benders Decomposition», *Management Science* 5, 822 - 844, 1974.
[32] M. Gerla and L. Kleinrock, «On the Topological Design of Distributed Computer Networks», *IEEE Transactions on Communications* COM-25, 1, 48 - 60, 1977.
[33] R.E. Gomory and T.C. Hu, «Multiterminal Network Flows», *SIAM Journal Applied Mathematics* 9, 551 - 570, 1961.
[34] R.E. Gomory and T.C. Hu, «Application of Generalized Linear Programming to Network Flows», *SIAM Journal on Applied Mathematics* 10, 260 - 283, 1962.
[35] R.E. Gomory and T.C. Hu, «Synthesis of a Communication Network», *SIAM Journal on Applied Mathematics* 12, 348 - 369, 1964.
[36] M. Gondran and M. Minoux, *Graphes et Algorithmes*, Eyrolles Paris, 1979. English translation, J. Wiley, 1984.
[37] C.C. Gonzaga and R.M. Persiano, «Planning the Expansion of a Power Transmission System. A Graph-Theoretical Approach», *6th International Conference on System Sciences, Hawai*, 393 - 396, 1973.
[38] P. Gray, «Exact solution of the Fixed Charge Transportation Problem», *Operations Research* 19, 1529 - 1538, 1971.
[39] S.L. Hakimi, «Steiner's Problem in Graphs and its Applications», *Networks* 1, 113 - 135, 1972.
[40] A.D. Hall, «An Overview of Economies of Scale in Existing Communications Systems», *IEEE Transactions on Systems, Man and Cybernetics* SMC-5, 1, 2 - 14, 1975.
[41] H.H. Hoang, «A Computational Approach to the Selection of an Optimal Network», *Management Science* 19, 488 - 498, 1973.
[42] H.H. Hoang, «Topological Optimization of Networks: a Nonlinear Mixed Integer Model Employing Generalized Benders Decomposition», *IEEE Transactions on Automatic Control* AC-27, 164 - 169, 1982.
[43] T.C. Hu, «Optimum Communication Spanning Trees», *SIAM Journal on Computing* 3, 188 - 195, 1974.
[44] D.S. Johnson, J.K. Lenstra and A.H.G. Rinnooy Kan, «The Complexity of the Network Design Problem», *Networks* 8, 279 - 285, 1978.
[45] J. Kennington, «Multicommodity Flows: a State-of-the-Art Survey of Linear Models and Solution Techniques», *Operations Research* 26, 209 - 236, 1978.
[46] A. Kershenbaum, «Computing Capacitated Minimal Spanning Trees Efficiently», *Networks* 4, 299 - 310, 1974.
[47] A. Kershenbaum and R.R. Boorstyn, «Centralized Teleprocessing Network Design», *Networks* 13, 279 - 293, 1983.
[48] H. Kobayashi, «Communication Network Design and Control Algorithms - a Survey», *Research Report RC 9233, IBM Thomas J. Watson Research Center*, 1982.
[49] E. Lawler, *Combinatorial Optimization: Networks and Matroids*, Holt, Rinehart and Winston, 1976.

[50] L.J. Leblanc, «An Algorithm for the Discrete Network Design Problem», *Transaction Science* 9, 283 - 287, 1975.
[51] T.L. Magnanti, P. Mireault and R.T. Wong, «Tailoring Benders Decompostion for Network Design», *Working paper OR 125 - 83, Operations Research Center, Massachusetts Institute of Technology*, 1983.
[52] T.L. Magnanti and R.T. Wong, «Network Design and Transportation Planning Models and Algorithms», *Transportation Science* 18, 1 - 55, 1984.
[53] M. Malek Zavarei and I.T. Frisch, «On the Fixed Cost Flow Problem», *International Journal of Control* 16, 897 - 902, 1972.
[54] P. Marcotte, «An Analysis of Heuristics for the Network Design Problem», *Publication #200, Centre de Recherche sur les Transports, Université de Montréal*, 1982.
[55] M. Minoux, «Planification à court et à moyen terme d'un réseau de Télécommunications», *Annales des Télécommunications* 29, 509 - 536, 1974.
[56] M. Minoux, «Multiflots dynamiques de coût actualisé minimal», *Annales des Télécommunications* 30, 51 - 58, 1975.
[57] M. Minoux, «Optimisation et planification des réseaux de Télécommunications», in *Optimization Techniques* G. Goos, J. Hartmanis and J. Cea, eds., Lecture Notes in Computer Science 40, Springer Verlag, 419 - 430, 1976.
[58] M. Minoux, «Multiflots de coût minimal avec fonctions de coût concaves», *Annales des Télécommunications* 31, 77 - 92, 1976.
[59] M. Minoux, «Algorithmes gloutons et algorithmes gloutons accélérés pour la résolution des grands problèmes combinatoires», *Bulletin de la Direction Etudes et Recherches EDF*, Série C, n. 1, 59 - 68, 1977.
[60] M. Minoux, «Accelerated Greedy Algorithms for Maximizing Submodular Set Functions», *Proceedings IFIP* (J. Stoer, ed.), Springer Verlag, 234 - 243, 1977.
[61] M. Minoux, «Optimum Synthesis of a Network with Nonsimultaneous Multicommodity Flow Requirements», in, *Studies on Graphs and Discrete Programming*, P. Hansen, ed., Annals of Discrete Mathematics 11, North Holland, 269 - 277, 1981.
[62] M. Minoux, «Accelerated Greedy Algorithms for Minimizing Convex and Concave Functions on Partially Ordered Sets», *Proceedings International Congress «Mathematics for Computer Science»*, Paris March 1982, AFCET Publisher, 1982.
[63] M. Minoux, *Programmation Mathématique: Théorie et algorithmes*, Dunod, Paris, 1983.
[64] M. Minoux, «Subgradient Optimization and Benders Decomposition For Large Scale Programming», *Proc. Internat Congress on Mathematical Programming, Rio*, in R.W. Cottle, M.L. Kelmanson and B. Korte, eds., *Mathematical Programming*, 271 - 288, North Holland, 1984.
[65] M. Minoux, «Network Synthesis and Optimum Network Design Problems: Models, Solution Methods and Applications», to appear, 1985.
[66] M. Minoux and A. Guerard, «Choix d'investissements en matériels de Télécommunications: une nouvelle approche de la notion de cout équivalent», *Annales des Télécommunications* 36, 602 - 612, 1981.
[67] M. Minoux and J.Y. Serreault, «Synthèse Optimale d'un réseau de Télécommunications avec contraintes de sécurité», *Annales des Télécommunications* 36, 211 - 230, 1981.
[68] M. Minoux and J.Y. Serreault, «Subgradient optimization and large scale programming: an application to network synthesis with security constraints», *RAIRO* 15, 185 - 203, 1981.
[69] M. Minoux and J.J. Strodiot, «Un algorithme exact pour les problèmes de multiflots de coût minimum avec fonctions de coût concaves», *Unpublished report, CNET*, 1982.
[70] T.L. Morin and R.E. Marsten, «Branck and Bound Strategies for Dynamic Programming», *Operations Research* 24, 611 - 627, 1976.
[71] M. Schwartz, *Computer Communications Network Design and Analysis*, Prentice Hall, Englewood Cliffs, 1977.
[72] A.J. Scott, «The Optimal Network Problem: some Computational Procedures», *Transportation Research* 3, 201 - 210, 1969.
[73] P.A. Steenbrink, *Optimization of Transport Networks*, J. Wiley & Sons, 1974.
[74] D.I. Steinberg, «The Fixed Charge Problem», *Naval Research Logistics Quarterly* 17, 217 - 236, 1970.
[75] D.T. Tang, L.S. Woo and L.R. Bahl, «Optimization of Teleprocessing Networks with Concentrators and Multiconnected Networks», *IEEE Transactions on Computers* C-27, 7, 594 - 604, 1978.
[76] H.M. Weingartner and D.M. Ness, «Methods for the Solution of Multidimensional 0-1 Knapsack Problems», *Operations Research* 15, 83 - 103, 1967.
[77] B. Yaged Jr, «Minimum Cost Routing for Static Network Models», *Network* 1, 139 - 172, 1971.
[78] B. Yaged Jr, «Minimum Cost Routing for Dynamic Network Models», *Networks* 3, 193 - 224, 1973.

[79] N. Zadeh, «On Building Minimum Cost Communication Networks», *Networks* 3, 315 - 331, 1973.
[80] N. Zadeh, «On Building Minimum Cost Communication Networks Over Time», *Networks* 4, 19 - 34, 1974.
[81] W.I. Zangwill, «Minimum Concave Cost Flows in Certain Networks», *Managements Science* 14, 429 - 450, 1968.

Michel Minoux
Université de Paris 9 - Dauphine
Lamsade
Place du Maréchal De Lattre de
Tassigny 75775 Paris
France

PARALLEL COMPUTER MODELS
AND COMBINATORIAL ALGORITHMS

Celso Carneiro RIBEIRO

1. Introduction

As pointed out by Hwang and Briggs [81], the computer industry has experienced four generations of development over the past four decades, coming from components consisting of relays and vacuum-tubes (1940 - 1950) to diodes and transistors (1950 - 1960), to small and medium scale integrated circuits (1960 - 1970) and to large and very large scale integrated systems more recently (from 1970 onwards). This evolution was characterized until some years ago by constant improvements in device speed and reliability, associated with reductions in hardware size and cost.

More recently there were significant changes in computer architecture and a new era in very high-performance computing is beginning, in which high-performance systems will be dominated by parallel architectures. Parallelism is a set of techniques that introduce concurrency in computer systems, enabling several units to be simultaneously active, in order to increase the computational throughput. It is important to notice that the idea of parallelism induces a structural change in computation theory and algorithm design. The different ways used for introducing parallelism into computer systems and the theoretical concepts underlying them should be examined.

Two different levels of parallelism should be distinguished. From the hardware point-of-view, there is parallelism if there are at least two processing units working simultaneously. Hence, concurrency of operation seeks to achieve better utilization of available hardware by overlapping activities which use disjointed parts of the computing system. Among the techniques used to introduce concurrency at the hardware level, there are pipelining, vector registers, replication of processors and interconnection networks, which are very often combined in practical assemblages.

From the sofware point-of-view, parallelism exists whenever two or more logical tasks can be performed simultaneously (and one is able to make use of this logical parallelism through the computer languages available for a given machine). Hence, another line to seek speed-up in computer systems is the design of parallel algorithms based on the logical parallelism inherent in each problem. The computer programs corresponding to these algorithms should be coded taking into account the architecture of the machine on which they will run, in order to maximize their efficiency.

The main objective of this work is to study parallelism in terms of how it affects combinatorial algorithm design and complexity. In section 2, some important issues concerning parallel computer models and architectures are reviewed: classification schemes of parallel machines (as well as examples of practical assemblages), communication issues in parallel processing, interconnection network models and systolic devices. Section 3 is concerned with the complexity theory and its extensions to synchronous parallel computation. The main concepts are the parallel computation thesis, solvability in polylog parallel time and log-space completeness for P. Finally, section 4 presents several examples of parallel combinatorial algorithms, ranging from sorting and searching to graph theory and NP-hard optimization problems.

2. Parallel Computer Models and Architectures

In this section some important issues in parallel computer models and architectures are studied: classification schemes of parallel architectures, communication issues in the design of parallel algorithms, network models used in the implementation of parallel architectures and systolic devices.

2.1. Parallel Computer Architectures

Several classification schemes of computer architectures have been proposed, based on different points-of-view; for surveys see Thurber and Wald [176] and Hockney and Jesshope [79], among others. The taxonomy proposed by Flynn [58, 59] is based only on the multiplicity of instruction and data streams. In spite of sometimes being quite ambiguous, it is still used because of its simplicity and some of the associated terminology has become part of the language of parallel computing. Four classes of machines are defined according to Flynn:

(i) *SISD (single instruction stream, single data stream)*: This class corresponds to the conventional sequential (or serial) computer architecture proposed by von Neumann: the instructions are executed sequentially, each instruction is performed at a time on a single set of data.

(ii) *SIMD (single instruction stream, multiple data stream)*: This class corresponds to machines where the same instruction is executed simultaneously on several sets of data. A high number of processing elements, each one with its own local storage, is controlled by a unique control unit. At a given moment, each processor is performing the same instruction on a different set of data from a distinct data stream. As Hockney and Jesshope [79] point out, the term SIMD became synonymous with an array of many processors working in lockstep under common control. The first computer with this type of architecture was the ILLIAC IV, with 64 processors organized as an 8×8 array; see Barnes et alii [12].

Other examples of SIMD machines are the ICL DAP (Distributed Array Processor) and the Burroughs BSP (Scientific Processor), described respectively in Reddaway [145] and Austin [9].

(iii) *MISD (multiple instruction stream, single data stream)*: This class corresponds to machines where a number of processors simultaneously perform different instructions on the same set of data. So far, it has received less attention and no practical realization seems to be available.

(iv) *MIMD (multiple instruction stream, multiple data stream)*: This class corresponds to the most general model of architecture. More frequent up to now are control-flow multiprocessor architectures (synchronous or asynchronous), in which the operations are executed in an order predetermined by a control convention used by the programmer, corresponding to a program stored in the memory. Data-flow and reduction machines have been proposed and designed. The latter are characterized by the fact that the operations are executed in an order determined by data interdependencies and resource availability. Data-flow machines carry out data-driven computations in which the operations are executed in an order determined by the availability of input data, while reduction machines carry out demand-driven computations in which the operations are executed in an order determined by the requirements for data. These architectures are based upon completely different principles with respect to the traditional control-flow architectures and will require completely different algorithmic techniques and programming languages; the interested reader is referred to Treleaven et alii [178], Chudik [35] and Sharp [161] among others.

Hockney [78] proposed a structural classification for control-flow MIMD systems, which can also be applied to SIMD models, defining two major categories of architectures. The first one is formed by the so-called switched systems, in which there is an identifiable and separate switch unit that connects together a number of processors and memory modules. Within this class there are shared memory and distributed memory architectures. In shared memory architectures a number of processors are connected via the switch unit to a number of independent memory modules, so that the memory is shared by all processors on an equal basis. Examples of shared memory MIMD systems are CRAY-XMP, CRAY-2 and CRAY-3, ETA-10 (which will be comprised of 8 CPUs each based on the architecture of the CDC CYBER 205) and Denelcor HEP (Heterogeneous Element Processor, comprised of 16 Process Execution Modules, each capable of having up to 50 user instruction streams; see e.g. Smith [168]). The characteristic of distributed memory systems is that the memory modules are attached directly to the processors, so that the role of the switch is to interconnect the processors and there are no memory modules connected directly to the switch. The second category is that of network systems. In these architectures a number of processors are connected through an interconnection network with a particular topology. The processing elements which form the nodes of the network may only communicate with neighbor processors directly connected to them. Several topologies for these interconnection networks have been proposed and some of them are reviewed in section 2.3; see e.g. Siegel [167].

Being a very simple classification, Flynn's scheme is sometimes ambiguous and is not always able to perfectly characterize all machines, e.g. pipelined computers. In a pipelined computer, concurrency is introduced by subdividing basic tasks into several subtasks. Each task is executed by a specialized hardware stage that operates concurrently with other stages in the pipeline, while inputs and results are streamed through the pipeline in such a way that successive tasks are executed in an overlapped fashion. As soon as a subtask is carried out, the stage responsible for its execution can accept and process a new set of inputs. Let us suppose that some task should be executed on N sets of data and that it can be pipelined into P stages, each one taking T time units for its execution. Then, the overall execution time in pipelined mode will be $(N + P - 1)T$, while in sequential mode it would be NPT, which represents savings of $O(P)$ in terms of computing time. Some pipelined computers are also characterized by the existence of vector registers that allow the feeding of pipelined functional units at a rate of one set of data per clock period. Hockney and Jesshope [79], Kogge [98] and Hwang and Briggs [81] present very detailed studies of pipelined computers. Among them we have CRAY-1 (and its successors), CDC CYBER 205, FPS AP-120B, TEXAS ASC, FACOM VP-200 and IBM 3090, described respectively in Russell [153], Kascic [93], Harte [72], Watson [194], Tamura et alii [172] and IBM [82].

Besides the examples of parallel architectures already mentioned, several other parallel machines have been proposed and designed, such a CDC STAR-100, SOLOMON I and SOLOMON II, PEPE (Parallel Element Processing Ensemble), STARAN, OMEN (Orthogonal Mini Embedment), MPP (Goodyear Massively Parallel Processor), C.mmp, Cm* and Cosmic Cube; the reader is referred to Thurber and Wald [176], Zakharov [198] and Hockney [78], where these and other machines are described in detail and the original references are quoted.

2.2. Communication Issues in Parallel Processing

Communication plays a very important role in parallel processing. Factors like the geometry of the array of processors and the way they are interconnected affect the performance and the effectiveness of parallel algorithms, due to the introduction of delays on inter-processor communication and data transmission. It is sometimes the time for data movement, rather than the computation time based on data dependencies, which limits execution speed.

According to the way in which information is transmitted among the processors, two general models of parallel architectures are distinguished: paracomputers and ultracomputers; see Schwartz [159], Lint and Agerwala [116] and Ausiello and Bertolazzi [8].

In a paracomputer, a very large number N of identical processors share a global storage area and at any time any number of processors can simultaneously access the common memory. Although the technology currently available prohibits the realization of a paracomputer, because of physical fan-in limitations, they are of great theoretical interest and can provide important complexity results concerning the limits of parallel algorithms and computations. A structured complexity theory for parallel computation is important, in order to make it possible to understand the range of problems for which good parallel algorithms may exist. The complexity theory of synchronous parallel computation is based on models of shared memory parallel machines which have been derived from the paracomputer, such as PRAM, WRAM and CREW-PRAM (see section 3.2).

In a realistic and physically realizable assemblage, one could not expect that any processing element could be connected to more than a fixed number k of other computer elements. A parallel computer model where N processors (each one with its own memory) communicate through an interconnection network with a particular topology should therefore be considered. This kind of architecture is called an ultracomputer. Within an ultracomputer the processors communicate through an interconnection network, while in a paracomputer they communicate through the memory. Many of these interconnection networks have been proposed, some designed for particular problems. In the next section some of the general purpose networks are reviewed; for complementary references the reader is referred to Haynes et alii [73], Valiant [186] and Vishkin [190].

2.3. Basic Network Models

The interconnection networks mentioned in the previous sections can be seen as undirected graph models, as described by Ausiello and Bertolazzi [8]: N identical processors are located at the nodes of a potentially infinite recursive graph structure and communicate through the edges of this graph. The practical use of this kind of structure is limited by wiring constraints, as well as by design and fabrication principles:

(i) The difficulty of performing interprocessor communication in logarithmic time (due to constraints on the capacity of communication lines).

(ii) The fan-in physical limitations which require that the number of neighbor processors be either constant or at most logarithmic in terms of the total number of processors.

(iii) The wiring constraints which do not allow (specially for VLSI implementations) more than two or three levels of wiring and which impose restrictions on the density and length of connection wires in a layout.

Taking these limitations into account, Preparata [139] suggests some requirements that should be satisfied by an architecture graph:

(i) *Degree-boundedness*: the degree d_1 of each node should be bounded by a small constant.

(ii) *Size-independence*: the structure of the deployed processors should be independent of the system size (i.e., the number N of processors).

(iii) *Regularity and modularity*: the layout of the graph should be highly regular (i.e., susceptible of a simple and compact description) and modular (i.e., obtainable by suitably combining layouts of analogous graphs of smaller sizes).

(iv) *Logarithmic diameter*: the maximum path length (number of edges) d_2 between any pair of nodes of the graph should grow at most logarithmically in the number N of processors, ensuring fast communication.

Of the many interconnection networks that have been proposed, the large-purpose ones described below (among others) exhibit these features, with minor exceptions:

(i) *Rectangular mesh*: The rectangular mesh consists of $N = m \cdot n$ processors located at the nodes of an $m \times n$ grid. At each node (i, j) of the grid there is a processor that can communicate with the processors located at nodes $(i \pm 1, j)$ and $(i, j \pm 1)$, provided they exist. In this very regular and modular structure all nodes have degree less than or equal to four, i.e. $d_1 = 4$. If the two-dimensional $n \times n$ square mesh connected network proposed by Unger [182] is considered, then the diameter $d_2 = 2(n-1) = O(\sqrt{N})$, which is greater than $O(\log N)$; see figure 1. The ILLIAC-IV computer has its 64 processors arranged in an 8×8 square mesh network. This type of interconnection network is very useful for finite difference approximations to various partial differential equation problems; see e.g. Lomax and Pulliam [117].

(ii) *Cube connected network*: This structure, proposed by Squire and Palais [169], corresponds to an n-dimensional hypercube with $N = 2^n$ processors located at its vertices and interconnections along the edges. In this case, $d_1 = d_2 = n = O(\log N)$. Seitz [160] presents the architecture and applications of the Cosmic Cube, a machine which is a hypercube with $N = 2^6$ processors hosted by a VAX 11/780; see also Hockney [78].

(iii) *Cube connected cycles network*: This structure, proposed by Preparata and Vuillemin [140], is obtained from the previous one by substituting each of the vertices of the cube by a cyclically connected set of n processors, so that the total number of processors is $N = 3 \cdot 2^n$; see figure 2. Each processor has at most two cycle connections and one edge connection. It has several desirable features: $d_1 = 3$, $d_2 = O(\log N)$, the modules are size-independent and its layout is highly regular and modular. This network has been shown to be optimal for several problems with respect to the area $\times$ time^2 complexity measure proposed by Thompson [174] for VLSI implementations; see Preparata and Vuillemin [140], Ausiello and Bertolazzi [8], Preparata [139], Leighton [111] and Leiserson [113]. Nevertheless, a severe limitation to its physical realization comes from the technological problems concerning the layout of non-planar structures.

(iv) *Perfect shuffle-exchange graph*: The foundations of this structure go back to Clos [36] and Benes [15], in the context of switching networks. It was later explored and improved by Stone [170]. There are $N = 2^n$ processors and for

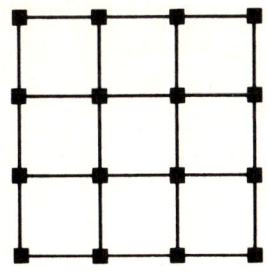

Fig. 1. Square mesh connected network ($n = 4, N = 16$).

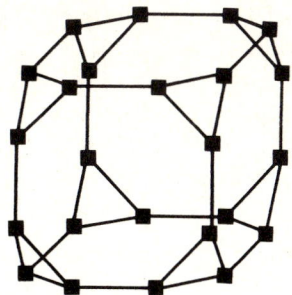

Fig. 2. Cube connected cycles network ($n = 3$, $N = 24$), from Kindervater and Lenstra [95].

each $i = 0, 1, 2, \ldots, N - 1$, processor P_i is connected to processors P_{i+1} (resp. P_{i-1}) if i is even (resp. odd), $P_{\sigma(i)}$ and $P_{\sigma^{-1}(i)}$, where

$$\sigma(i) = \begin{cases} 2i, & \text{if } 0 \leqslant i < N/2 \\ 2i + 1 - N, & \text{if } N/2 \leqslant i < N \end{cases}$$

$$\sigma^{-1}(i) = \begin{cases} i/2, & \text{if } i \text{ is even} \\ (i - 1)/2 + N/2, & \text{if } i \text{ is odd,} \end{cases}$$

all sums being taken modulo N; see figure 3. Schwartz [159] shows that since any arbitrary permutation of the contents of the processors can be performed in time $O(\log N)$, the shuffle interconnection has very important applications in some algorithms having a recursive, «divide and conquer» characteristic. This architecture is also characterized by $d_1 = 3$ and $d_2 = O(\log N)$. In spite of its lack of layout regularity and modularity, the perfect shuffle has also been shown to be optimal with respect to the area × time² complexity measure for VLSI implementations; see Preparata [138], Leighton [111] and Leiserson [113], among others.

(v) *Binary tree*: In this kind of architecture, the processors correspond to intermediary nodes and leaves of trees. Mead and Conway [191] describe an n-level binary tree architecture, in which $N = 2^n - 1$ processors are available and each processor controls two subprocessors. Bentley and Kung [17] propose a different architecture consisting of two $(n + 1)$-level binary trees with common processors located at their leaves. The first tree (an out-tree) is used to send data down to the processors located at the leaves and the second tree (an in-tree) combines the results already obtained. There are 2^n common processors

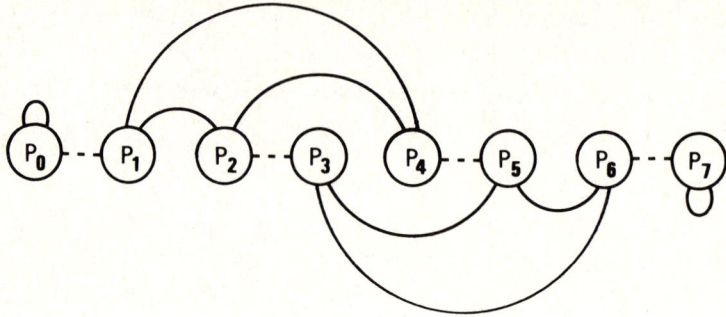

Fig. 3. Perfect shuffle interconnection network ($n = 3, N = 8$).

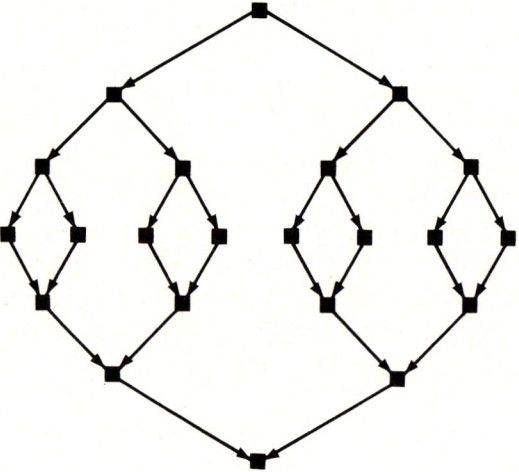

Fig. 4. Binary tree network with common leaves ($n = 3, N = 22$).

and each tree has $2^n - 1$ additional processors, hence $N = 3 \cdot 2^n - 2$. In this case, as in the previous one, $d_1 = 3$ and $d_2 = O(\log N)$; see figure 4. The combination of binary tree and mesh connected networks gives rise to the orthogonal trees network. As described by Preparata [139], a two-dimensional orthogonal mesh-of-trees has the following structure. Given an $n \times n$ square mesh with edges removed, for each row (and column) a binary tree is built having the vertices of that row (column) as its leaves, as illustrated in figure 5 for $n = 4$. The inter-row (column) area is used to embed $2n(n-1)$ non-leaf processors corresponding to the root and internal nodes of the associated row (column) tree, hence there are $N = 3n^2 - 2n$ processors; see Nath et alii [134] for details.

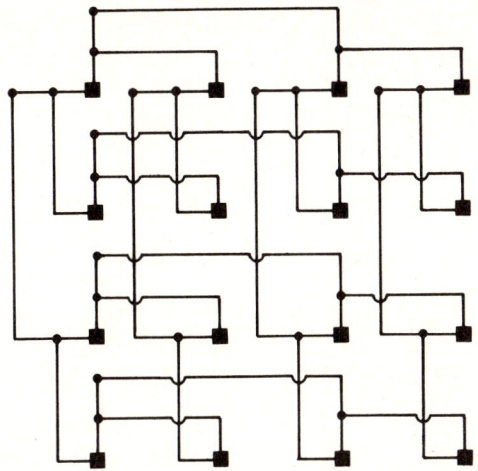

Fig. 5. Orthogonal trees network ($n = 4, N = 40$), from Nath et alii [134].

2.4. Systolic Arrays

Systolic arrays are highly parallel synchronous architectures suitable for the implementation of highly synchronized computations, characterized by the following properties; see Kung [102] and Hwang and Briggs [81]:

(i) Each systolic architecture is appropriate for a special class of problems.

(ii) The architecture is based on a small number of different types of processors (or cells), each capable of performing some simple operation.

(iii) The processors are locally and regularly connected, in order to form an array or a tree.

(iv) The flows of inputs and results are pipelined through the interconnection network during the computations, in synchronous parallel mode.

(v) Communication of data occurs only between neighboring processors and control signals propagate through the array like data.

(vi) Communication with the outside world occurs only at the boundary cells.

As a consequence, they are very well suited for reliable and inexpensive implementation using many identical components. Systolic arrays assume many different structures, depending on the type of algorithm that will be implemented on them. Figure 6 shows some of these structures and table 1 illustrates their potential usage. Two-dimensional systolic arrays are attracting a great deal of attention even though their implementation on VLSI chips still faces many practical difficulties. Three-dimensional arrays in which multi-layer devices are used have also been recently proposed. One of the first commercialy available systolic machines seems to be the GAPP (Geometric Arithmetic Parallel Processor) designed by NCR. It consists of 72 processors organized as a 6 x 12 two-dimensional array (each processor communicates with its four neighbors) and

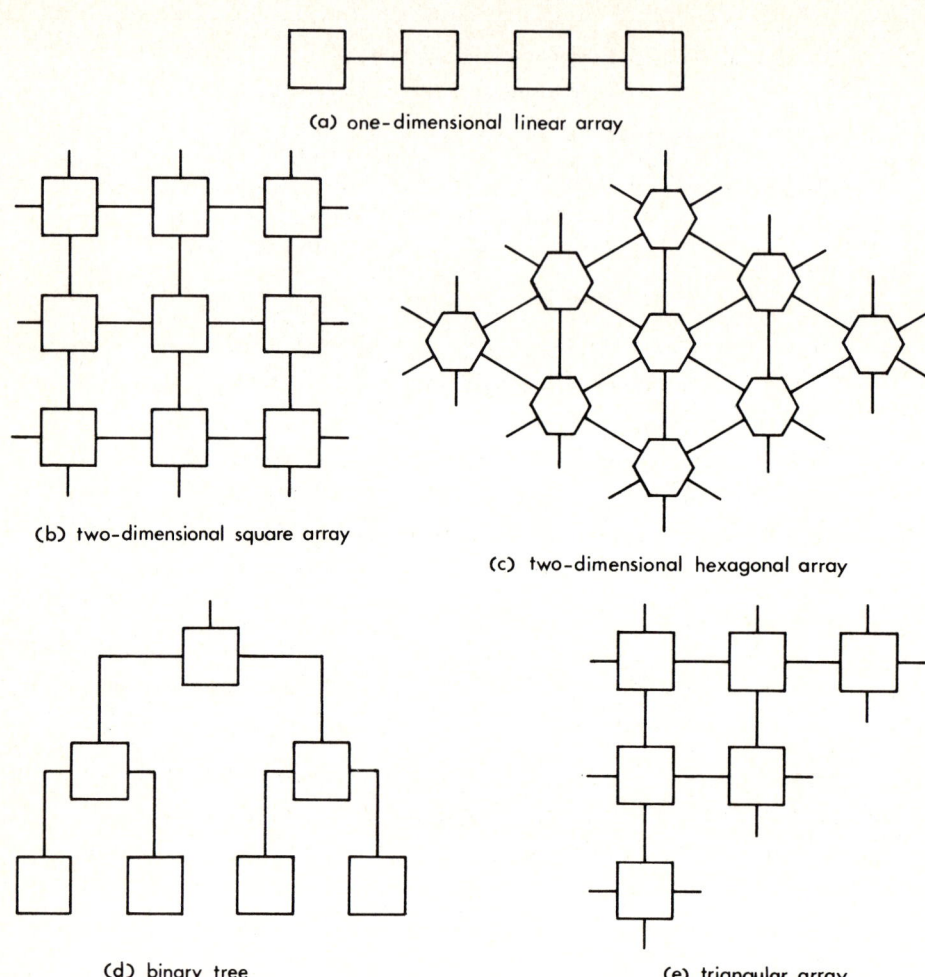

Fig. 6. Some systolic array configurations: (a) one-dimensional linear array, (b) two-dimensional square array, (c) two-dimensional hexagonal array, (d) binary tree and (e) triangular array (from Hwang and Briggs [81]).

several of such blocks can be placed side by side in order to obtain a more powerful machine; see Breteuil [27] for details.

Some examples of systolic architectures and algorithms are presented below:

Example 1. Matrix-Vector Multiplication

The problem of multiplying an $n \times n$ matrix $A = (a_{ij})$ with an n-vector $x = (x_i)$ is addressed here. The elements in the product $y = (y_1, \ldots, y_n)$ can be computed by the following recurrences, as described by Mead and Conway [121]:

Table 1.
Some applications of systolic arrays and their desired VLSI structures.

Processor array structure	Applications
1-D linear arrays	Convolution, discrete Fourier transform, solution of triangular linear systems, matrix-vector multiplication, odd-even transposition sort.
2-D square arrays	Dynamic programming, graph algorithms involving adjacency matrices.
2-D hexagonal arrays	Matrix multiplication, LU decomposition, transitive closure.
Trees	Searching algorithms, parallel function evaluation.
Triangular arrays	Inversion of triangular matrices.

$$y_i^1 = 0$$
$$y_i^{k+1} = y_i^k + a_{ik} x_k \qquad (k = 1, 2, \ldots, n)$$
$$y_i = y_i^{n+1}.$$

These recurrences can be computed by pipelining vectors x and y through $2n - 1$ linearly connected processors of the type described in figure 7, each one with three input and three output gates.

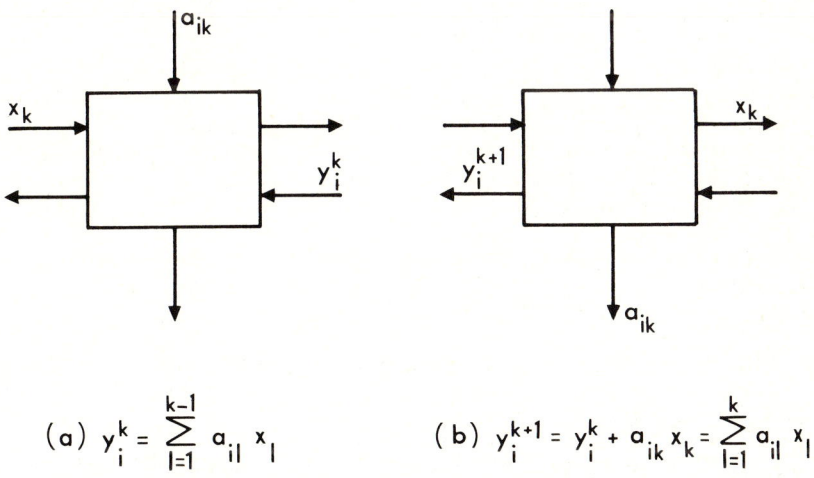

Fig. 7. Processor for matrix-vector multiplication at (a) the beginning and (b) the end of a cycle.

As illustrated in figure 8, vector y (which is initially zero) moves to the left, while vector x moves to the right and matrix A moves down, all movements being pipelined and synchronized.

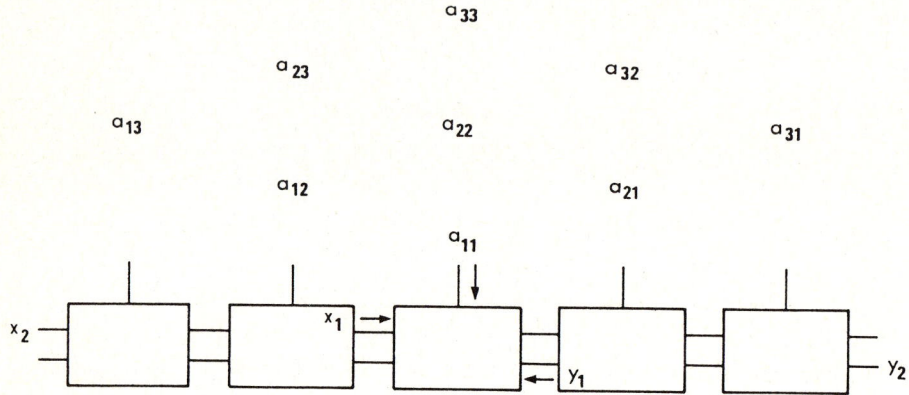

Fig. 8. Linearly connected array for matrix-vector multiplication.

Example 2. Matrix Multiplication

Multiplying two $n \times n$ matrices $A = (a_{ij})$ and $B = (b_{ij})$ can be done in $O(n)$ time with different types of mesh connected networks. The first type, illustrated in figures 9 and 10, is based on processing elements with two input gates, two output gates and one register each. Matrix A is pipelined through the rows of the mesh, while matrix B is pipelined through its columns. At each cycle the processor located at vertex (i, j) of the mesh receives elements a_{ik} and b_{kj} as inputs, multiplies them and adds the result to the contents of its register. After $2n - 1$ cycles each processing element (i, j) will contain the value

$$c_{ij} = \sum_{k=1}^{n} a_{ik} b_{kj}.$$

The second type of mesh connected network is more suited to band matrix multiplication and is based on a kind of hexagonal processor shown in figure 11. In this case, the partial sums obtained at each cycle are also pipelined through the mesh and the overall computations also take $O(n)$ time; see Mead and Conway [121] for details.

The reader interested in systolic algorithms is referred to the following references: Rote [151], where it is shown how the Gauss-Jordan elimination algorithm for the algebraic path problem can be implemented on a hexagonal systolic array with a quadratic number of processors in linear time; André et alii [5],

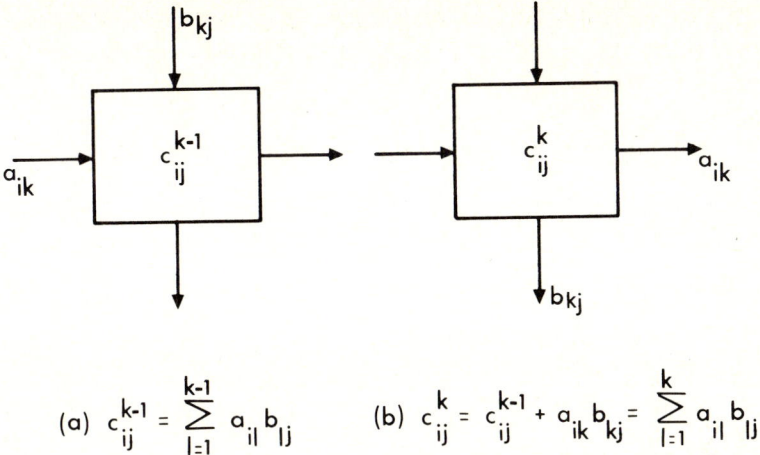

Fig. 9. Processing element for matrix multiplication in a mesh connected network (first type) at (a) the beginning and (b) the end of a cycle.

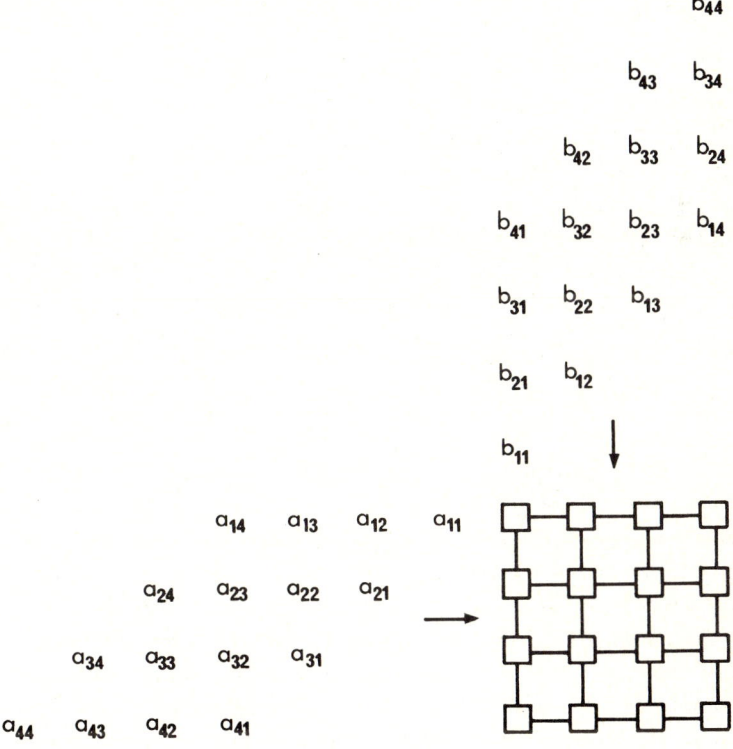

Fig. 10. Matrix multiplication on a mesh connected network.

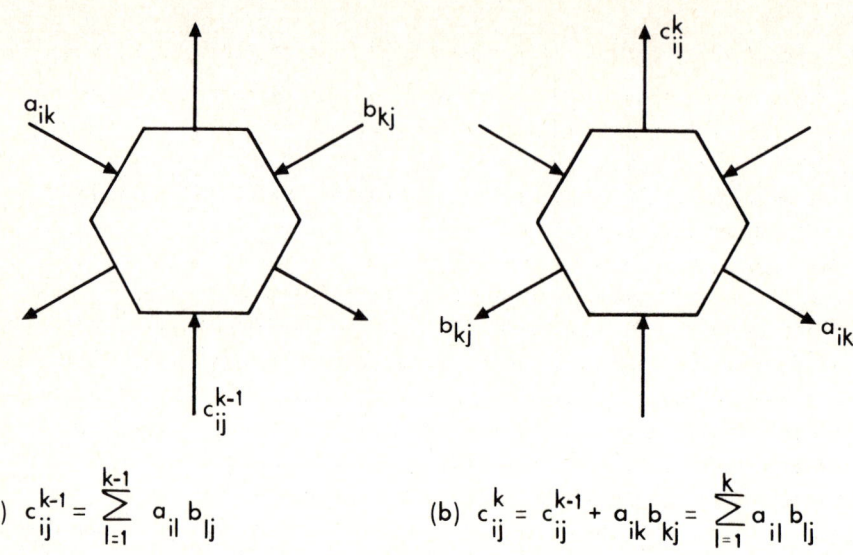

Fig. 11. Hexagonal processor for matrix multiplication (second type) at (a) the beginning and (b) the end of a cycle.

where a systolic architecture is studied for the word detection problem; Mead and Conway [121], where systolic algorithms are proposed for several other problems on square and hexagonal mesh-connected networks (LU decomposition of a matrix, resolution of triangular linear systems, computation of convolutions, finite impulse response filters and discrete Fourier transforms) and binary trees (matrix multiplication and the clique and color-cost problems); Hwang and Briggs [81], for an in-depth study of systolic techniques (including reconfigurable processor arrays that can host several different systolic array organizations).

3. An Overview of Complexity Theory of Parallel Computation

As Kindervater and Lenstra [95] point out, beyond the basic distinction between solvability in polynomial time and completeness for NP (which are the most important issues in sequential computation), many other concepts have been studied for parallel computation. Among them are the very important notions of solvability in polylog parallel time and log-space completeness for P. After reviewing the main concepts from complexity theory, the model of a parallel random access machine is discussed and an introduction to the main complexity results for parallel computation is presented; the reader is referred to Cook [39] for a complete study on the complexity theory of synchronous parallel computation.

3.1. A Quick Overview of the Complexity Theory for Sequential Computation

In this section the main notions and results from complexity theory are reviewed; the reader is referred to Garey and Johnson [62] for an exhaustive study, to Cook [40] for a very clear overview of this theory, where its history and current lines of research are discussed, and e.g. to Papadimitriou and Steiglitz [136] for a very didactical explanation.

Associated with every optimization problem one can always define a decision problem of the type «given a problem instance I (i.e. a set of data), is there an associated structure S satisfying a certain property Q?», i.e., a question that can be answered by «yes» or «no». A problem instance is said to be feasible if it leads to an answer «yes». The size of an instance is defined as the number of bits needed to encode the data under any reasonable encoding scheme. The running time of an algorithm can be seen as the number of elementary operations required for its solution. An algorithm is said to be polynomial if its running time is bounded by a polynomial function of the problem size. Since one can always solve the optimization problem and check whether its solution satisfies property Q or not, the decision problem is never harder than the corresponding optimization problem. Under very general and realistic assumptions, it can also be shown that the optimization problem can be solved efficiently whenever the decision problem can. Since the decision problem is never harder than the original optimization problem, any negative results proved about the complexity of the decision problem will apply to the optimization problem as well. For this reason, complexity theory is based on the analysis of decision problems rather than optimization problems.

The class of decision problems solvable by polynomial time algorithms is denoted by P, i.e., a problem is in class P if an algorithm exists that determines in polynomial time, for any instance, whether the answer is «yes» or «no». Since polynomial time algorithms are largely accepted as efficient, P can be seen as the class of easy and well-solvable decision problems for which efficient algorithms exist. NP denotes the class of decision problems for which algorithms exist that can always check in polynomial time if some given candidate leads to a «yes» answer. By their definition, it can easily be seen that $P \subseteq NP$.

A decision problem A transforms polynomially to B (which is denoted by $A \propto B$) if a polynomial transformation T exists so that the answer for problem A given any instance I is «yes» if and only if the answer for problem B given the instance $T(I)$ is «yes». If $A \propto B$ and there is a polynomial time algorithm for B, there is also a polynomial time algorithm for A. It turns out that there is a subset of problems in NP which are the hardest problems, in the sense that if a polynomial time algorithm for one of the problems of this subset is found, it will apply to all other problems in NP. Hence, a decision problem is said to be NP-complete if it belongs to NP and all other problems in NP polynomially transform to it. The term NP-hard is sometimes associated with the optimization version of an

NP-complete problem. The interested reader is referred to Garey and Johnson [62], Papadimitriou and Steiglitz [136] and Johnson and Papadimitriou [89] for in-depth studies of many other complexity classes.

Up to this point, complexity has been considered only in terms of computing time. In practice, the amount of storage (space) required for solving a problem is also important. Using the model of the Turing Machine, Garey and Johnson [62] show that all problems solvable in polynomial time can be solved in polynomial space. PSPACE is defined as the class of decision problems solvable by polynomial space algorithms. A problem is said to be PSPACE-complete if it belongs to PSPACE and all other problems in PSPACE transform polynomially to it. Hence, a PSPACE-complete problem could belong to P if and only if P = PSPACE. As Garey and Johnson [62] point out, the fact that a problem is PSPACE-complete is an even stronger indication that it is intractable than if it were NP-complete, since we could have P $\neq$ PSPACE even if P = NP. Using the model of a non-deterministic Turing Machine, one could define analogously the class NPSPACE, but it can be shown that PSPACE = NPSPACE; see Savitch [158] and Garey and Johnson [62]. As a consequence, PSPACE-completeness is the strongest type of completeness result currently available.

3.2. Parallel Random Access Machines

A very basic issue in complexity theory is to find the most appropriate model of a computer. Different models could lead to different computations and the number of steps executed by the same algorithm would be different, depending on the model of machine considered. The widely accepted model of a RAM (Random Access Machine) consists of a read-only input tape, a write-only output tape, a program and a memory; see e.g. Shepherdson and Sturgis [162] and Aho et alii [1] for a detailed description of the properties and characteristics of the RAM.

Fortune and Wyllie [60] proposed the model of a PRAM (Parallel Random Access Machine) for parallel computations, which is an extension of the RAM model. In a PRAM there is a potentially infinite sequence of global registers, plus a potentially infinite set of N identically-programmed processors. Each processor has a set of local registers, and at any time it can refer to either its local registers or global ones: simultaneous reads are allowed, while simultaneous writes are not (i.e., the system crashes in case of an access conflict due to two processors trying to write in the same global register at the same time). At the beginning of the computations the input is loaded into the global registers and only the first processor is active. At each time unit any active processor can perform one of a set of primitive operations, among them the operation of activating a new processor (that will run in parallel with those already in operation from that time on, until it is turned-off). The first processor is always active and the computation is considered terminated when it halts.

Since any active processor can access any global register, the PRAM model coincides with an N-processor paracomputer (see section 2.2) in which $O(\log N)$ time is required to activate the N processors. This $O(\log N)$ activation time would be realistic in any reasonable model, due to delays introduced in practice by fan-in limitations. Other models of shared memory parallel machines derived from the concept of a paracomputer have been proposed in the literature, like the WRAM and the CREW-PRAM; see e.g. Shiloach and Vishkin [163] and Borodin and Hopcroft [26]. The basic differences between each of these models are concerned with how they deal with read and write conflicts of memory access.

3.3. Complexity Theory of Synchronous Parallel Computation

An overview of the main results derived from the complexity theory of synchronous parallel computation is presented here. The reader is referred to Johnson [88] and Kindervater and Lenstra [95] for very clear presentations on which this section is largely based, as well as to Cook [39] for a more in-depth survey.

Several authors have mentioned that there is evidence that parallel time and space are equivalent up to a polynomial factor. This parallel computation thesis is stated by Chandra et alii [32] and Goldschlager [67] as: time-bounded parallel machines are polynomially related to space-bounded sequential machines. This conjecture states that the class of decision problems solvable in time $T(n)^{O(1)}$ by a machine with unbounded parallelism (i.e., polynomial in $T(n)$, where n is the problem size) is the same class of decision problems solvable in space $T(n)^{O(1)}$ by a sequential machine. However, Blum [25] proved that the parallel computation thesis is not sustained when, for the parallel computer model considered, an arbitrarily large though finite number of processors can be activated in one parallel step and each memory cell of an arbitrarily large though finite memory can be accessed by any processor. As pointed out by Blum, these properties are fulfilled for many models (see e.g. Shiloach and Vishkin [163] and Borodin and Hopcroft [26]), but are not satisfied for the models where $O(\log N)$ time is needed to activate N processors; see Van Emde Boas [188] for a survey of several cases of parallel computer models and time bounds for which this thesis is consistent.

The parallel computation thesis holds, in particular, in the case where $T(n) = n^{O(1)}$ (i.e., a polynomial function of the problem size) and the computer model considered is a PRAM. This model is quite powerful: the class of decision problems that a PRAM could solve in polynomial time would be precisely PSPACE, including the very difficult NP-complete and PSPACE-complete problems. However, if the more realistic case of a PRAM with a polynomial number of processors (instead of unbounded parallelism) is considered, the class of decision problems solvable in polynomial time is exactly P. Under this restrictive but realistic assumption, all that can be obtained with parallelism is the speeding-up of the solution of problems in P: hard problems (for which it is unlikely that polynomial time algorithms could ever be found) remain as hard as they were without parallelism, in terms of theoretical complexity.

In turns out that many problems in P can be solved in parallel time

$O((\log n)^{O(1)})$, i.e., time that is polynomially bounded by the logarithm of the problem size, which is called polylog parallel time. Among them are the problems of maximum finding, partial sums, sorting, preemptive scheduling, scheduling fixed jobs, maximum flow in a planar graph and linear programming with a fixed number of variables; for details about polylog parallel time algorithms for these problems see Kindervater and Lenstra [95], Dekel and Sahni [47, 48], Muller and Preparata [127], Johnson and Venkatesan [90] and Meggido [122]. Based on the parallel computation thesis, the class **POLYLOGSPACE** of decision problems that can be solved in polylog sequential space is defined as corresponding to the class of decision problems solvable in polylog parallel time. Since any realistic computer model needs at least $O(\log N)$ time to activate N processors, it is very unlikely that sublogarithmic algorithms could ever be found. In this sense, the class **POLYLOGSPACE** can be considered as formed by the easiest problems in P. Section 4 shows several examples of polylog parallel time algorithms for combinatorial problems.

A problem is said to be log-space complete for P if it belongs to P and all problems in P can be reduced to it by means of some transformation computable using logarithmic work space. Examples of log-space complete problems are: solvability of a path system, circuit value, linear programming and maximum flow; see Cook [38], Ladner [103], Goldschlager [66], Dobkin et alii [54], Valiant [185] and Goldschlager et alii [68]. If any log-space complete problem could ever be solved in polylog (sequential) space, so would be all problems in P. Since this assumption does not seem to hold, log-space complete problems are unlikely to be in **POLYLOGSPACE** and, by the parallel computation thesis, are unlikely to be solvable in polylog parallel time. In this sense, the log-space complete problems could be seen as the hardest problems in P.

Other approaches to the study of the complexity of parallel and distributed algorithms exist; see e.g. Santoro [154] and Lavallée and Lavault [108].

4. Combinatorial Algorithms

While many publications and numerical results are available concerning the use of parallel algorithms for problems arising in partial differential equations, linear algebra, non-linear optimization, fast Fourier transform, image processing and power system analysis, the same is not true in the field of combinatorial algorithms, mainly with respect to computational results obtained with the use of parallel machines.

In this section, parallel algorithms for several combinatorial problems are studied; the reader is referred to Kindervater and Lenstra [96] for a complete commented survey of the literature up to 1983. Basically, four classes of problems are considered: sorting and related problems (searching, merging, data transmission and permuting), graph theory problems, well-solvable and NP-hard optimization problems. Also included in this survey of parallel algorithms are distributed algorithms, suited to asynchronous MIMD machines in which no common memory is available, so that the processors (connected by means of an interconnection network) communicate by exchanging messages.

4.1. Sorting, Merging and Data Permutation

Some references on sorting and related problems are presented below; the majority of them are quoted and commented by Kindervater and Lenstra [96]:

(i) Sorting networks: Batcher [13], Muller and Preparata [127], Ajtai et alii [3], Siegel [166], Leighton [112] and Bilardi and Preparata [23].

(ii) Algorithms for shared memory SIMD and MIMD computers: Even [57], Gavril [63], Valiant [183], Todd [177], Hirschberg [75], Preparata [138], Barlow et alii [11], Shiloach and Vishkin [163], Reischuk [148], Borodin and Hopcroft [26], Kruskal [99], Aigner [2], Akl [4], Reischuk [149] and Cole and Yap [37].

(iii) Algorithms for mesh-connected networks: Orcutt [135], Thompson and Kung [175], Nassimi and Sahni [128, 130], Kumar and Hirschberg [101] and Lang et alii [106].

(iv) Algorithms for other interconnection networks and architectures: Baudet and Stevenson [14], Valiant and Brebner [187], Lev et alii [114], Nassimi and Sahni [131, 132], Valiant [184], Reif and Valiant [147], Nath et alii [134], Stout [171] and Gottlieb and Kruskal [69].

The reader is referred to Bitton et alii [24] for a survey on parallel sorting algorithms, where a taxonomy of parallel sorting is proposed, encompassing a broad range of sorting algorithms. Some implementations of sorting algorithms on different types of parallel architectures are presented below, in order to illustrate the interrelationships between algorithms, implementations and architectures.

Example 1. Implementation of Batcher's bitonic sorting algorithm on a perfect shuffle interconnection

Batcher [13] proposed a sorting algorithm that can sort N numbers in $O(\log^2 N)$ time when embedded in a sorting network consisting of $O(N \log^2 N)$ comparison-exchange modules. Stone [170] shows how a single rank of $N/2$ modules is all that is necessary to perform Batcher's bitonic sort when a perfect shuffle interconnection network is used.

The central idea of Batcher's algorithm is that the network shown in figure 12 can sort a bitonic sequence of $N = 2^k$ numbers. A sequence of N real numbers $a_0, a_1, a_2, \ldots, a_{N-1}$ is bitonic (i) if there is an index i such that the sequence $a_0, a_1, a_2, \ldots, a_i$ is monotonically increasing and the sequence $a_i, a_{i+1}, a_{i+2}, \ldots, a_{N-1}$ is monotonically decreasing or (ii) if the sequence can be shifted cyclically so that condition (i) is satisfied. This network consists basically of $N/2$ comparison-exchange modules that perform a comparison of a_i and $a_{i+N/2}$ for $i = 0, 1, 2, \ldots, N/2 - 1$ and place the lower (resp. higher) value of the two

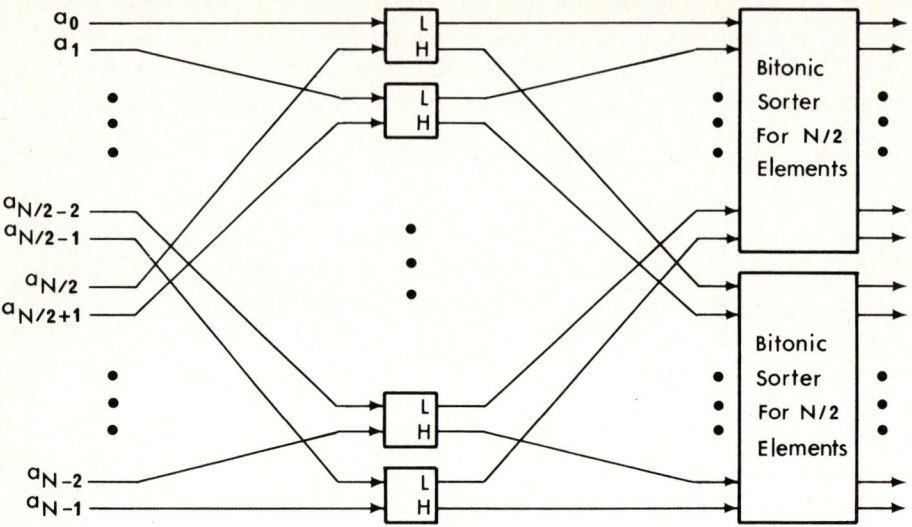

Fig. 12. The structure of a bitonic sorter for N elements.

inputs at the output gate labelled L (resp. H). Batcher proved that after performing the comparison-exchange operations (i) the subsequences consisting of the first $N/2$ numbers and the last $N/2$ numbers are both bitonic and (ii) every number in the first subsequence is not greater than any number in the second subsequence. A complete bitonic sorter can be devised based on a recursive construction technique which uses the basic structure presented in figure 12 (the last stage of the network consists of a bitonic sorter for two numbers, which is simply a comparison-exchange element). Then, a complete sorter can be constructed by (i) successively sorting and merging smaller sequences into larger ones until a bitonic sequence of size N is obtained and (ii) inputing this bitonic sequence to the bitonic sort network already described.

Using the fact that every pair of items that enter a comparison-exchange module differ only by a single bit in their binary representation, Stone [170] shows that the bitonic sort algorithm can be implemented with a single rank of $N/2$ comparison-exchange modules connected by means of a shuffle network, as illustrated in figure 13. It is also shown that if a mask is used to control the behavior of the compare-exchange modules, reversing the outputs of appropriate modules at precise iterations, the first phase of the algorithm (sorting and merging smaller sequences until a bitonic sequence is obtained) can also be implemented on the same network. The total number of comparison-exchange steps is $k(k+ + 1)/2$ and the number of shuffles of the data is $k(k-1)$, hence the algorithm takes $O(\log^2 N)$ time.

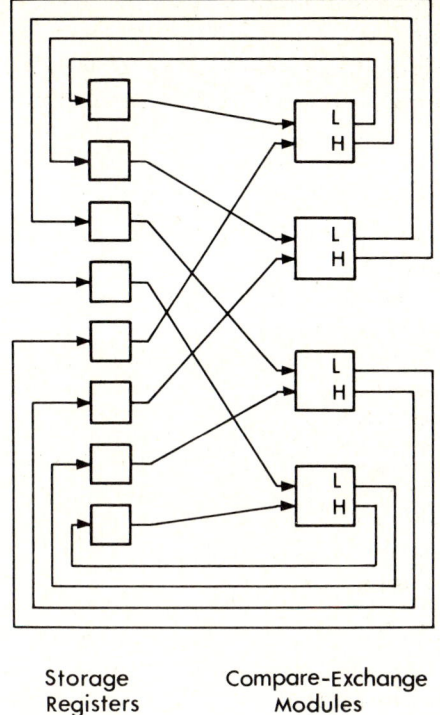

Storage　　　Compare-Exchange
Registers　　　Modules

Fig. 13. Sorting processor for performing Batcher's bitonic sort algorithm.

Example 2. Implementation of the odd-even transposition sort on a mesh-connected network

Given N numbers $a_0, a_1, a_2, \ldots, a_{N-1}$ to be sorted in ascending order, the odd-even transposition sort works as follows; see Knuth [97]:

Step 0: Set $k \leftarrow 1$.

Step 1: For all $i = 0, 1, 2, \ldots, \lfloor N/2 - 1 \rfloor$ such that $a_{2i+2} < a_{2i+1}$, set $t \leftarrow a_{2i+2}$, $a_{2i+2} \leftarrow a_{2i+1}$ and $a_{2i+1} \leftarrow t$.

Step 2: For all $i = 0, 1, 2, \ldots, \lfloor N/2 - 1 \rfloor$ such that $a_{2i+1} < a_{2i}$, set $t \leftarrow a_{2i+1}$, $a_{2i+1} \leftarrow a_{2i}$ and $a_{2i} \leftarrow t$.

Step 3: Set $k \leftarrow k + 1$. If $k \leq N/2$ go back to step 1. Otherwise, stop.

Assume now that $N = n^2$ numbers should be sorted on an $n \times n$ square mesh-connected network where each processor is connected to all its neighbors and wrap-around connections between processors at the perimeter do not exist. Observe that, in the worst case, elements initially loaded at opposite corner processors will have to be transposed during the sort and at least $4(n-1)$ unit-distance routing steps will be needed. Hence, $O(n)$ is a lower bound for the time-complexity of any algorithm designed for sorting n^2 numbers on an $n \times n$ mesh-connected network.

As Thompson and Kung [175] state, the processors of the mesh may be indexed by any function that is a one-to-one mapping from $1, 2, \ldots, n \times 1, 2, \ldots, n$ onto $0, 1, 2, \ldots, N-1$. The N items to be sorted are initially loaded in the N processors and the sorting problem is defined to be the problem of moving the jth smallest element to the processor indexed by $j-1$ for all $j = 1, 2, 3, \ldots, N$. Three different ways of indexing the processors (row-major indexing, shuffled row-major indexing and snake-like row-major indexing) are illustrated in figure 14. Orcutt [135], Thompson and Kung [175] and Nassimi and Sahni [128], among others, present adaptations of Batcher's bitonic sort for sorting $N = n^2$ elements in $O(N)$ time on $n \times n$ mesh-connected networks using different indexing schemes.

It is shown now how the odd-even transposition sort can be implemented in $O(N)$ time for sorting $N = n^2$ elements on an $n \times n$ mesh-connected network with the processors indexed in snake-like row-major order (which can be seen as a linear array; see figure 14). Step 2 is a «cheap comparison-interchange operation»: the contents of even and odd processors located at the same row of the mesh are compared and eventually interchanged; this can be done in $O(1)$ parallel time. Step 1 is an «expensive comparison-interchange operation»: the processors which have their contents interchanged are sometimes located at different rows; this can also be done in $O(1)$ parallel time, but more time is needed for data routing. Since $N/2$ steps are necessary, the overall algorithm takes $O(N)$ time.

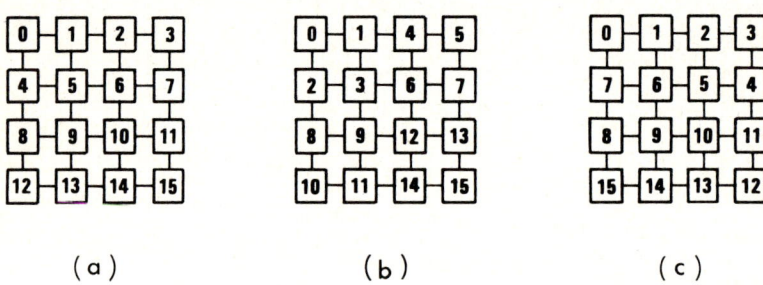

Fig. 14. (a) row major-indexing, (b) shuffled row-major indexing and (c) snake-like row-major indexing.

Example 3. Implementation of an enumeration sort on a SIMD machine with shared memory where simultaneous writes are prohibited

Enumeration sort methods are based on counting the number of items which should precede each item; see Knuth [97]. This kind of algorithm consists basically of three distinct steps: (i) counting (computation of the relative ranks), (ii) ranking (computation of the position of each item) and (iii) permutation (routing each element to the processor corresponding to its rank). The algorithm proposed by Muller and Preparata [127] is studied below, assuming that the sequence $a_1, \ldots, a_N$ to be sorted is such that $a_i \neq a_j$ for $i \neq j$. The notation

$$\text{par } [p \leqslant i \leqslant q] \, s_i$$

is introduced to denote that the statements s_i are to be executed in parallel for all values of the index i in the given range, each one corresponding to a different processor. Their algorithm can be stated as follows; see Kindervater and Lenstra [95]:

Step 1 (counting): par $[1 \leqslant i, j \leqslant N]$ $r_{ij} \leftarrow 1$, if $a_i \leqslant a_j$; $r_{ij} \leftarrow 0$, otherwise.

Step 2 (ranking): par $[1 \leqslant j \leqslant N]$ $p_j \leftarrow \text{sum} \{r_{ij} \mid 1 \leqslant i \leqslant N\}$.

Step 3 (routing): par $[1 \leqslant j \leqslant N]$ $a_{p_j} \leftarrow a_j$.

Step 1, corresponding to the computation of the relative ranks r_{ij}, can be performed in $O(1)$ time with N^2 processors or in $O(\log N)$ time with $N^2/\log N$ processors. For the computation of the positions p_j the reader is referred to Dekel and Sahni [47] and Kindervater and Lenstra [95], where a partial sums algorithm is described: step 2 can be performed in $O(\log N)$ time with $N^2/\log N$ processors. Finally, the data routing step can be performed in $O(1)$ time using only N processors. Hence, the algorithm requires $O(\log N)$ time and $N^2/\log N$ processors. Nassimi and Sahni [132] study the implementation of an enumeration sort on a cube connected network in $O(\log n)$ time using $N^2 = 2^{2n}$ processors.

4.2. Graph Theory and Path Problems

The reader is referred to Kindervater and Lenstra [96] for a commented survey of the literature up to 1983 and to Ivanov and Shevchenko [84] for a survey of the russian literature. Some complexity results concerning the implementation of parallel algorithms for graph theory and path problems on synchronous SIMD and MIMD architectures with shared memory are presented in table 2. These results are complemented by table 3, where some examples of graph algorithms on interconnection networks, asynchronous MIMD models and pipelined vector processors are presented. The reader is referred to the references, as well as to Quinn and Deo [144], for more details on the algorithms, implementations and architectures considered. Other references are Crane [42], one of the first papers about parallel graph algorithms; Hirschberg [74], Goldschlager [65] and Wyllie [196], where some complexity results for graph problems are given; Wisniewski and Sameh [195], where the single source shortest path problem is solved through the resolution of systems of the form $x = Ax + b$ in the regular algebra of Carré, by solution methods (mostly known from linear algebra) that are parallelized; Tsin [179], where algorithms for the bridge-connectivity and biconnectivity problems are presented and it is shown how they can be implemented on any parallel computer model on which an ordinary matrix multiplication algorithm

Table 2.
Parallel algorithms for some graph problems for synchronous shared memory computer models (n = number of vertices, m = number of arcs, d = diameter of the graph, SR = simultaneous reads allowed, SW = simultaneous writes allowed).

Problem	Parallel time complexity	Number of processors	Reference and model of machine
Minimum spanning tree	$O(\log^2 n)$	$n^2/\log n$	Savage [155], SIMD-SR
	$O((n^2/p)\log n)$	$p \leq n$	Deo and Yoo [50], MIMD-SR
	$O(\log^2 n)$	n^2	Savage and Ja'Ja' [157], SIMD-SR
	$O(\log^2 n)$	n^2	Nath and Maheshwari [133], SIMD
	$O(\log^2 n)$	$n^2/\log^2 n$	Chin et alii [34], SIMD-SR
	$O(\log n)$	n^5	Kucera [100], SIMD-SR-SW
Transitive closure	$O(\log^2 n)$	$n^2/\log n$	Hirschberg et alii [76], SIMD-SR
	$O(\log^2 n)$	$n^2/\log^2 n$	Chin et alii [34], SIMD-SR
	$O(\log n)$	n^4	Kucera [100], SIMD-SR-SW
Shortest paths	$O(\log^2 n)$	$n^3/\log n$	Savage [155], SIMD-SR
	$O(\log n)$	n^4	Kucera [100], SIMD-SR-SW
Connected components	$O(\log^2 n)$	n^3	Reghbati and Corneil [146], SIMD-SR
	$O(\log^2 n)$	$n^2/\log n$	Hirschberg et alii [76], SIMD-SR
	$O(\log n \log d)$	$n^3/\log n$	Savage and Ja'Ja' [157], SIMD-SR
	$O(\log^2 n)$	$m + n \log n$	Savage and Ja'Ja' [157], SIMD-SR
	$O(\log n)$	$n + 2m$	Shiloach and Vishkin [164], MIMD-SR-SW
	$O(\log^2 n)$	n^2	Nath and Maheshwari [133], SIMD
	$O(\log^2 n)$	$n^2/\log^2 n$	Chin et alii [34], SIMD-SR
	$O(\log n)$	n^4	Kucera [100], SIMD-SR-SW
Biconnected components	$O(\log^2 n)$	$n^3/\log n$	Savage and Ja'Ja' [157], SIMD-SR
	$O(\log^2 n)$	$n^2/\log^2 n$	Tsin and Chin [181], SIMD-SR
	$O(\log n)$	$n + m$	Tarjan and Vishkin [173], SIMD-SR-SW
Triconnected components	$O(\log^2 n)$	n^4	Ja'Ja' and Simon [86], SIMD-SR
Weakly-connected components	$O(\log^2 n)$	$n^2/\log^2 n$	Chin et alii [34], SIMD-SR
Breadth-first and depth-first search	$O(n + m/p)$	p	Eckstein and Alton [56], MIMD-SR
Bridge finding	$O(\log^2 n)$	$n^2 \log n$	Savage and Ja'Ja' [157], SIMD-SR
	$O(\log^2 n)$	$n^2/\log^2 n$	Tsin and Chin [181], SIMD-SR
Spanning forest	$O(\log^2 n)$	$n^2/\log^2 n$	Chin et alii [34], SIMD-SR
	$O(\log^2 n)$	$n^2/\log^2 n$	Tsin and Chin [181], SIMD-SR
Fundamental cycles	$O(\log^2 n)$	n^3	Savage and Ja'Ja' [157], SIMD-SR
	$O(\log^2 n)$	$n^2/\log n$	Tsin and Chin [181], SIMD-SR
Separation vertices	$O(\log^2 n)$	$n^2/\log^2 n$	Tsin and Chin [181], SIMD-SR
Bisection width in trees	$O(\log^3 n)$	n^2	Goldberg and Miller [64], SIMD-SR-SW
Planarity testing	$O(\log^2 n)$	n^4	Ja'Ja' and Simon [86], SIMD-SR
Topological sort of acyclic graphs	$O(\log n)$	n^4	Kucera [100], SIMD-SR-SW
Strong orientation of undirected graphs	$O(\log^2 n)$	n^3	Atallah [6], SIMD-SR
	$O(\log n)$	n^3	Atallah [6], SIMD-SR-SW
	$O(\log n)$	$n + m$	Vishkin [191], SIMD-SR-SW
	$O(\log^2 n)$	$n^2/\log^2 n$	Tsin [180], SIMD-SR

Table 3.
Parallel algorithms for some graph problems for interconnection networks, asynchronous MIMD machines and vector processors.

Problem	Architectures	References
All shortest paths	systolic array (hexagonal mesh) MIMD (Denelcor HEP) cube connected, perfect shuffle systolic array (hexagonal mesh) vector processor (CRAY-1)	Levitt and Kautz [115] Deo et alii [49] Dekel et alii [44] Rote [151] Ribeiro [150]
Shortest path tree	MIMD (Denelcor HEP) cube connected, perfect shuffle binary tree, systolic array, MIMD (asynchronous shared memory) MIMD (full interconnection) vector processor (CRAY-1)	Deo et alii [49] Dekel et alii [44] Mateti and Deo [120] Chandy and Misra [33] Ribeiro [150]
Minimum spanning tree	systolic array (square mesh) binary tree MIMD (full interconnection) perfect shuffle, orthogonal tree two-dimensional square mesh systolic array (linear) MIMD (asynchronous shared memory)	Levitt and Kautz [115] Bentley [16] Humblet [80], Parker and Samadi [137], Gallager et alii [61], Lavallée and Roucairol [110] Nath and Maheshwari [133] Awerbuch and Shiloach [10] Savage [156] Lavallée [107]
Enumeration of spanning trees	MIMD (Neptune System)	Mai and Evans [118]
Transitive closure	two-dimensional square mesh systolic array (hexagonal mesh) vector processor (CDC CYBER 205)	Guibas et alii [70], Van Scoy [189] Rote [151] Courteille and Fraisse [41]
Topological sort and critical paths	cube connected, perfect shuffle	Dekel et alii [44]
Connected components	k-dimensional mesh perfect shuffle, orthogonal tree two-dimensional square mesh k-dimensional mesh systolic array (linear)	Nassimi and Sahni [129] Nath and Maheshwari [133] Awerbuch and Shiloach [10] Hambrush [71] Savage [156]
Radius, diameter, center and medians	cube connected, perfect shuffle	Dekel et alii [44]
Bridges, articulation points and shortest cycle	two-dimensional square mesh	Atallah and Kosaraju [7]

exists; Ja'Ja' [85], where general lower bound techniques are developed to determine the VLSI complexity of several graph problems; Vishkin and Wigderson [192], where a new technique for proving lower bounds for parallel computation is introduced, enabling the obtention of non-trivial tight lower bounds for shared memory models of parallel computation that allow several processors to have simultaneous access to the same memory location; Bertossi and Bonuccelli [19], where polylog parallel time algorithms are given for finding maximum cliques, maximum independent sets, minimum clique covers and minimum dominating sets of interval graphs, as well as hamiltonian circuits and the minimum bandwidth of proper interval graphs.

Two implementations of parallel graph algorithms are described below. In both cases, the tree-structured searching machine originally devised by Bentley and Kung [17] for searching and sorting problems is used; see figure 15. This structure contains three types of processors, represented by circles (which broadcast data), squares (which store data and compute) and triangles (which combine their inputs in a suitable way). The input and ouput nodes (processors) are connected.

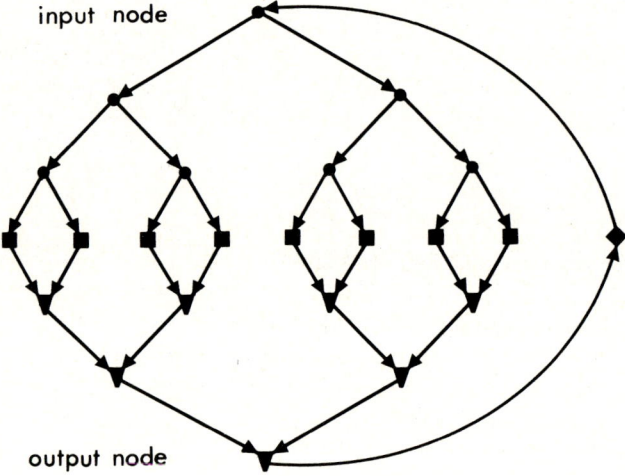

Fig. 15. Structure of the tree machine.

Example 1. Minimum spanning tree

Given a complete undirected graph G with vertex set $V = \{1, 2, \ldots, N\}$ and edge lengths c_{ij}, the algorithm proposed by Prim [141] obtains the minimum spanning tree of G after N iterations in $O(N^2)$ time. At the end of iteration k, let V^k be the set of nodes already spanned and T^k the set of edges used to span them. At iteration $k + 1$, set $V^{k+1} \leftarrow V^k \cup \{\alpha\}$ and $T^{k+1} \leftarrow T^k \cup \{(\alpha, \beta)\}$, where α and β are such that

$$c_{\alpha\beta} = \min_{\substack{i \notin V^k \\ j \in V^k}} \{c_{ij}\}.$$

Bentley [16] studies a parallel version of this algorithm, using a binary tree machine with N square processors (suppose that $N = 2^q - 1$ for some integer q). The algorithm is initialized with $V^1 = \{1\}$ and $T^1 = \phi$. At the end of iteration k, each active processor P_i (such that $i \notin V^k$) stores the values β_i and d_i given by

$$d_i = c_{i\beta_i} = \min_{j \in V^k} \{c_{ij}\}.$$

For each $k = 1, 2, \ldots, N - 1$ the following computations are performed:

(i) Apply a minimum finding procedure to obtain the node α such that d_α is the minimum of the values d_i among all active processors, i.e.,

$$d_\alpha = \min_{i \notin V^k} \{d_i\}.$$

(ii) Let $\beta = \beta_\alpha$, $V^{k+1} \leftarrow V^k \cup \{\alpha\}$ and $T^{k+1} \leftarrow T^k \cup \{(\alpha, \beta)\}$. Next, send α from the input node in order to turn-off P_α (forever) and each P_i with $d_i \leq c_{i\alpha}$ (temporarily for the rest of this iteration).

(iii) For all active processors P_i recompute their contents ($\beta_i \leftarrow \alpha$ and $d_i \leftarrow c_{i\alpha}$), turn-on all processors P_i temporarily turned-off, set $k \leftarrow k + 1$ and go back to step (i).

Hence, data is broadcast three times from the input node to the output node during each iteration. Since each of these phases can be performed in $O(\log N)$ time and $N - 1$ iterations are necessary, the overall computation takes $O(N \log N)$ time.

Example 2. Shortest path from a single source

Given a complete directed graph G with vertex set $V = \{1, 2, \ldots, N\}$ and positive arc lengths c_{ij}, the algorithm proposed by Dijkstra [53] solves the shortest path problem from a single source in $O(N^2)$ time. Mateti and Deo [120] describe a parallel implementation of Dijkstra's algorithm on the tree machine shown in figure 15. Each square processor P_i contains the length d_i of the shortest path already obtained from the source to node i and the initialization is such that $d_i = +\infty$ for all $i = 1, 2, \ldots, N$. Let V^k be the set formed by all vertices for which the shortest path from the source is already known, with $V^1 = \{\text{source}\}$. For each $k = 1, 2, \ldots, N - 1$ the following computations are performed:

(i) Input to the root a pair of numbers (x, d_x) consisting of the (last) vertex that was included in V^k (at the end of the previous iteration) and the length of the shortest path from the source to this node. The pair (source, 0) is inputed for $k = 1$.

(ii) Broadcast the input to the square processors through the circle processors.

(iii) Each active square processor P_i receiving (x, d_x) as input performs the following computations: if $x \neq i$ set $d_i \leftarrow \min\{d_i, d_x + c_{xi}\}$ and broadcast the pair (i, d_i); otherwise turn-off P_i.

(iv) Each triangle processor receives (x, d_x) from the left and (y, d_y) from the right and broadcasts (z, d_z), where $z = x$ if $d_x < d_y$ and $z = y$ otherwise.

(v) The node α that should be included in V^k is obtained at the output node. Set $V^{k+1} \leftarrow V^k \cup \{\alpha\}$, $k \leftarrow k + 1$ and broadcast the pair (α, d_α) to the input node.

Each iteration takes $O(\log N)$ time and $N - 1$ iterations are necessary, thus the overall computation takes $O(N \log N)$ time, as for the minimum spanning tree algorithm.

4.3. Well-Solvable Combinatorial Optimization Problems

Even though this class is not as rich as the previous one in terms of applications of parallelism, efficient algorithms have been proposed for some scheduling problems, network flows and linear programming.

(a) *Scheduling problems*

Dekel and Sahni [45, 46, 47, 48] study polylog parallel time algorithms for many scheduling problems. Their algorithms are based on binary trees which provide the basis for computing the partial sums of n numbers in $O(\log n)$ time with $O(n/\log n)$ processors and for finding a maximum matching in n-vertex convex bipartite graphs in $O(\log^2 n)$ time using $O(n)$ processors; the reader is referred to Kindervater and Lenstra [95] for a very clear presentation of some of these algorithms, as well as to Kindervater and Lenstra [96] for a summary of the main results concerning their complexity.

(b) *Network flow problems*

Goldschlager et alii [68] show that the maximum flow problem is log-space complete for P, using a log-space transformation from the monotone circuit value problem. Shiloach and Vishkin [165] present an algorithm for solving the maximum flow problem that is closely related to the sequential methods due to Dinic and Karzanov. Their algorithm is devised for a synchronous shared memory machine (simultaneous reads allowed, simultaneous writes allowed provided the same value is written) and has parallel time complexity $O((n^3/p) \log n)$ when $p \leqslant n$ processors are available. Megiddo [123] points out that the efficiency of sequential algorithms for one problem may be improved by exploiting the parallelism in other problems. It is shown, for instance, how parallel sorting algorithms turn out to be useful for cost-effective resource allocation, and how parallel all-shortest-path algorithms serve for the minimum ratio cycle problem. Sometimes, the study of parallel algorithms for a given problem may also help to devise and understand the behavior of sequential algorithms. As an example, the algorithm proposed by Shiloach and Vishkin [165] induces a new rather simple $O(n^3)$ algorithm for the maximum flow problem. Johnson and Venkatesan [90] consider the model of a synchronous shared memory machine (simultaneous reads allowed, no simultaneous writes allowed) and show that the maximum flow in planar directed n-vertex networks can be computed in $O(\log^3 n)$ time using $O(n^4)$ processors or in $O(\log^2 n)$ time using $O(n^6)$ processors (it is also shown that $O(\log^2 n)$ time and $O(n^4)$ processors suffice for planar undirected networks). More recently, Janiga and Koubek [87] showed that the maximum flow problem in planar graphs can be solved with only $O(n^4)$ processors ($O(n^3)$ suffice when the graph is known to be planar beforehand) in $O(\log^2 n)$ time.

Bertsekas and El-Baz [22] study distributed asynchronous relaxation methods for convex network flow problems, well suited for implementation on parallel machines. In their pure form these algorithms modify the dual variables (node prices) one at a time, using only local node information while aiming to improve the dual cost. They can be shown to converge for problems with strictly convex arc costs, even if relaxation at each node is carried out asynchronously with out-of-date price information from neighboring nodes. In a more recent paper, Bertsekas [21] describes a distributed algorithm for solving the classical linear assignment problem. This algorithm employs exclusively pure relaxation steps whereby the prices of sources and sinks are changed individually on the basis of only local (neighboring) node price information. It can be implemented in an asynchronous manner and seems to be efficient for problems with small cost ranges.

(c) *Linear programming and decomposition*

Using a log-space transformation starting from the unit resolution problem (see Jones and Laaser [91]) and in conjunction with Kachian's algorithm, Dobkin et alii [54] show that linear programming is log-space complete for P. Other results concerning linear programming can be found in Kamdoum [92], where it is shown that each pivot step of the simplex method can be executed p times faster when p processors are available (for small values of p, when compared to the number of variables and constraints), and Megiddo [122], where a parallel implementation of a linear programming method is proposed, running in $O(\log^n m)$ time when the number n of variables is fixed (m is the number of constraints). Decomposition schemes using parallel computation are studied in Dutta et alii [55], where a SIMD implementation of the simplex method is presented and a parallel architecture for implementing the Dantzig-Wolfe decomposition algorithm is proposed, and Ho [77], where three approaches for the introduction of parallel computation in linear programming are compared: parallel simplex paths, parallel block pricing and Dantzig-Wolfe decomposition. A parallel successive overrelaxation method for linear programming is proposed by Mangasarian and DeLeone [119], suitable for a multiprocessor such as the Madison Crystal multicomputer. The Crystal multicomputer consists of twenty VAX 11/750, each with 2 megabytes of memory, linked via a 10 megabit/second token ring. Parallel algorithms for large-scale traffic assignment and generalized networks, along with computational results on Crystal, are presented by Meyer [124].

4.4. Enumeration Methods and NP-Hard Optimization Problems

In this section, parallel implementations of enumeration schemes (such as dynamic programming and branch and bound) and other results for some NP-hard optimization problems are discussed.

Dynamic programming is a technique very well suited for implementation on systolic arrays and synchronous MIMD (or SIMD) machines, since a regular sequence of many highly similar and quite simple instructions is performed. Casti et alii [31] seems to be the first paper where a parallel implementation of the dynamic programming technique is proposed. A mesh-connected architecture is considered and different schemes are studied, corresponding to parallel state, decision and stage algorithms. Al-Dabass [43] studies the efficiency of the algorithms proposed in the previous paper, considering their implementation on a master-slave architecture. Bertolazzi and Pirozzi [18] make some assumptions on the general problem to which the previous algorithms are applied, in such a way that communication steps no longer depend on the size of the problem, making it possible to propose more efficient architectures in which communication can be carried out in constant time. Two classes of problems which can be solved by parallel algorithms with a reduction of complexity are proposed. Some applications are discussed, concerning the problems of optimal parenthesization and extraction from a picture of one line of fixed length which is optimal according to some given figure of merit; see respectively Guibas et alii [70] and Montanari [126] for more details about these problems. The reader is also referred to Bertsekas [20], where asynchronous distributed algorithms for solving dynamic programming problems are discussed. The class of problems considered is very broad and includes shortest path problems and finite and infinite horizon stochastic optimal control problems.

Parallel dynamic programming approaches have been considered for the resolution of the 0-1 knapsack problem, formulated as to

$$\text{maximize} \quad \sum_{i=1}^{n} c_i x_i$$

$$\text{subject to} \quad \sum_{i=1}^{n} a_i x_i \leqslant b$$

$$x_i \in \{0, 1\} \quad i = 1, 2, \ldots, n,$$

where b and a_i are positive integers, for $i = 1, 2, \ldots, n$. The first of such approaches was the one proposed by Casti et alii [31], in which the states of each stage are handled in parallel. This algorithm is described by Kindervater and Lenstra [95] as:

$$\text{par } [0 \leqslant z \leqslant b] \quad c(0, z) \leftarrow 0$$

for $j = 1$ to n do:

$$\text{par } [0 \leqslant z < a_j] \quad c(j, z) \leftarrow c(j-1, z)$$

$$\text{par } [a_j \leqslant z \leqslant b] \quad c(j, z) \leftarrow \max \{c(j-1, z), c(j-1, z - a_j) + c_j\}$$

where $c(j, z)$ denotes the value of the optimal solution of the subproblem

$$\text{maximize} \sum_{i=1}^{j} c_i x_i$$

$$\text{subject to} \sum_{i=1}^{j} a_i x_i = z$$

$$x_i \in \{0, 1\} \quad i = 1, 2, \ldots, j.$$

It can be easily seen that this scheme requires $O(n)$ time when $b + 1$ processors are available. Ribeiro [150] uses the same idea to develop a vector implementation for the resolution of this problem on a CRAY-1 computer. The numerical results obtained show that maximum efficiency is attained and computer times are reduced by a significant factor through vectorization. A theoretical result is obtained by Yao [197], who considers the complexity of solving the knapsack problem on a parallel computer with real arithmetic and branching operations. A time-processor trade-off constraint is derived and, in particular, it is shown that an exponential number of processors has to be used if the problem is to be solved in $O(\sqrt{n})$ time.

In contrast to dynamic programming, branch and bound methods are more suited to implementation in a distributed way on asynchronous MIMD machines. The subproblems that should be solved are distributed and each processor deals with a subproblem. Bounds and feasible solutions are either communicated through the processors or stored in a shared global memory (which can be used for processor communication). Desai [51] proposes a staged MIMD system to solve 0-1 integer programs using implicit enumeration. El-Dessouki and Huen [52] study a distributed branch and bound scheme in which each part (subtree) of the search tree is associated with one of the processors after the first branching operation. The processors communicate only to broadcast new feasible solutions and bounds or to redistribute the subproblems that remain to be solved. Depth-first search is used by each processor for exploring the subtree associated with it. Another approach is presented by Imai and Fukumura [83] for shared memory MIMD machines. Each processor selects one unresolved subproblem (following a global depth-first strategy) and either finds that it is a leaf of the search tree or creates all the successors of the corresponding node. For both of these approaches numerical results obtained through simulation are presented. Burton et alii [29] consider distributed branch and bound algorithms for execution on cube connected networks of processors. Burton et alii [30] and Lai and Sahni [104, 105] analyse anomalies arising in the context of distributed branch and bound algorithms: it is shown that it is quite possible for a parallel branch and bound algorithm using n_2 processors to take more time than one using n_1 processors even though $n_1 < n_2$; furthermore, it is also possible to achieve speedups greater than n_2/n_1 (experimental results for the 0-1 knapsack and traveling salesman problems are presented). Lavallée and Roucairol [109] review the previous approaches, compare two general schemes for implementing distributed branch and bound algorithms and specificate both of them using CSP. The first scheme is called vertical parallelization, in which each processor deals with one of the subtrees generated after the first branching operation (as proposed by El-Dessouki and Huen [52]), seeming to be more suited to computer networks or multiprocessors with slow intercommunication. The second one is a horizontal parallelization scheme, in which each processor is associated with one node of the search tree and expands all of its descendants (as proposed by Imai and Fukumura [83]), more suited to architectures with a large number of processors and efficient message passing patterns. A branching scheme based on the Bellmore-Malone criterion is proposed, in order to better use the capabilities of the parallel processors. Roucairol [152] studies the adaptation of the second parallelization scheme (proposed in the precedent paper) for asynchronous shared memory MIMD machines such as the CRAY-XMP. The distribution of the subproblems to be solved is done by the use of a shared list which contains information about every node that should be expanded. This list is kept sorted according to some criterion corresponding to the selection rule (depth-first search, breadth-first search, best-first search). This parallel branch and bound scheme is applied to the resolution of quadratic assignment problems and numerical results obtained on a simulator of the CRAY-XMP are reported.

A different approach is adopted by Wah and Ma [193], who propose the design of an architecture for implementing parallel branch and bound algorithms. The critical element in this architecture is the selection network, which selects subproblems to be evaluated by the parallel processors based on a kind of best-first search. In a (synchronous) parallel branch and bound algorithm using best-first search, a set of subproblems equal in size to the number of processors must be selected in each iteration. Provided that these subproblems are those with the smallest lower bounds, they do not have to be selected in a sorted order. Selection is carried out in parallel and it is shown that an unidirectional ring network is cost-effective for this operation, as well as in connecting the memory controllers so that the workload of the system is balanced. The other major components of this architecture are: secondary storage (which provides virtual memory support for excess subproblems which can not be stored in the subproblem memory controllers due to memory limitations), secondary storage redistribution network (which allows a subproblem memory controller to store and access subproblems in any set or subset of the secondary storage modules), parallel processors (which are general purpose computers for partitioning subproblems and evaluating lower bounds), global data register (which is accessible to all memory controllers and contains the value of the best feasible solution found) and subproblem memory controllers (which are designed to manage the local list of subproblems and communicate with other controllers through the selection network). This architecture is simulated by a simulation model written in C language and implemented on a VAX 11/780 computer with virtual storage. Numerical results, concerning the resolution of vertex covering problems, are given.

Some results concerning the traveling salesman problem are available. Pruul [142] describes a parallel implementation of the subtour elimination algorithm for an asynchronous MIMD machine, in which each processor performs its own depth-first search. The algorithm is simulated on a sequential machine. Mohan [125] shows how it can be solved on Cm*, a multiprocessor system, using two parallel search programs based on branch and bound algorithms. One of these programs is synchronous and has a master-slave architecture, while the other is asynchronous and has an egalitarian architecture. Their absolute execution times and speedups are discussed. Quinn and Deo [143] describe an implementation of the farthest-insertion heuristic for the Euclidean traveling salesman problem to run on the Denelcor HEP with p processors in $O(n^2/p + np)$ time. Kindervater [94] presented the parallel complexity of several different heuristics for the traveling salesman problem, showing that the nearest-addition heuristic (and others based on the same principles) can be implemented in polylog parallel time $O(\log^2 n)$, while the problem corresponding to the nearest-neighbor heuristic is log-space complete (by a reduction from the circuit value problem).

Other results available for NP-hard combinatorial optimization problems are those concerning the maximum cardinality clique and color cost problems, studied by Browning [28]. A binary tree based algorithm (each node of the tree is a processor with general computing capability) with time complexity $O(n^2)$ with $2^n - 1$ processors is proposed for the maximum cardinality clique problem on n-vertex undirected graphs. The color cost problem is an adaptation of the k-colorability problem. Given an undirected graph with n nodes and a set of n colors, each with an associated cost, find a minimum cost coloring of the graph such that no nodes sharing an edge are colored with the same color. The algorithm proposed by Browning runs in $O(n^2)$ time on a binary tree machine with $2n^n - 1$ processors.

Acknowledgements

This work was sponsored by FINEP-Financiadora de Estudos e Projetos, under research contract number 6.2.84.0416.00. The author wishes to thank Ivan Lavallée, for many constructive comments and important remarks; Miguel Menasche, for the reading of the final draft of this monograph and some fruitful discussions; William Koelsch and Pamela Schmitt, for improving the readability of this work; and Angela Jaconianni, for the carefully prepared illustrations. The author is also grateful to Henri Maître for the warm reception at the Image Laboratory of the ENST (Paris) where this work was finished, as well as to IBM Brasil for providing financial support which enabled the travel between Rio de Janeiro and Paris.

References

[1] A.V. Aho, J.E. Hopcroft and J.D. Ullman, *The Design and Analysis of Computer Algorithms*, Addison-Wesley, Reading, 1974.
[2] M. Aigner, «Parallel Complexity of Sorting Algorithms», *Journal of Algorithms* 3, 79 - 88, 1982.
[3] M. Ajtai, J. Komlos and E. Szemeredi, «An $O(n \log n)$ Sorting Network», *Proceedings of the 15th Annual Symposium on Theory of Computing*, 1 - 9, 1983.
[4] S.G. Akl, «An Optimal Algorithm for Parallel Selection», *Information Processing Letters* 19, 47 - 50, 1984.
[5] F. André, P. Frison and P. Quinton, «Algorithmes Systoliques: de la Théorie à la Pratique», *Rapport de Recherche no. 214,* INRIA, Rennes, France, 1983.
[6] M.J. Atallah, «Parallel Strong Orientation of an Undirected Graph», *Information Processing Letters* 18, 37 - 39, 1984.
[7] M.J. Atallah and S.R. Kosaraju, «Graph Problems on a Mesh-Connected Processor Array», *Proceedings of the 14th Annual ACM Symposium on Theory of Computing*, 345 - 353, 1982.
[8] G. Ausiello and P. Bertolazzi, «Parallel Computer Models: An Introduction», *Proceedings of the Symposium on Vector and Parallel Processors*, Rome, Italy, 1982.
[9] J.H. Austin Jr., «The Burroughs Scientific Processor», in C.R. Jesshope and R.W. Hockney, eds., *Infotech State of The Art Report: Supercomputers - Vol. 2*, 1 - 31, Infotech Intl. Ltd., Maidenhead, 1979.
[10] B. Awerbuch and Y. Shiloach, «New Connectivity and MSF Algorithms for Ultracomputer and PRAM», *Proceedings of the 1983 International Conference on Parallel Processing*, 175 - 179, 1983.
[11] R.H. Barlow, D.J. Evans and J. Shanehchi, «A Parallel Merging Algorithm», *Information Processing Letters* 13, 103 - 106, 1981.
[12] G.H. Barnes, R.M. Brown, M. Kato, D.J. Kuck, D.L. Slotnick and R.A. Stokes, «The ILLIAC-IV Computer», *IEEE Transactions on Computers* C - 17, 746 - 757, 1968.
[13] K.E. Batcher, «Sorting Networks and Their Applications», *Proceedings of the AFIPS Spring Joint Computer Conference* 32, 307 - 314, 1968.
[14] G. Baudet and D. Stevenson, «Optimal Sorting Algorithms for Parallel Computers», *IEEE Transactions on Computers* C-27, 84 - 87, 1978.
[15] V.E. Benes, *Mathematical Theory of Connecting Networks and Telephone Traffic*, Academic Press, New York, 1965.
[16] J.L. Bentley, «A Parallel Algorithm for Constructing Minimum Spanning Trees», *Journal of Algorithms* 1, 51 - 59, 1980.
[17] J.L. Bentley and H.T. Kung, «A Tree Machine for Searching Problems», *Proceedings of the 1979 International Conference on Parallel Processing*, 257 - 266, 1979.
[18] P. Bertolazzi and M. Pirozzi, «Parallel Algorithms for Dynamic Programming Problems», *Proceedings of the Symposium on Vector and Parallel Processors*, Rome, Italy, 1982.
[19] A.A. Bertossi and M.A. Bonuccelli, «Some Parallel Algorithms on Interval Graphs», in P. Bertolazzi and F. Luccio, eds., *VLSI: Algorithms and Architectures*, 69 - 78, North-Holland, Amsterdam, 1985.
[20] D.P. Bertsekas, «Distributed Dynamic Programming», *IEEE Transactions on Automatic Control* 27, 610 - 616, 1982.
[21] D.P. Bertsekas, «A Distributed Asynchronous Relaxation Algorithm for the Assignment Problem», *Proceedings of the 24th IEEE Conference on Decision and Control* (to appear).
[22] D.P. Bertsekas and D. El-Baz, «Distributed Asynchronous Relaxation Methods for Convex Network Flow Problems», *LIDS Report* P-1417, Massachusetts Institute of Technology, Cambridge, 1984.
[23] G. Bilardi and F. Preparata, «A Minimum VLSI Network for $O(\log N)$ Time Sorting», *IEEE Transactions on Computers* C-34, 336 - 343, 1985.
[24] D. Bitton, D.J. DeWitt, D.K. Hsiao and J. Menon, «A Taxonomy of Parallel Sorting», *ACM Computing Surveys* 16, 287 - 318, 1984.

[25] N. Blum, «A Note on the Parallel Computation Thesis», *Information Processing Letters* 17, 203-205, 1983.
[26] A. Borodin and J.E. Hopcroft, «Routing, Merging and Sorting on Parallel Models of Computation», *Proceedings of the 14th Annual ACM Symposium on Theory of Computing*, 338-344, 1982.
[27] H. Breteuil, «Une machine qui a du coeur: le processeur systolique Gapp de NCR», *Minis et Micros* 235, 67-73, 1985.
[28] S.A. Browning, «Algorithms for the Tree Machine», in C. Mead and L. Conway, authors, *Introduction to VLSI Systems*, 295-312, Addison-Wesley, Reading, 1980.
[29] F.W. Burton, G.P. McKeown, V.J. Rayward-Smith and M.R. Sleep, «Parallel Processing and Combinatorial Optimization», in L.B. Wilson, C.S. Edwards and V.J. Rayward-Smith, eds., *Combinatorial Optimization III*, 19-36, University of Stirling, 1982.
[30] F.W. Burton, M.N. Huntbach, G.P. McKeown and V.J. Rayward-Smith, «Parallelism in Branch-and-Bound Algorithms», *Report CSA/3/1983*, University of East Anglia, Norwich, 1983.
[31] J. Casti, M. Richardson and R. Larson, «Dynamic Programming and Parallel Computers», *Journal of Optimization Theory and Applications* 12, 423-438, 1973.
[32] A.K. Chandra, D.C. Kozen and L.J. Stockmeyer, «Alternation», *Journal of the ACM* 28, 114-133, 1981.
[33] K.M. Chandy and J. Misra, «Distributed Computation on Graphs: Shortest Path Algorithms», *Communications of the ACM* 25, 833-837, 1982.
[34] F.Y. Chin, J. Lam and I-N. Chen, «Efficient Parallel Algorithms for Some Graph Problems», *Communications of the ACM* 25, 659-665, 1982.
[35] J. Chudik, «Data Flow Computer Architecture», in J. Miklosko and V.E. Kotov, eds., *Algorithms, Software and Hardware of Parallel Computers*, 323-358, Springer-Verlag, Berlin, 1984.
[36] C. Clos, «A Study of Nonblocking Switching Networks», *Bell Systems Technical Journal* 32, 406-424, 1953.
[37] R. Cole and C.K. Yap, «A Parallel Median Algorithm», *Information Processing Letters* 20, 137-139, 1985.
[38] S.A. Cook, «An Observation on Time-Storage Trade Off», *Journal of Computer and System Sciences* 9, 308-316, 1974.
[39] S.A. Cook, «Towards a Complexity Theory of Synchronous Parallel Computation», *Enseignement Mathématique* 27, 99-124, 1981.
[40] S.A. Cook, «An Overview of Computational Complexity», *Communications of the ACM* 26, 401-408, 1983.
[41] F. Courteille and P. Fraisse, «Vectorisation d'Algorithmes de Calcul de Fermeture Transitive», *Actes du Séminaire le Calcul Parallèle et ses Applications aux Télécommunications*, Note Technique CNET NT/PAA/TIM/MTI/1510, 79-122, Issy-les-Moulineaux, 1985.
[42] B.A. Crane, «Path Finding with Associative Memory», *IEEE Transactions on Computers* C-17, 691-693, 1968.
[43] D. Al-Dabass, «Two Methods for the Solution of the Dynamic Programming Algorithm on a Multiprocessor Cluster», *Optimal Control Applications and Methods* 1, 227-238, 1980.
[44] E. Dekel, D. Nassimi and S. Sahni, «Parallel Matrix and Graph Algorithms», *SIAM Journal on Computing* 10, 657-675, 1981.
[45] E. Dekel and S. Sahni, «A Parallel Matching Algorithm for Convex Bipartite Graphs and Applications to Scheduling», *Technical Report 81-3*, Computer Science Department, University of Minnesota, Minneapolis, 1981.
[46] E. Dekel and S. Sahni, «A Parallel Matching Algorithm for Convex Bipartite Graphs», *Proceedings of the 1982 International Conference on Parallel Processing*, 178-184, 1982.
[47] E. Dekel and S. Sahni, «Binary Trees and Parallel Scheduling Algorithms», *IEEE Transactions on Computers* C-32, 307-315, 1983.
[48] E. Dekel and S. Sahni, «Parallel Scheduling Algorithms», *Operations Research* 31, 24-49, 1983.
[49] N. Deo, C.Y. Pang and R.E. Lord, «Two Parallel Algorithms for Shortest Path Problems», *Proceedings of the 1980 International Conference on Parallel Processing*, 244-253, 1980.
[50] N. Deo and Y.B. Yoo, «Parallel Algorithms for the Minimum Spanning Tree Problem», *Proceedings of the 1981 International Conference on Parallel Processing*, 188-189, 1981.
[51] B.C. Desai, «The BPU, a Staged Parallel Processing System to Solve the Zero-One Problem», *Proceedings of ICS'78*, 802-817, 1978.
[52] O.I. El-Dessouki and W.H. Huen, «Distributed Enumeration on Between Computers», *IEEE Transactions on Computers* C-29, 818-825, 1980.
[53] E.W. Dijkstra, «A Note on Two Problems in Connexion with Graphs», *Numerische Mathematik* 1, 269-271, 1959.

[54] D. Dobkin, R.J. Lipton and S. Reiss, «Linear Programming is Log-Space Hard for P», *Information Processing Letters* 8, 96 - 97, 1979.
[55] A. Dutta, H.J. Siegel and A.B. Whinston, «On the Application of Parallel Architectures to a Class of Operations Research Problems», *RAIRO Recherche Opérationnelle* 17, 317 - 341, 1983.
[56] D.M. Eckstein and D.A. Alton, «Parallel Searching of Non-Sparse Graphs», *Technical Report 77-02*, Department of Computer Science, University of Iowa, Iowa City, 1977.
[57] S. Even, «Parallelism in Tape Sorting», *Communications of the ACM* 17, 202 - 204, 1974.
[58] M.J. Flynn, «Very High-Speed Computing Systems», *Proceedings of the IEEE* 54, 1901 - 1909, 1966.
[59] M.J. Flynn, «Some Computer Organizations and Their Effectiveness», *IEEE Transactions on Computers* C-21, 948 - 960, 1972.
[60] S. Fortune and J. Wyllie, «Parallelism in Random Access Machines», *Proceedings of the 10th Annual ACM Symposium on Theory of Computing*, 114 - 118, 1978.
[61] R.G. Gallager, P.A. Humblet and P.M. Spira, «A Distributed Algorithm for Minimum-Weight Spanning Trees», *ACM Transactions on Programming Languages and Systems* 5, 66 - 77, 1983.
[62] M.R. Garey and D.S. Johnson, *Computers and Intractability: A Guide to the Theory of NP-Completeness*, W.H. Freeman and Company, San Francisco, 1979.
[63] F. Gavril, «Merging with Parallel Processors», *Communications of the ACM* 18, 588 - 591, 1975.
[64] M. Goldberg and Z. Miller, «A Parallel Algorithm for Bisection Width in Trees», *XII International Symposium on Mathematical Programming*, Cambridge, 1985.
[65] L.M. Goldschlager, «Synchronous Parallel Computation», *Technical Report 114*, University of Toronto, Toronto, Canada, 1977.
[66] L.M. Goldschlager, «The Monotone and Planar Circuit Value Problems are Log-Space Complete for P», *SIGACT News* 9 (2), 25 - 29, 1977.
[67] L.M. Goldschlager, «A Universal Interconnection Pattern for Parallel Computers», *Journal of the ACM* 29, 1073 - 1086, 1982.
[68] L.M. Goldschlager, R.A. Shaw and J. Staples, «The Maximum Flow Problem is Log-Space Complete for P», *Theoretical Computer Science* 21, 105 - 111, 1982.
[69] A. Gottlieb and C.P. Kruskal, «Complexity Results for Permuting Data and Other Computations on Parallel Processors», *Journal of the ACM* 31, 193 - 209, 1984.
[70] L.J. Guibas, H.T. Kung and C.D. Thompson, «Direct VLSI Implementation of Combinatorial Algorithms», *Caltech Conference on VLSI*, 509 - 525, 1979.
[71] S.E. Hambrush, «VLSI Algorithms for the Connected Component Problem», *SIAM Journal on Computing* 12, 354 - 365, 1983.
[72] J. Harte, «The FPS AP-120B Array Processor», in C.R. Jesshope and R.W. Hockney, eds., *Infotech State of the Art Report: Supercomputers - Vol. 2*, 183 - 203, Infotech Intl. Ltd., Maidenhead, 1979.
[73] L.S. Haynes, R.L. Lau, D.P. Siewiorek and D.W. Mizell, «A Survey of Highly Parallel Computing», *IEEE Computer* 15(1), 9 - 24, 1982.
[74] D.S. Hirschberg, «Parallel Algorithms for the Transitive Closure and the Connected Component Problems», *Proceedings of the 8th Annual ACM Symposium on Theory of Computing*, 55 - 57, 1976.
[75] D.S. Hirschberg, «Fast Parallel Sorting Algorithms», *Communications of the ACM* 21, 657 - 661, 1978.
[76] D.S. Hirschberg, A.K. Chandra and D.V. Sarwate, «Computing Connected Components on Parallel Computers», *Communications of the ACM* 22, 461 - 464, 1979.
[77] J.K. Ho, «Decomposition of Linear Programs Using Parallel Computation», *XII International Symposium on Mathematical Programming*, Cambridge, 1985.
[78] R.W. Hockney, «MIMD Computing in the USA-1984», *Parallel Computing* 2, 119 - 136, 1985.
[79] R.W. Hockney and C.R. Jesshope, *Parallel Computers*, Adam Hilger Ltd., Bristol, 1981.
[80] P.A. Humblet, «A Distributed Algorithm for Minimum Weight Directed Spanning Trees», *Technical Report LIDS-P-1149*, Massachusetts Institute of Technology, Cambridge, 1981.
[81] K. Hwang and F.A. Briggs, *Computer Architecture and Parallel Processing*, McGraw-Hill Book Company, New York, 1984.
[82] IBM Corporation, *The IBM 3090 Processor*, 1985.
[83] M. Imai and T. Fukumura, «A Parallelized Branch-and-Bound Algorithm Implementation and Efficiency», *Systems-Computers-Controls* 10, 62 - 70, 1979.
[84] E.A. Ivanov and V.P. Shevchenko, «Parallel Computations on Graphs», *Cybernetics* 20, 418 - 425, 1985.
[85] J. Ja'Ja', «The VLSI Complexity of Selected Graph Problems», *Journal of the ACM* 31, 377 - 391, 1984.

[86] J. Ja'Ja' and J. Simon, «Parallel Algorithms in Graph Theory: Planarity Testing», *SIAM Journal on Computing* 11, 314 - 328, 1982.

[87] L. Janiga and V. Koubek, «A Note on Finding Minimum Cuts in Directed Planar Networks by Parallel Computations», *Information Processing Letters* 21, 75 - 78, 1985.

[88] D.S. Johnson, «The NP-Completeness Column: An Ongoing Guide (seventh edition)», *Journal of Algorithms* 4, 189 - 203, 1983.

[89] D.S. Johnson and C.H. Papadimitriou, «Computational Complexity», in E. Lawler, J.K. Lenstra, A.H.G. Rinnooy Kan and D.B. Shmoys, eds., *The Traveling Salesman Problem*, 37 - 85, Wiley, New York, 1985.

[90] D.B. Johnson and S.M. Venkatesan, «Parallel Algorithms for Minimum Cuts and Maximum Flows in Planar Networks», *Proceedings of the 23rd Annual IEEE Symposium on Foundations of Computer Science*, 244 - 254, 1982.

[91] N.D. Jones and W.T. Laaser, «Complete Problems for Deterministic Polynomial Time», *Theoretical Computer Science* 3, 105 - 117, 1976.

[92] B. Kamdoum, «Speeding up the Primal Simplex Algorithm on Parallel Computer», *SIGMAP Bulletin* 31, 19 - 23, 1982.

[93] M.J. Kascic Jr., «Vector Processing on the CYBER-200», in C.R. Jesshope and R.W. Hockney, eds., *Infotech State of the Art Report: Supercomputers - Vol. 2*, 237 - 270, Infotech Intl., Maidenhead, 1979.

[94] G.A.P. Kindervater, «The Complexity of Traveling Salesman Heuristics on Parallel Computers», *XII International Symposium on Mathematical Programming*, Cambridge, 1985.

[95] G.A.P. Kindervater and J.K. Lenstra, «An Introduction to Parallelism in Combinatorial Optimization», *Report OS-R8501*, Department of Operations Research and System Theory, Centre for Mathematics and Computer Science, Amsterdam, 1985.

[96] G.A.P. Kindervater and J.K. Lenstra, «Parallel Algorithms», in M. O'hEigeartaigh, J.K. Lenstra and A.H.G. Rinnooy Kan, eds., *Combinatorial Optimization*, 106 - 128, Wiley, Chichester, 1985.

[97] D.E. Knuth, *The Art of Computer Programming - Vol. 3: Sorting and Searching*, Addison-Wesley, Reading, 1973.

[98] P.M. Kogge, *The Architecture of Pipelined Computers*, McGraw-Hill Book Company, New York, 1981.

[99] C.P. Kruskal, «Results in Parallel Searching, Merging and Sorting», *Proceedings of the 1982 International Conference on Parallel Processing*, 196 - 198, 1982.

[100] L. Kucera, «Parallel Computation and Conflicts in Memory Access», *Information Processing Letters* 14, 93 - 96, 1982.

[101] M. Kumar and D.S. Hirschberg, «An Efficient Implementation of Batcher's Odd-Even Merge Algorithm and Its Application in Parallel Sorting Schemes», *IEEE Transactions on Computers* C-32, 254 - 264, 1983.

[102] H.T. Kung, «Why Systolic Architectures?», *IEEE Computer* 15(1), 37 - 46, 1982.

[103] R.E. Ladner, «The Circuit Value Problem is Log-Space Complete for P», *SIGACT News* 7 (1), 18 - 20, 1975.

[104] T.-H. Lai and S. Sahni, «Anomalies in Parallel Branch-and-Bound Algorithms», *Proceedings of the 1983 International Conference on Parallel Processing*, 183 - 190, 1983.

[105] T.-H. Lai and S. Sahni, «Anomalies in Parallel Branch-and-Bound Algorithms», *Communications of the ACM* 27, 594 - 602, 1984.

[106] H.W. Lang, M. Schimmler, H. Schmeck and H. Schröder, «A Fast Sorting Algorithm for VLSI», in J. Díaz, ed., *Proceedings of the 10th Colloquium on Automata, Languages and Programming*, Lecture Notes in Computer Science 154, 408 - 419, Springer-Verlag, Berlin, 1983.

[107] I. Lavallée, «Un Algorithme Parallèle Efficace pour Construire un Arbre de Poids Minimal dans un Graphe», *RAIRO Recherche Opérationnelle* 19, 57 - 69, 1985.

[108] I. Lavallée and C. Lavault, «Algorithmique Parallèle et Distribuée», *Rapport de Recherche no. 471*, INRIA, Rocquencourt, France, 1985.

[109] I. Lavallée and C. Roucairol, «Parallel Branch and Bound Algorithms», *MASI Research Report*, EURO VII, Bologna, Italy, 1985.

[110] I. Lavallée and G. Roucairol, «A Fully Distributed (Minimal) Spanning Tree Algorithm», *Information Processing Letters* (to appear).

[111] T. Leighton, *Complexity Issues in VLSI-Optimal Layouts for the Shuffle Exchange Graph and Other Networks*, The MIT Press, Boston, 1983.

[112] T. Leighton, «Tight Bounds on the Complexity of Parallel Sorting», *IEEE Transactions on Computers* C-34, 344 - 354, 1985.

[113] C.E. Leiserson, *Area-Efficient VLSI Computation*, The MIT Press, Boston, 1983.

[114] G. Lev, N. Pippenger and L.G. Valiant, «A Fast Parallel Algorithm for Routing in Permutation

Networks», *IEEE Transactions on Computers* C-30, 93 - 100, 1981.

[115] K.N. Levitt and W.H. Kautz, «Cellular Arrays for the Solution of Graph Problems», *Communications of the ACM* 15, 789 - 801, 1972.

[116] B. Lint and T. Agerwala, «Communication Issues in the Design and Analysis of Parallel Algorithms», *IEEE Transactions on Software Engineering* 7, 174 - 188, 1981.

[117] H. Lomax and T.H. Pulliam, «A Fully Implicit, Factored Code for Computing Three-Dimensional Flows on the ILLIAC-IV», in G. Rodrigue, ed., *Parallel Computations*, 217 - 250, Academic Press, New York, 1982.

[118] S.-W. Mai and D.J. Evans, «A Parallel Algorithm for the Enumeration of the Spanning Trees of a Graph», *Parallel Computing* 1, 275 - 286, 1984.

[119] D.L. Mangasarian and R. DeLeone, «A Parallel Successive Overrelaxation (SOR) Algorithm for Linear Programming», *XII International Symposium on Mathematical Programming*, Cambridge, 1985.

[120] P. Mateti and N. Deo, «Parallel Algorithms for the Single Source Shortest Path Problem», *Computing* 29, 31 - 49, 1982.

[121] C. Mead and L. Conway, *Introduction to VLSI Systems*, Addison-Wesley, Reading, 1980.

[122] N. Megiddo, «Poly-log Parallel Algorithms for LP with an Application to Exploding Flying Objects», unpublished manuscript, 1982.

[123] N. Megiddo, «Applying Parallel Computation Algorithms in the Design of Serial Algorithms», *Journal of the ACM* 30, 852 - 865, 1983.

[124] R.R. Meyer, «Large-Scale Parallel Optimization on the Crystal Multicomputer», *XII International Symposium on Mathematical Programming*, Cambridge, 1985.

[125] J. Mohan, «Experience with Two Parallel Programs Solving the Traveling Salesman Problem», *Proceedings of the 1983 International Conference on Parallel Processing*, 191 - 193, 1983.

[126] V. Montanari, «On the Optimal Detection of Curves in Noisy Pictures», *Communications of the ACM* 14, 335 - 345, 1971.

[127] D.E. Muller and F.P. Preparata, «Bounds to Complexities of Networks for Sorting and for Switching», *Journal of the ACM* 22, 195 - 201, 1975.

[128] D. Nassimi and S. Sahni, «Bitonic Sort on a Mesh-Connected Parallel Computer», *IEEE Transactions on Computers* C-28, 2 - 7, 1979.

[129] D. Nassimi and S. Sahni, «Finding Connected Components and Connected Ones on a Mesh-Connected Parallel Computer», *SIAM Journal on Computing* 9, 744 - 757, 1980.

[130] D. Nassimi and S. Sahni, «An Optimal Routing Algorithm for Mesh-Connected Parallel Computers», *Journal of the ACM* 27, 6 - 29, 1980.

[131] D. Nassimi and S. Sahni, «Data Broadcasting in SIMD Computers», *IEEE Transactions on Computers* C-30, 101 - 106, 1981.

[132] D. Nassimi and S. Sahni, «Parallel Permutation and Sorting Algorithms and a New Generalized Connection Network», *Journal of the ACM* 29, 642 - 667, 1982.

[133] D.D. Nath and S.N. Maheshwari, «Parallel Algorithms for the Connected Components and Minimal Spanning Tree Problems», *Information Processing Letters* 14, 7 - 11, 1982.

[134] D.D. Nath, S.N. Maheshwari and P.C.P. Bhatt, «Efficient VLSI Networks for Parallel Processing Based on Orthogonal Trees», *IEEE Transactions on Computers* C-32, 569 - 581, 1983.

[135] S.E. Orcutt, «Implementation of Permutation Functions on Illiac IV-Type Computers», *IEEE Transactions on Computers* C-25, 929 - 936, 1976.

[136] C.H. Papadimitriou and K. Steiglitz, *Combinatorial Optimization: Algorithms and Complexity*, Prentice Hall, Englewood Cliffs, 1982.

[137] D.S. Parker Jr. and B. Samadi, «Distributed Minimal Spanning Tree Algorithms», in G. Pujolle, ed., *Performance of Data Communication Systems and Their Applications*, 45 - 52, North-Holland, Amsterdam, 1981.

[138] F.P. Preparata, «New Parallel-Sorting Schemes», *IEEE Transactions on Computers* C-27, 669 - 673, 1978.

[139] F.P. Preparata, «Algorithm Design and VLSI Architectures», *Proceedings of the Symposium on Vector and Parallel Processors*, Rome, Italy, 1982.

[140] F.P. Preparata and J. Vuillemin, «The Cube-Connected Cycles: A Versatile Network for Parallel Computation», *Communications of the ACM* 24, 300 - 309, 1981.

[141] R.C. Prim, «Shortest Connection Networks and Some Generalizations», *Bell System Technical Journal* 36, 1389 - 1401, 1957.

[142] E.A. Pruul, «Parallel Processing and a Branch-and-Bound Algorithm», *M.Sc. Thesis*, Cornell University, Ithaca, 1975.

[143] M.J. Quinn and N. Deo, «A Parallel Approximate Algorithm for the Euclidean Traveling Salesman

Problem», *Report CS-83-105*, Computer Science Department, Washington State University, Pullman, 1983.
[144] M.J. Quinn and N. Deo, «Parallel Graph Algorithms», *ACM Computing Surveys* 16, 319 - 348, 1984.
[145] S.F. Reddaway, «The DAP Approach», in C.R. Jesshope and R.W. Hockney, eds., *Infotech State of the Art Report: Supercomputers - Vol. 2*, 311 - 329, Infotech Intl. Ltd., Maidenhead, 1979.
[146] E. Reghbati and D.G. Corneil, «Parallel Computations in Graph Theory», *SIAM Journal on Computing* 7, 230 - 237, 1978.
[147] J.H. Reif and L.G. Valiant, «A Logarithmic Time Sort for Linear Size Networks», *Proceedings of the 15th Annual ACM Symposium on Theory of Computing*, 10 - 16, 1983.
[148] R. Reischuk, «A Fast Probabilistic Parallel Sorting Algorithm», *Proceedings of the 22nd Annual IEEE Symposium on Foundations of Computer Science*, 212 - 219, 1981.
[149] R. Reischuk, «Probabilistic Parallel Algorithms for Sorting and Selection», *SIAM Journal on Computing* 14, 396 - 409, 1985.
[150] C.C. Ribeiro, «Performance Evaluation of Vector Implementations of Combinatorial Algorithms», *Parallel Computing* 1, 287 - 294, 1984.
[151] G. Rote, «A Systolic Array Algorithm for the Algebraic Path Problem», *Bericht 84-43*, Technische Universität Graz, Institut für Mathematik, Graz, Austria, 1984.
[152] C. Roucairol, «A Parallel Branch and Bound Algorithm for the Quadratic Assignment Problem», *School on Combinatorial Optimization*, Rio de Janeiro, Brazil, 1985.
[153] R.M. Russell, «The CRAY-1 Computer System», *Communications of the ACM* 21, 63 - 72, 1978.
[154] N. Santoro, «On the Message Complexity of Distributed Problems», *Research Report*, Distributed Computing Group, Carleton University, Ottawa, 1982.
[155] C. Savage, «Parallel Algorithms for Graph Theoretic Problems», *Ph.D. Thesis*, University of Illinois, 1977.
[156] C. Savage, «A Systolic Design for Connectivity Problems», *IEEE Transactions on Computers* C-33, 99 - 104, 1984.
[157] C. Savage and J. Ja'Ja', «Fast, Efficient Parallel Algorithms for Some Graph Problems», *SIAM Journal on Computing* 10, 682 - 691, 1981.
[158] W.J. Savitch, «Relationships between Nondeterministic and Deterministic Tape Complexities», *Journal of Computer and System Sciences* 4, 177 - 192, 1970.
[159] J.T. Schwartz, «Ultracomputers», *ACM Transactions on Programming Languages and Systems* 2, 484 - 521, 1980.
[160] C.L. Seitz, «The Cosmic Cube», *Communications of the ACM* 28, 22 - 33, 1985.
[161] J.A. Sharp, *Data Flow Computing*, Ellis Horwood Limited, Chichester, 1985.
[162] J.C. Shepherdson and H.E. Sturgis, «Computability of Recursive Functions», *Journal of the ACM* 10, 217 - 255, 1963.
[163] Y. Shiloach and U. Vishkin, «Finding the Maximum, Merging and Sorting in a Parallel Computation Model», *Journal of Algorithms* 2, 88 - 102, 1981.
[164] Y. Shiloach and U. Vishkin, «An $O(\log n)$ Parallel Connectivity Algorithm», *Journal of Algorithms* 3, 57 - 67, 1982.
[165] Y. Shiloach and U. Vishkin, «An $O(n^2 \log n)$ Parallel MAX-FLOW Algorithm», *Journal of Algorithms* 3, 128 - 146, 1982.
[166] A.R. Siegel, «Minimum Storage Sorting Networks», *IEEE Transactions on Computers* C-34, 355 - 361, 1985.
[167] H.J. Siegel, «A Model of SIMD Machines and a Comparison of Various Interconnection Networks», *IEEE Transactions on Computers* C-28, 907 - 917, 1979.
[168] B.J. Smith, «A Pipelined, Shared Resource MIMD Computer», *Proceedings of the 1978 International Conference on Parallel Processing*, 6 - 8, 1978.
[169] J.S. Squire and S.M. Palais, «Programming and Design Considerations of a Highly Parallel Computer», *Proceedings of the AFIPS Spring Joint Computer Conference* 23, 395 - 400, 1963.
[170] H.S. Stone, «Parallel Processing with the Perfect Shuffle», *IEEE Transactions on Computers* C-20, 153 - 161, 1971.
[171] Q.F. Stout, «Sorting, Merging, Selecting, and Filtering on Tree and Pyramid Machines», *Proceedings of the 1983 International Conference on Parallel Processing*, 214 - 221, 1983.
[172] H. Tamura, S. Kamiya and T. Ishigai, «FACOM VP-100/200: Supercomputers with Ease of Use», *Parallel Computing* 2, 87 - 107, 1985.
[173] R.E. Tarjan and U. Vishkin, «An Efficient Parallel Biconnectivity Algorithm», *SIAM Journal on Computing* 14, 862 - 874, 1985.
[174] C.D. Thompson, «Area-Time Complexity for VLSI», *Proceedings of the 11th Annual ACM Symposium on Theory of Computing*, 81 - 88, 1979.

[175] C.D. Thompson and H.T. Kung, «Sorting on a Mesh-Connected Computer», *Communications of the ACM* 20, 263 - 271, 1977.

[176] K.J. Thurber and L.D. Wald, «Associative and Parallel Processors», *ACM Computing Surveys* 7, 215 - 255, 1975.

[177] S. Todd, «Algorithm and Hardware for a Merge Sort Using Multiple Processors», *IBM Journal of Research and Development* 22, 509 - 517, 1978.

[178] P.C. Treleaven, D.R. Brownbridge and R.P. Hopkins, «Data-Driven and Demand-Driven Computer Architecture», *ACM Computing Surveys* 14, 93 - 143, 1982.

[179] Y.H. Tsin, «Bridge-connectivity and Biconnectivity Algorithms for Parallel Computer Models», *Proceedings of the 1983 International Conference on Parallel Processing*, 180 - 182, 1983.

[180] Y.H. Tsin, «An Optimal Parallel Processor Bound in Strong Orientation of an Undirected Graph», *Information Processing Letters* 20, 143 - 146, 1985.

[181] Y.H. Tsin and F.Y. Chin, «Efficient Parallel Algorithms for a Class of Graph Theoretic Problems», *SIAM Journal on Computing* 13, 580 - 599, 1984.

[182] S.H. Unger, «A Computer Oriented Toward Spatial Problems», *Proceedings IRE* 46, 1744 - 1750, 1958.

[183] L.G. Valiant, «Parallelism in Comparison Problems», *SIAM Journal on Computing* 4, 348 - 355, 1975.

[184] L.G. Valiant, «A Scheme for Fast Parallel Communication», *SIAM Journal on Computing* 11, 350 - 361, 1982.

[185] L.G. Valiant, «Reducibility by Algebraic Projections», *Enseignement Mathématique* 28, 253 - 268, 1982.

[186] L.G. Valiant, «Parallel Computation», in *Foundations of Computer Science IV - Distributed Systems: Part I, Algorithms and Complexity*, 35 - 48, Mathematical Centre Tract 158, Centre for Mathematics and Computer Science, Amsterdam, 1983.

[187] L.G. Valiant and G.J. Brebner, «Universal Schemes for Parallel Communication», *Proceedings of the 13th Annual ACM Symposium on Theory of Computing*, 263 - 277, 1981.

[188] P. Van Emde Boas, «The Second Machine Class: Models of Parallelism», in J. Van Leeuwen and J.K. Lenstra, eds., *Parallel Computers and Computations*, CWI Syllabus, Centre for Mathematics and Computer Science, Amsterdam, 1985.

[189] F.L. Van Scoy, «The Parallel Recognition of Classes of Graphs», *IEEE Transactions on Computers* C-29, 563 - 570, 1980.

[190] U. Vishkin, «Synchronous Parallel Computation - A Survey», *Preprint*, Courant Institute, New York University, 1983.

[191] U. Vishkin, «On Efficient Parallel Strong Orientation», *Information Processing Letters* 20, 235 - 240, 1985.

[192] U. Vishkin and A. Wigderson, «Trade-Offs between Depth and Width in Parallel Computation», *SIAM Journal on Computing* 14, 303 - 314, 1985.

[193] B.W. Wah and Y.W. Ma, «MANIP-A Multicomputer Architecture for Solving Combinatorial Extremum-Search Problems», *IEEE Transactions on Computers* C-33, 377 - 390, 1984.

[194] W.J. Watson, «The TI-ASC - A Highly Modular and Flexible Super Computer Architecture», *AFIPS Proceedings* FJCC, 221 - 228, 1972.

[195] J.A. Wisniewski and A.H. Sameh, «Parallel Algorithms for Network Routing Problems and Recurrences», *SIAM Journal on Algebraic and Discrete Methods* 3, 379 - 394, 1982.

[196] J.C. Wyllie, «The Complexity of Parallel Computations», *Ph.D. Thesis*, Cornell University, Ithaca, 1979.

[197] A.C.-C. Yao, «On Parallel Computation for the Knapsack Problem», *Journal of the ACM* 29, 898 - 903, 1982.

[198] V. Zakharov, «Parallelism and Array Processing», *IEEE Transactions on Computers* C-33, 45 - 78, 1984.

Celso Carneiro Ribeiro
Catholic University of Rio de Janeiro
Department of Electrical Engineering
Gávea-Caixa Postal 38063
Rio de Janeiro 22452
Brazil

PROBABILISTIC ANALYSIS OF ALGORITHMS

Alexander H.G. RINNOOY KAN

1. Introduction

Suppose that two thieves meet on a regular basis to divide the proceeds of their joint effort. Each stolen object has a specific dollar value and has to be assigned to one of the two. For obvious reasons, they are interested in a quick and fair partitioning scheme.

In spite of its apparent simplicity, the above combinatorial problem is not easy to solve if we insist on an optimal solution, i.e., one in which the difference between the values assigned to each thief is as small as possible. As is the case with most practical problems, this problem too is known to belong to the class of *NP-complete* problems. This implies that any optimization method for its solution could be expected to perform very poorly on some occasions: more formally, its *worst case running time* is likely to grow *exponentially* with *problem size*.

Hence, in choosing a solution scheme, the thieves will be forced into an unpleasant trade-off between two features of algorithmic quality: the *computational effort* (the smaller running time, the better) on one hand and the *computational result* (the smaller deviation from optimality, the better) on the other hand. Complexity theory indicates that we cannot insist on a simultaneous *absolute guarantee* for both, i.e., on a fast (polynomially bounded) running time as well as a zero deviation from the optimal solution value.

One possible way out of this dilemma is to change the perspective on the analysis by no longer demanding an *absolute* guarantee. For practical purposes an algorithm that, with respect to both effort and result, does well in the majority of cases or even on average might be perfectly acceptable.

Probability theory provides the natural setting for such an *analysis of algorithms*. This analysis starts from a specification of what an average problem instance would look like, in terms of a *probability distribution* over the *class of all instances*. The running time and the solution value of a particular algorithm are then considered as *random variables*, whose behaviour can be studied and evaluated. This approach can therefore be viewed as the analytical counterpart to the familiar *experimental* analysis in which an algorithm is tried out on a set of supposedly representative test problems and evaluated statistically. Here we obtain the rigor of mathematical analysis at the expense of a certain naivete, in that only relatively straightforward solution methods can be analyzed probabilistically in full detail.

Although the *probabilistic analysis of algorithms* has only recently become an active research area, it has already generated an impressive number of publications. A concise survey of this area would require the prior introduction of many techniques from probability theory and could hardly do justice to the diversity of ideas and approaches that one finds in the literature. Fortunately, a recent annotated bibliography (Karp et al. [32]) provides an up to date survey of the available articles and publications. In view of the existence of this source of detailed information, no attempt at completeness will be made in what follows below. Rather, the nature of the analysis and of the results will be illustrated by some typical examples.

In Section 2, we consider the problem of the two thieves in more detail. It is, of course, none other than the well know PARTITION problem in which one seeks to minimize the size of the largest share. In Section 3, we review some representative result that are known for problems defined in the *Euclidean* plane. In Section 4, we examine the fertile area of optimization problems defined on *graphs and networks*. In Section 5, we indicate some possibilities for future work in this very lively research area.

We conclude this introduction by a short digression on modes of *stochastic convergence*, clearly an essential concept if we want to analyze the notion of a random variable such as the error of a heuristic going to 0 with increasing problem size.

Almost sure convergence of a sequence of random variables $\underline{y}_n$ to a constant c by definition means that $\Pr\{\lim_{n \to \infty} \underline{y}_n = c\} = 1$; it is a strong form of stochastic convergence and implies the weaker *convergence in probability*, which stands for $\lim_{n \to \infty} \Pr\{|\underline{y}_n - c| > \epsilon\} = 0$ for all $\epsilon > 0$. The reverse implication holds under the additional assumption that, for all $\epsilon > 0$,

$$\sum_{n=1}^{\infty} \Pr\{|\underline{y}_n - c| > \epsilon\} < \infty. \tag{1}$$

Similarly, convergence of $\underline{y}_n$ to c *in expectation*, i.e.

$$\lim_{n \to \infty} |E(\underline{y}_n) - c| = 0 \tag{2}$$

also implies convergence in probability, with the reverse implication holding under additional boundedness assumptions on $\underline{y}_n$.

2. The partition problem

Perhaps the simplest possible way to solve the PARTITION problem of the two thieves is to allow each thief to choose a particular item in turn until they have all been assigned. If the j-th item has value a_j ($j = 1, \ldots, n$), then this amounts to ordering the items according to decreasing a_j values $a^{(n)} \geq a^{(n-1)} \geq \ldots \geq a^{(1)}$; one thief receives $a^{(n)} + a^{(n-2)} + \ldots$, the other $a^{(n-1)} + a^{(n-3)} + \ldots$.

This is clearly a fast heuristic method that may, however, produce a very inequitable result: in the worst case, the first thief may receive up to 50 percent more than the optimal partition would grant him. (Take $a_1 = 2$, $a_{2i} = a_{2i+1} = 2^{-i}$ ($i \geq 1$). How about its average case behaviour? To answer that question, we specify a probability distribution over all problem instances by assuming that robberies are so frequent and haphazard that the a_j can be thought of as independent draws from a uniform distribution on, say, $[0, 1]$. (Actually, many of the results mentioned below hold under much more general assumptions, though independence is always required.)

Under the uniform assumption, the optimal solution value $\underline{Z}_n^{\text{OPT}}$ of the partition problem (i.e., the smallest possible size of the larger share) turns out to be almost surely (a.s.) *asymptotic to* the lower bound $(\Sigma_{j=1}^n \underline{a}_j)/2$:

$$\frac{\underline{Z}_n^{\text{OPT}}}{\left(\sum_{j=1}^n \underline{a}_j\right)/2} \to 1 \quad \text{(a.s.)}. \tag{3}$$

This result provides a first example of *asymptotic probabilistic value analysis*: for n large enough, the optimal solution value $\underline{Z}_n^{\text{OPT}}$ can be guessed with increasing (relative) accuracy. What about the heuristic solution value $\underline{Z}_n^H$, i.e. the size of the share under the heuristic scheme proposed above? We know that

$$\underline{Z}_n^H = \frac{1}{2}\sum_{j=1}^n \underline{a}_j + \frac{1}{2}[(\underline{a}^{(n)} - \underline{a}^{(n-1)}) + (\underline{a}^{(n-2)} - \underline{a}^{(n-3)}) + \ldots], \tag{4}$$

and it is not hard to show that the difference between $\underline{Z}_n^H$ and $(\Sigma_{j=1}^n \underline{a}_j)/2$ a.s. converges to $1/4$, so that also

$$\frac{\underline{Z}_n^H}{\left(\sum_{j=1}^n \underline{a}_j\right)/2} \to 1 \quad \text{(a.s.)} \tag{5}$$

and hence

$$\frac{\underline{Z}_n^H}{\underline{Z}_n^{\text{OPT}}} \to 1 \quad \text{(a.s.)}. \tag{6}$$

This implies that the heuristic is *asymptotically optimal*: its *relative error* (i.e., its percentage deviation from the optimum) $(\underline{Z}_n^H - \underline{Z}_n^{\text{OPT}})/\underline{Z}_n^{\text{OPT}}$ a.s. goes to 0.

Hence, a probabilistic analysis leads to a much more optimistic conclusion than a worst case oriented one.

What about the *absolute error* $Z_n^H - Z_n^{OPT}$ or the *absolute difference* between the two shares? Neither of these two quantities goes to 0 for the above heuristic, so there is room for improvement. A slightly more sophisticated scheme would be to allow each thief in turn to select items until the value of his share exceeds the value of the current share of his colleague. Even from a worst case point of view, this is a much more reasonable approach: the larger share can never exceed its smallest possible size by more than 16 2/3 percent (Graham [20]). (For a worst case example, take e.g., $a_1 = a_2 = 3$, $a_3 = a_4 = a_5 = 2$.) In a probabilistic sense, the difference is even more impressive. Of course, the relative error again goes to 0, but the absolute difference between the shares $\underline{d}_n^H$ (which is an upper bound on the absolute error) also satisfies

$$\underline{d}_n^H \to 0 \quad \text{(a.s.)} \tag{7}$$

(Frenk and Rinnooy Kan [14]). To prove this result, one observes that

$$\underline{d}_n^H \leq \max\{\underline{d}_{n-1}^H - \underline{a}^{(1)}, \underline{a}^{(1)}\} \tag{8}$$

which, after repeated application, yields that

$$\underline{d}_n^H \leq \max_{1 \leq k \leq n} \left\{ \underline{a}^{(k)} - \sum_{j=1}^{k-1} \underline{a}^{(j)} \right\} \leq$$

$$\leq \underline{a}^{(\lceil \delta n \rceil)} + \max \left\{ \underline{a}^{(n)} - \sum_{j=1}^{[\delta n]} \underline{a}^{(j)}, 0 \right\} \tag{9}$$

for all $\delta > 0$. The first term converges a.s. to δ and can therefore be made arbitrarily small; for any fixed δ, the second term converges a.s. to 0 since $\sum_{j=1}^{[\delta n]} \underline{a}^{(j)} = O(n)$ and $\underline{a}^{(n)}$ is $o(n)$ for every distribution with finite first moment.

The two results presented so far demonstrate the importance of the theory of *order statistics* for the analysis of heuristics that involve the sorting of numbers; *priority rules* generally fall into this class.

Can we do still better? One weakness of a result such as (7) is its *asymptotic nature*, i.e. its validity only for *sufficiently large* values of n. At the very least, one would like to know the rate at which $\underline{d}_n^H$ converges to 0. It can be shown (Frenk and Rinnooy Kan [16]) that

$$\limsup_{n \to \infty} \frac{\underline{d}_n^H}{\log \log n / n} < \infty \quad \text{(a.s.)} \tag{10}$$

and a simple argument shows that the *rate of convergence* for this heuristic has to be at least $1/n$ in expectation (Karp [32]). This rate is good but not quite good enough: indeed, it is also known that d_n^{OPT}, the smallest possible absolute difference, satisfies

$$\limsup_{n \to \infty} \frac{d_n^{OPT}}{n^2 2^{-n}} < \infty \quad \text{(a.s.)} \tag{11}$$

(Karmarkar et al. [29]). Hence, the exponential effort that may be required for the computation of the optimal partition is at least rewarded by an exponential decrease to 0 of the difference between the two shares. Can this also be achieved a.s. in polynomial time?

The answer to this question is unknown, but the previous heuristic can again be improved upon by essentially assigning two items at a time and compensating for their difference. More precisely, in the first iteration $a^{(n)}$ would be assigned to one thief, and $a^{(n-1)}$ to the other. The two items would then be replaced by a single item of value $a^{(n)} - a^{(n-1)}$ and the process would be repeated on the new set of $n-1$ items until only one item remains; its value represents the difference between the two shares. A simple backtracking procedure establishes the partition in terms of the original items.

In the worst case, this heuristic method is not better than the previous one. The probabilistic analysis of its performance is quite difficult: as on so many other occasions, each step in the algorithm conditions the probability distribution encountered in the succeeding steps in a complicated fashion. Since the difference of two independent uniformly distributed random variables follows a triangular distribution, there is no *distributional invariance* throughout the steps of this method and yet such invariance is an essential prerequisite for a successful analysis. One way to overcome this kind of obstacle is to change the algorithm so that (with high probability) the value produced will not be affected but its modified behaviour can be analyzed rigorously. In this particular case, the new version of algorithm works roughly as follows. In iteration m, it deals with numbers in an interval $[0, \alpha_m]$, which is first partitioned into subintervals of size α_{m+1}. The original differencing method is then applied to random pairs of numbers taken from the set S_i found in the i-th subinterval ($i = 1, \ldots, \alpha_m/\alpha_{m+1}$), until a set $S_i' \subset [0, \alpha_{m+1}]$ is obtained and at most one of the numbers originally in S_i remains. The original differencing method is then used on the set of remaining numbers to reduce them to a single number in $[0, \alpha_{m+1}]$.

Now, before the method can be applied recursively to $[0, \alpha_{m+1}]$, the issue of distributional invariance has to be addressed. To do so, we assume (inductively) that $S = \cup S_i$ was divided into two subsets, a subset G of «good» points that can be assumed to come from a uniform distribution over $[0, \alpha_m]$ and a subset B of «bad» points. During the application of the differencing method to the sets S_i the numbers entering S_i' are labeled good only if they are obtained as the difference of two good numbers in S_i. As a result, we know that the good numbers in $S' = \cup S_i'$ follow a distribution related to the triangular one. In a final step, a subset of these numbers is relabeled «bad» so that the remaining good ones are again uniformly distributed over $[0, \alpha_{m+1}]$. All that remains to be shown is that for appropriate choices of α_m, enough good numbers remain for the method to reach the number of iterations that is required for a good result.

Through this approach, it could be established in Karmarkar and Karp [28] that

$$\limsup_{n\to\infty} \frac{\underline{d}_n^H}{n^{-\log n}} < \infty \quad \text{(a.s.).} \tag{12}$$

This, in $O(n \log n)$ time this method guarantees a rate of convergence that is superpolynomial, yet subexponential. It is tempting to conjecture that this is best possible for a polynomial time heuristic!

We have dealt with this simple example in some detail, since it exhibits many of the ingredients typically encountered in a probabilistic analysis:
— a *combinatorial problem* that may be difficult to solve to optimality (the PARTITION problem is NP-complete);
— a *probability distribution* over all problem instances, that generates the problem data as realizations of *independent and identically distributed* (i.i.d.) random variables (the $\underline{a}_j$ are independent and uniform over [0, 1]);
— a *probabilistic value analysis* that yields a description of the asymptotic optimal solution value as a simple function of the problem data ($\underline{Z}_n^{\text{OPT}}/(n/4) \to 1$ (a.s.));
— simple, fast heuristics whose *relative error* or *absolute error* may decrease to 0 in some stochastic sense or may be otherwise well behaved;
— a *rate of convergence analysis* that allows further differentiation among the heuristics.

The state of the art for a particular problem class can conveniently be monitored by means of the above concepts. Consider, for example, the MULTIKNAPSACK problem $\max\{\Sigma_j c_j x_j | \Sigma_j a_{ij} x_j \leq b_i \ (i = 1, \ldots, m), \ x_j \in \{0, 1\} \ (j = 1, \ldots, n)\}$, which is a generalization of PARTITION (take $m = 1$, $c_j = a_{1j}$ $(j = 1, \ldots, n)$, $b_1 = (\Sigma_{j=1}^n a_{1j})/2$).

Let us assume that the $\underline{c}_j$ and $\underline{a}_{ij}$ are i.i.d. uniform on [0, 1] and that $b_i = n\beta_i$ is constant. As above, we are interested in the optimal solution value as a function of $\beta = (\beta_1, \ldots, \beta_m)$ and in heuristics whose error vanishes asymptotically with high probability.

The analysis of this problem in Meanti et al. [45] is of interest in that exploits the close relationship (in probability) between certain difficult nonconvex combinatorial optimization problems such as MULTIKNAPSACK and their convex LP relaxations obtained by replacing the constraints $x_j \in \{0, 1\}$ by $0 \leq x_j \leq 1$. It is easy to see that the absolute difference between the solution values of these two problems is bounded by m, so that the relative error that we make by focusing on the LP relaxation goes to 0. But the LP relaxation (or, rather, its dual) is much easier to analyze: its value is given by

$$\min_\lambda \underline{L}_n(\lambda) \tag{13}$$

with $\lambda = (\lambda_1, \ldots, \lambda_m)$ and

$$\underline{L}_n(\lambda) = \sum_{i=1}^m \lambda_i b_i + \max\left\{\Sigma_j \left(\underline{c}_j - \sum_{i=1}^m \lambda_i \underline{a}_{ij}\right) x_j \mid 0 \leq x_j \leq 1\right\} \quad (14)$$

where the maximization problem in (14) is solved by setting

$$\underline{x}_j(\lambda) = \begin{cases} 1 & \text{if } \underline{c}_j - \sum_{i=1}^m \lambda_i \underline{a}_{ij} \geq 0 \\ 0 & \text{otherwise.} \end{cases} \quad (15)$$

Results from convex analysis can then be used to establish that the optimal solution value Z_n^{OPT} satisfies

$$\frac{\underline{Z}_n^{OPT}}{n} \to \min_\lambda L(\lambda) \quad \text{(a.s.)}, \quad (16)$$

where

$$L(\lambda) = \sum_{i=1}^m \lambda_i \beta_i + \frac{1}{n} E\underline{c}^T \underline{x}(\lambda) - \frac{1}{n} \sum_{i=1}^n \lambda_i E\underline{a}_i^T \underline{x}(\lambda) \quad (17)$$

is a convex, twice differentiable function with a unique minimum that can actually be computed in closed form in some simple case (e.g., for $m = 1$).

Not surprisingly, successful heuristics for this problem also have a strong LP flavor (cf. Frieze and Clarke [17]). A natural one to consider is the *generalized greedy heuristic* in which $x_j's$ are set equal to 1 in order of nonincreasing ratio's $c_j/(\Sigma_{i=1}^m \lambda_i a_{ij})$. If the λ_i are chosen to be equal to the values minimizing the right hand side of (17), then the relative error of this greedy method goes to 0 a.s. A heuristic whose absolute error vanishes asymptotically is not known, however, and further analysis of the model reveals puzzling differences between the minimization and maximization version of MULTIKNAPSACK that stil have to be resolved. None the less, the probabilistic analysis of the model yields surprisingly high returns.

What are the strong and the weak points of the approach? One one hand, the algorithmic insights nicely complement the more traditional worst case analysis, with an emphasis that is not so much on an exact *guarantee* that a certain running time or a certain error will not be exceeded, as on an *explanation* why the algorithm may perform so much better in practice than the worst case analysis would seem to suggest. On the other hand, the results are obtained under probabilistic assumptions that can always be questioned, and are usually only valid for «sufficiently large» problem sizes, with little indication of how large they might have to be. Rate of convergence results somewhat compensate for this latter deficiency.

Apart from its contribution to error analysis, a probabilistic value analysis may, however, yield additional benifits. An estimate of the optimal solution value can, for instance, be used in a *branch and bound procedure* to replace weak upper or lower bounds, at the possible expense of sacrificing optimality but at the gain of a significant improvement in running time (Derigs [10]). Or it may be used in a *two stage stochastic programming* problem, where the first stage decision determines the value of some parameters of the second stage problem. A probabilistic value analysis may then reveal how the second stage solution value depends on these parameters so that they can be given optimal values in the first stage (Stougie [53]).

As a fist step towards a classification of results in probabilistic analysis, let us note that the PARTITION problem and its generalization, the MULTIKNAPSACK problem, both have a problem input that consists of *numbers*. The probabilistic analysis of algorithms for these problems usually assumes that these are independently generated from a fixed distribution. To close this section, we briefly review some typical results that were obtained for other problems in this category.

The MULTIMACHINE SCHEDULING problem is an appropriate starting point, since it can be viewed as yet another generalization of PARTITION: it is the problem to distribute jobs with processing times $a_1, \ldots, a_n$ among m identical machines so as to minimise the maximal sum of processing times assigned to any machine (the *makespan*); the PARTITION problem corresponds to the case that $m = 2$.

The second heuristic proposed for PARTITION can be viewed as a special case of the LPT (Largest Processing Time) heuristic for MULTIMACHINE SCHEDULING, in which jobs are assigned to the earliest available machine in order of nonincreasing a_j. If the a_j are i.i.d. uniform on $[0, 1]$, the proof technique based on order statistics can again be applied to show that the absolute error of the LPT heuristic goes to 0 a.s.; the optimal solution value is asymptotic to $n/(2m)$. The order in which jobs are assigned to machines turns out to be really essential: an arbitrary list scheduling heuristic will have a relative error that goes to 0 a.s. but an absolute error that does not. The Karp-Karmarkar heuristic can be extended to the case of arbitrary m to improve the rate of convergence to optimality from $1/n$ for the LPT rule to $n^{-\log n}$. We refer to Coffman et al. [8] for additional references.

The famous BIN PACKING problem is in a sense dual to MULTIMACHINE SCHEDULING: here, the makespan is fixed (say, equal to 1) and the objective now is to find the minimum number of bins (machines) into which the items (jobs) of size $a_1, \ldots, a_n$ can be packed. The probabilistic analysis of this problem is again usually carried out under the assumption that the jobs are i.i.d. uniform on $[0, 1]$. It has yielded many beautiful results. To give one example, consider the heuristic that inspects the items in order of decreasing a_j and matches each item a_k with the largest unassigned item a_l satisfying $a_k + a_l \leq 1$. To analyze the performance of this heuristic, we consider the set of all a_j on the interval [0, 1] and replace each item a_k larger than $1/2$ by $1 - a_k$, marking it with a «+»; the items smaller than $1/2$ get a «−» sign. The heuristic now amounts to pairing each «+» with largest «−» to its left; the number of poorly filled bins (i.e., bins with only one item) is related to the excess of «+»s over «−»s. But the sequence of «+»s and «−»s can be viewed as generated by flips of a coin, and results from the theory of random walks can be invoked to show that the expected total number of bins used is $n/2 + O(\sqrt{n})$; since the expected optimal number of bins is known to be $n/2 + \Omega(\sqrt{n})$, the relative error of this heuristic goes to 0 in expectation (Frederickson [12], Knödel [36]).

This analysis reveals an interesting connection between *matching problems* on the interval [0, 1/2] and bin packing methods. In a similar vein, a connection can be established between *on line* bin packing methods (i.e., methods in which items arrive in arbitrary order and have to be irrevocably assigned to a bin right away; the previous method is *off line*) and matching problems on a *square*, where one of the dimensions corresponds to item size and the other dimension represents time, i.e., it indicates the order in which the items arrive for packing. In this case, feasibility dictates that each point can only be matched to a *right upward* neighbour; this matching problem was studied in a different context in Karp et al. [33], and the clever extension and refinement of these results to on-line bin packing can be found in Shor [50]. Further references can be found in the useful survey by Coffman et al. [8] that was already quoted above.

A final number problem that deserves to be mentioned is LINEAR PROGRAMMING, not because it is a hard combinatorial problem (the Khachian method solves it in polynomial time) but because probabilistic analysis played such a vital role in understanding the excellent average performance of the *simplex method*. The history of the analysis illustrates the importance of an appropriate *probabilistic model*: ultimately, the concept of a random polytope as being generated by m fixed hyperplanes in $\mathbb{R}^n$ and m coin flips to determine the direction of the corresponding inequalities reduced the computation of the average number of simplex pivots to a combinatorial counting question. Within this model, various simplex variants admit of a quadratic upper bound on the expected number of iterations (including those in Phase I), which takes us very close to the behaviour observed in practice (Haimovich [22], Adler et al. [1], Adler and Megiddo [2], Todd [54]).

In the next section, we turn to problems with a geometric flavor, whose probabilistic analysis involves random sets of points in the Euclidean plane.

3. Euclidean problems

In this section we are concerned with problems whose input includes a set of n

points in the Euclidean plane. The most famous problem of this type is surely the TRAVELLING SALESMAN problem, which is to find the shortest tour connecting the n points. It has a venerable history, of which its probabilistic analysis forms one of the most recent chapters (see Lawler et al. [10]).

To carry out such an analysis, one usually assumes the points to be uniformly distributed over a fixed region, e.g., the unit (1×1) square. Under such an assumption, it is not difficult to arrive at an intuitive probabilistic value analysis for the TRAVELLING SALESMAN problem. For n large, the length of an optimal tour through a 2×2 square with $4n$ points will be approximately 4 times as large as the optimal tour in a unit square with n points. Scaling back the 2×2 square to a unit one, we conclude that the optimal tour length Z_n^{OPT} is likely to grow proportionally to $\sqrt{n}$. Indeed, a heuristic from Few [11] shows that its value is bounded deterministically from above by $\sqrt{2n}$, and it is not hard to show that there exists a positive constant c such that it is a.s. bounded by $c\sqrt{n}$ from below.

The actual convergence argument is much more difficult and was first provided in Beardwood et al. [5]: as expected,

$$\frac{Z_n^{OPT}}{\sqrt{n}} \to \beta \quad \text{(a.s.)} \tag{18}$$

where β is a constant that has been estimated empirically to be about 0.765.

The proof of this result involves a technique useful in a broader context: rather than viewing the problem on a fixed set of n points, uniformly distributed on a square, they are assumed to be generated by a *Poisson process* of intensity 1. The advantage is that point sets in disjoint regions are now fully independent; the disadvantage that an expression for, say, the expected routelength $F(t)$ in a square $[0, t] \times [0, t]$ has to be converted back into a result for the expected length EZ_n^{OPT} of a tour through n points, using the relation

$$F(t) = \sum_{n=1}^{\infty} tEZ_n^{OPT} e^{-t^2} \frac{(t^2)^n}{n!}. \tag{19}$$

Certain *Tauberian theorems* allow one to do precisely that.

In the specific case of the TRAVELLING SALESMAN problem, $F(t)$ is computed through the use of a heuristic that embodies the intuitive insight mentioned above. The square $[0, t] \times [0, t]$ is divided into m^2 equal size subsquares; a tour in each subsquare has expected lenght $F(t/m)$. These tours can be linked to form a feasible solution to the original problem by adding segments of length $O(tm)$, and hence

$$F(t) \leq m^2 F(t/m) + O(tm). \tag{20}$$

It is not hard to see that this implies that $F(t)/t^2$ converges to a constant β and to conclude from (19) that $EZ_n^{OPT} = \beta\sqrt{n}$. The argument is completed by proving the variance of Z_n^{OPT} to be $O(1)$ through the *Efron-Stein* inequality; Chebyshev's inequality and the Borel-Cantelli lemma then yield (18) (Karp & Steele [34]).

Given the result of the probabilistic value analysis, it now becomes attractive to search for heuristics whose absolute error is $o(\sqrt{n})$; their relative error then goes to 0 almost surely. As in the case of other Euclidean problems, *partitioning heuristics* do precisely that. Generally, in these heuristics the square is appropriately partitioned into subregions (e.g. rectangles), subproblems defined by the points in each subregion are analyzed separately, and a feasible solution to the problem as a whole is composed out of the separate results.

For the TRAVELLING SALESMAN problem, one partitioning approach is to execute an alternating sequence of horizontal and vertical *cuts* through the point with current median vertical and horizontal coordinate respectively, until the resulting rectangles contain no more than $\sqrt{\log n}$ points. Each of these subproblems is solved to optimality by some enumerative technique (say, dynamic programming, which takes $O(n^\epsilon)$ time per rectangle, and hence $O(n^{1+\epsilon})$ time overall, for any $\epsilon > 0$). The resulting tours define a connected graph with even degree at each point; the *Euler walk* that visits each edge of this graph can be converted into a tour of no greater length by eliminating multiple visits to one point. The difference between the length of this tour and the optimal one can be shown to be of the same order as the total perimeter of the rectangles generated, which is easily seen to be $o(\sqrt{n})$ in this case. Thus, the relative error of the heuristic goes to 0 a.s. (Karp [30], Karp and Steele [34]).

Not much is known about the rate of convergence to optimality of this heuristic, nor is any heuristic known whose absolute error goes to 0 asymptotically. The partitioning approach, however, has been generalized to a variety of other Euclidean problems. Consider, for example, the CAPACITATED ROUTING problem, where the points have to be equally distributed among salesmen, each of whom can visit no more than q customers before returning to a common depot. If we assume that the i-th customer is at distance r_i from the depot (r_i i.i.d., with finite first moment $E\underline{r}$), then it is not hard to prove that the optimal solution value satisfies

$$Z_n^{OPT} \geq \max\left\{T_n^{OPT}, 2\frac{\sum_{i=1}^{n} r_i}{q}\right\} \qquad (21)$$

where T_n^{OPT} is the length of a single travelling salesman tour through the n customers. A tour partitioning heuristic, in which T_n is optimally divided into $[n/q]$ consecutive segments containing q customers each, can be shown to yield a value no more than the sum of the two terms on the right hand side of (21), so that — in view of (18) — Z_n^{OPT} is a.s. asymptotic to $2nE\underline{r}/q$. As a byproduct of this result, one obtains that the tour partitioning heuristic (of the «*route first, cluster second*» type) has relative error going to 0 a.s.. A similar result holds for certain region partitioning heuristics («*cluster first, route second*»): as indicated above, their absolute error is dominated by the total perimeter of the subregions generated, which in this case is $O(\sqrt{n/q})$ and hence vanishes asymptotically relative to $\underline{Z}_n^{OPT}$.

All these results presuppose that q is a constant that does not grow with n; if it does, than the results hold as long as $q = o(\sqrt{n})$. Above this threshold, Z_n^{OPT} behaves as in (18), since at that point the total cost of moving among the groups of customers, T_n, starts to dominate the total cost $2\Sigma_{i=1}^n r_i/q$ of reaching these groups from the depot (cf. (21)).

The ubiquitous presence of partitioning techniques in probabilistic Euclidean analysis points in the direction of a common generalization, which was indeed provided in Steele [51]. There, (18) is generalized to problems of arbitrary dimension, provided that the objective function is *subadditive* on sets of points (and that a few other technical conditions are satisfied). The TRAVELLING SALESMAN problem is one example, the MATCHING problem (find the segments linking each point to a neighbour, of minimum total length) is another one: see Papadimitriou [46] for a probabilistic value analysis and a heuristic that together establish the optimal solution value to be a.s. asymptotic to $\epsilon\sqrt{n}$, with $\epsilon \in [0.25, 0.40106]$.

We close this section by discussing an interesting class of problems that cannot quite be handled by Steele's techniques, i.e. LOCATION problems, in which k depots have to be located so as to minimize the average distance between each point to its closest depot (the k-MEDIAN problem) or the maximum of these distances (the k-CENTER problem). The probabilistic value analysis for both problems leads to surprisingly similar results, provided that $k = O(n/\log n)$ (the case $k = \alpha n$ is partially open; see Hochbaum and Steele [25]). Both optimal solution values are asymptotically proportional to $1/\sqrt{k}$, albeit for different constants of proportionality.

The analysis leading to these results is of special interest, since it relies heavily on the similarity between the original (*discrete*) problem for large n and the *continuous* problem is which customer demand is not concentrated in a finite set of points but spread uniformly and continuously over the entire region. A simple partitioning heuristic in which a depot is located in each of k nonempty subsquares of size $1/\sqrt{k}$ by $1/\sqrt{k}$ already provides an $O(1/\sqrt{k})$ upper bound on both optimal solution values. An asymptotically optimal heuristic, however, is only obtained by partitioning the region into regular *hexagons* (the *honeycomb* heuristic), with the constants of proportionality being determined by the optimal solution value of the continuous problem with $k = 1$ over one such hexagon. This heuristic actually solves the continuous problem to optimality, and a detailed error analysis shows that, for n sufficiently large, its relative error in the discrete case becomes vanishingly small (Haimovich [23], Zemel [56]).

4. Graphs and networks

We now turn to the rich area of combinatorial optimization problems defined on graphs and networks. One of the reasons for the wide variety of probabilistic results for this class of problems is the existence of a substantial theory dealing with *random graphs*. There are two popular definitions of this concept: $G_{n,p}$ is defined to be the (undirected) graph for n vertices for which each of the $n(n-1)/2$ edges occurs independently with equal probability p; G_n^N is defined by assuming that each of the $\binom{n(n-1)/2}{N}$ undirected graphs on n vertices occurs with equal probability (see Bollobas [6] for survey of the theory). Especially for *structural graph optimization* problems, in which we are interested in graph properties that depend only on the node-edge incidence structure, random graphs provide a natural probability distribution over the set of all problem instances of size n.

Continuing in the spirit of the previous two sections, we again refer to Karp et al. [33] for a list of recent references in this area and review only a few typical probabilistic analyses of heuristics for NP-complete problems, whose expected performance compares favorably with the limits set by worst case analysis. In doing so, we (reluctantly) exclude many beautiful results on problems of CONNECTIVITY and MATCHING that can be solved in worst case polynomial time.

A typical example of a difficult structural problem is the CLIQUE problem of finding a *complete subgraph* of G that has maximal size $\omega(G)$. To carry out a probabilistic value analysis of $\omega(G_{n,p})$ for fixed p, we observe that the expected number of cliques of size k in such a graph is equal to $\binom{n}{k} p^{k(k-1)/2}$. We would expect the maximal clique size k to occur when the expected number is approximately equal to 1, i.e., when (from Stirling's approximation of $k!$)

$$\frac{1}{\sqrt{2\pi k}} \left(\frac{nep^{(k-1)/2}}{k} \right)^k \approx 1. \tag{14}$$

The left hand side of (14) decreases very rapidly as k increases and passes through the value 1 where

$$\frac{nep^{(k-1)/2}}{k} = 1 \tag{15}$$

i.e. when

$$k = 2 \log_{1/p} n + 2 \log_{1/p} e - 2 \log_{1/p} k + 1 \tag{16}$$

so that, approximately, $k \approx k(n,p)$ with

$$k(n,p) = 2 \log_{1/p} n - 2 \log_{1/p} \log_{1/p} n + 2 \log_{1/p} (e/2) + 1. \tag{17}$$

This estimate turns out to be very sharp indeed. In Matula [43], it is proved that, for all $\epsilon > 0$,

$$\lim_{n \to \infty} \Pr\{[k(n,p) - \epsilon] \leq \underline{Z}_n^{\text{OPT}} = \omega(G_{n,p}) \leq [k(n,p) + \epsilon]\} = 1 \qquad (18)$$

so that, for large enough n, the size of the largest clique can be predicted to be one of two consecutive integers with high probability of success.

This precise probabilistic value analysis again encourages the search for a fast heuristic whose absolute error compares favorably to $2 \log_{1/p} n$. Consider, for instance, the *sequential greedy algorithm*, which consider the vertices of G in arbitrary order and adds a vertex to the current clique if it is adjacent to all its members. For an analysis of the performance of this method, one observes that the expected number of trials to increase the clique size from j to $j + 1$ is $1/p^j$, so that we might guess the ultimate clique size Z^H to satisfy

$$\sum_{j=0}^{Z^H - 1} 1/p^j = \frac{1 - 1/p^{Z^H}}{1 - 1/p} = n \qquad (19)$$

i.e.,

$$Z^H \approx \log_{1/p} n. \qquad (20)$$

A more precise analysis shwos that, indeed, this greedy approach a.s. yields a clique of size $(1/2 - \epsilon) \underline{Z}_n^{\text{OPT}}$ (Grimmett and McDiarmid [21]). Thus, the relative error does not go to 0, but is almost surely close to 50 percent. (There is, by the way, no known polynomial time heuristic with any constant worst case bound on the relative error.)

The above result has immediate implications for the problem to find the INDEPENDENT SET in G of maximal size; it coincides with the maximal size clique in the complement of G. Again, the sequential greedy approach, which picks up each successive vertex that is not adjacent to any member of the current independent set, produces an independent set whose size is a.s. close to 50 percent of the optimal value. The COLORING problem, which is to partition the vertices of G into the smallest possible number $\chi(G)$ of independent sets, is much harder to analyze: the asymptotic optimal solution value $\underline{Z}_n^{\text{OPT}} = \chi(G_{n,p})$ is known for $p = 1/2$ (Korsunov [38]), though that (Russian) announcement has not been verified. The heuristic method, which greedily finds an independent set as above, deletes it and repeats on the remaining graph does poorly (McDiarmid [44]) but good enough to get within a factor of $2 + \epsilon$ a.s. (Grimmett and McDiarmid [21]).

The other class of structural graph problems for which probabilistic analysis has been successful is the HAMILTONIAN CIRCUIT problem of searching for a simple cycle containing all vertices. The emphasis here is more on conditions under which such a cycle exists (e.g., a.s. in G_n^N when $N = (1/2) n \log n +$ $+ (1/2) n \log \log n + cn$ (Komlos and Szemeredi [37]) and less on the development of fast heuristics that, with high probability, would be successful in finding such a cycle if it existed. However, the heuristic principle of *extension* and *rotation* has been applied to this class of problems with considerable success (Posa [48], Angluin and Valiant [3]). The general idea is as follows. Given a path of vertices $(v_0, \ldots, v_k)$, one of the neighbours of v_k, say w, is sampled randomly and the edge $\{v_k, w\}$ is added to the path. If w is not in the path, it is adjoined and the method is applied to w. If $w = v_l$ ($0 \leq l \leq k-1$), then the edge $\{v_l, v_{l+1}\}$ is removed from the path and the method is applied to v_{l+1}. If N exceeds the threshold by a sufficient amount (e.g., $N = cn \log n$, for large enough c) this method will be successful with high probability.

We now turn briefly to *number problems* on *weighted graphs*, i.e., graphs with weights on the edges, an area which mixes features addressed in Section 2 with the theory of random graphs. Here, most results refer to problems that admit of a worst case polynomially bounded algorithm. A typical example is provided by the LINEAR ASSIGNMENT problem of minimizing $\Sigma_i c_{i\pi(i)}$ over all permutations π. If the c_{ij} are i.i.d. with distribution F, then a probabilistic value analysis can be arrived at by viewing the problem as a weighted matching problem on a directed bipartite graph, with weights $\underline{b}_{ij}$ on edge (i,j) and $\underline{d}_{ij}$ on edge (j,i) such that $\underline{c}_{ij} \stackrel{d}{=} \min\{\underline{b}_{ij}, \underline{d}_{ji}\}$. If we now remove all edges except the s outgoing ones of minimal weight at each vertex and disregard edge orientations in the resulting bipartite graph, a result by Walkup shows that a perfect matching will be present with probability $1 - O(1/n)$ if $s = 2$ and $1 - O(2^{-n})$ if $s > 2$. Hence, Z_n^{OPT} is essentially determined by the small order statistics of F. In the case that the $\underline{c}_{ij}$ are uniform on $[0, 1]$, this yields that $E\underline{Z}_n^{OPT} \leq 3$ (Walkup [55]); generally, $E\underline{Z}_n^{OPT}$ is asymptotic to $nF^{-1}(1/n)$ (Frenk & Rinnooy Kan [15]).

Many other problems in this category have also been succesfully analyzed. For instance, in Frieze [18], the MINIMUM SPANNING TREE problem is studied under the assumption that the graph is complete and the edge weights are i.i.d. uniform on $[0, 1]$; the expected optimal solution value is equal to $\Sigma_{k=1}^{\infty} 1/k^3 \approx$ ≈ 1.2202. We also refer to Perl [17] for an analysis of the SHORTEST PATH problem; further references can be found in Karp et al. [33].

An NP-complete problem that belongs in this category is the ASYMMETRIC TRAVELLING SALESMAN problem, defined on a complete directed graph. We refer to Steele [52] for a probabilistic value analysis for a Euclidean variant of this problem. The optimal solution value is asymptotic to $\zeta\sqrt{n}$ in expectation; see also Karp [31] for a heuristic that patches the subcycles appearing in the linear assignment relaxation together, to achieve a relative error going to 0 in expectation. Parhaps the most peculiar result has been obtained for its generalization, the QUADRATIC ASSIGNMENT problem

$$\max\left\{\sum_{i=1}^{n}\sum_{j=1}^{n}\sum_{k=1}^{n}\sum_{l=1}^{n} c_{ij} d_{kl} x_{ik} x_{jl} \,\Big|\, \sum_{i=1}^{n} x_{ij} = 1, \qquad (j = 1, \ldots, n);\right.$$

$$\sum_{j=1}^{n} x_{ij} = 1 \ (i = 1, \ldots, n), \quad x_{ij} \in \{0, 1\}\bigg\}.$$

In Burkhard and Fincke [7] and in Frenk et al. [13], it is shown that for this problem with $\underline{c}_{ij}$ and $\underline{d}_{kl}$ i.i.d., the ratio of the best and the worst possible solution value tends to 1 in probability. It shows an unexpected side benefit of probabilistic analysis, in that it clearly indicates how not to generate test problems for an empirical analysis of heuristic solution methods!

5. Concluding remarks

In this final section, we first discuss a limitation of the preceding survey in that the probabilistic behaviour discussed there was only caused by factors extraneous to the algorithm itself. The algorithm itself could also contribute to the randomness of its outcome by containing *random steps*, i.e., steps whose result depends partially on a random mechanism.

A very early example of an *algorithm that flips coins* is the quicksort method for SORTING. In each step, the algorithm picks a number a_j from the set $a_1, \ldots, a_n$ to be sorted; it divides the set into the subset smaller than a_j and the subset larger than a_j, and repeats recursively on each subset of cardinality greater than two. The worst case and best case running time of the method are easily seen to be $O(n^2)$ and $O(n \log n)$ respectively; the average running time is $O(n \log n)$ under a variety of distributional assumptions but also under the assumption that for a *fixed input* the splitting number a_j is chosen randomly from the set of candidates. Note that the behaviour of the algorithm is now a random variable independent of any distributional assumption, avoiding the always controversial issue of what a random instance of a particular problem looks like.

The formal study of randomized algorithms is far from complete, and in particular the real power of randomization remains a mysterious issue; for instance, it is not clear to what extent (if any) the class of problems that can be solved in randomized polynomial time (i.e. fast with high reliability) strictly includes the class of problems that can be solved in worst case polynomial time. A recent annotated bibliography (Maffioli et al. [42]) provides a useful survey of the area. We restrict ourselves once again to a discussion of a few examples that highlight the variety of results produced so far.

Historically, PRIMALITY TESTING was the first successful application of randomization. In the algorithm in Rabin [49], a number is submitted to k tests and declared to be prime if it passes all of them, with the probability of it being composite none the less being equal to 2^{-k}. Such an algorithm is called a *Monte Carlo method*, in contrast to a *Las Vegas method* in which the algorithm never produces an incorrent answer, but may – with small probability – produce no answer at all. The method for GRAPH ISOMORPHISM in Babai [4] is of this nature. The two examples above are of special interest in that they concern two problems whose computational complexity (polynomially solvable or NP-complete) is still unknown.

Generally, randomization may produce a speed-up of a polynomial computation at the expense of a little uncertainty, as for instance in the case of a fast BIPARTITE MATCHING algorithm in Ibarra and Moran [26] with a small probability of error, or an $O(|V|^2)$ expected time algorithm for CIRCUIT COVERING (Itai and Rodeh [27]) that with small probability will not terminate. For NP-hard problems, there are other potential benefits. *Statistical inference* has been suggested as a way to estimate the optimal solution value from a sample of (supposedly independent) local minima, i.e., feasible solutions whose value cannot be further improved by a *local improvement method* (Dannenbring [9]). And the fashionable *simulated annealing* approach can be viewed as a randomized version of such a local improvement heuristic, in which neighbouring solutions that decrease the quality of the current feasible solution are also accepted, albeit with a small probability that is appropriately decreasing over time (Kirkpatrick et al. [35]).

It should be clear by now that the area of probabilistic analysis still harbors many interesting research challenges. The purpose of the preceding sections has, again, not been to provide an exhaustive review, but to provide some typical examples that convey the flavour of this area. They have ranged from the very complete insight we have into various solution methods for the PARTITION problem to the less satisfactory state of the art for the CLIQUE and the COLORING problem. Clearly, a lot of problems and a lot of algorithms await investigation. It is not hard to formulate open questions for probabilistic analysis; so far, however, it has turned out to be quite hard to come up with satisfactory answers for any but the simplest heuristics.

A particularly fascinating possibility is the development of a complexity theory that would lead to a class of problems for which solution to optimality in polynomial expected time is as unlikely as the equality of P and NP. A first step in that direction can be found in Levin [41], where a TILING problem is introduced, together with a probability distribution over its problem instances, such that any other problem with a (mildly restricted type of) probability distribution is reducible to the TILING problem.

To establish completeness for other problems in this class is a major challenge of considerable interest. After all, the reasonable average behaviour of enumerative methods (and the remarkable success of a nonenumerative method based on computations in an integer lattice (Lagarias and Odlyzko [39])) to solve some NP-complete problems and the apparent impossibility to find such algorithms for other NP-complete problems, still defy theoretical explanation!

Acknowledgements

This paper was written during a visit to the Department of Industrial Engineering and Operations Research and the School of Business Administration of the University of California at Berkeley, with partial support from a NATO Senior Scientist Fellowship. I gratefully acknowledge various useful discussions with Dick Karp and Dorit Hochbaum.

References

[1] I. Adler, R.M. Karp and R. Shamir, «A Simplex Variant Solving an $m \times d$ Linear Program in $0(\min(m^2, d^2))$ Expected Number of Pivot Steps», *Report UCB CSD 83/158, Computer Science Division, University of California, Berkeley*, 1983.
[2] I. Adler and N. Megiddo, «A Simplex Algorithm Whose Average Number of Steps is Bounded Between Two Quadratic Functions of the Smaller Dimension», *Technical Report, University of California, Berkeley*, 1983.
[3] D. Angluin and L.G. Valiant, «Fast Probabilistic Algorithms for Hamiltonian Circuits and Matchings», *Journal of Computer and System Science* 19, 155 - 193, 1979.
[4] L. Babai, «Monte Carlo Algorithms in Graph Isomorphism Testing», *Rapport de Recherches DMS 79 - 10, Département de Mathématique et de Statistique, Université de Montréal*, 1979.
[5] J. Beardwood, J.H. Halton and J.M. Hammersley, «The Shortest Path through Many Points», *Proceedings of the Cambridge Philosophical Society* 55, 299 - 327, 1959.
[6] B. Bollobás, «Lectures on Random Graphs» (to appear).
[7] R.E. Burkard and U. Fincke, «Probabilistic Asymtotic Properties of Quadratic Assignment Problem», *Zeitschrift fuer Operations Research* 27, 73 - 81, 1982.
[8] E.G. Coffman, Jr., M.R. Garey and D.S. Johnson, «Approximation Algorithms for Bin-Packing – An Updated Survey», *Bell Laboratories, Murray Hill, NJ*, 1983.
[9] D. Dannenbring, «Procedures for Estimating Optimal Solution Values for Large Combinatorial Problems», *Management Science* 23, 1273 - 1283, 1977.
[10] U. Derigs, Private communication, 1984.
[11] L. Few, «The Shortest Path and the Shortest Road through n Points», *Mathematika* 2, 141 - 144, 1955.
[12] G.N. Frederickson, «Probabilistic Analysis for Simple One – and Two – Dimensional Bin Packing Algorithms», *Information Processing Letters* 11, 156 - 161, 1980.
[13] J.B.G. Frenk, M. van Houweninge and A.H.G. Rinnooy Kan, «Asymptotic Properties of the Quadratic Assignment Problem», *Mathematics of Operations Research* 10, 100 - 116, 1982.
[14] J.B.G. Frenk and A.H.G. Rinnooy Kan, «The Asymptotic Optimality of the LPT Heuristic», *Mathematics of Operations Research* (to appear).
[15] J.B.G. Frenk and A.H.G. Rinnooy Kan, «Order Statistics and the Linear Assignment Problem», *Econometric Institute, Erasmus University, Rotterdam*, 1984.
[16] J.B.G. Frenk and A.H.G. Rinnooy Kan, «On the Rate of Convergence to Optimality of the LPT Rule», *Discrete Applied Mathematics* (to appear).
[17] A.M. Frieze and M.R.B. Clarke, «Approximation Algorithms for the m-dimensional 0-1 Knapsack Problem», *European Journal of Operational Research* 15, 100 - 109, 1984.
[18] A.M. Frieze, «On the Value of a Random Minimum Spanning Tree Problem», *Technical report, Department of Computer Science and Statistics, Queen Mary College, University of London*, 1982.
[19] B. Golden, Chapter 5. in E.L. Lawler, J.K. Lenstra, A.H.G. Rinnooy Kan and D.B. Shmoys, eds., *The Traveling Salesman Problem*, Wiley, Chichester, 1985.
[20] R.L. Graham, «Bounds on Multiprocessing Timing Anomalies», *SIAM Journal on Applied Mathematical* 17, 263 - 269, 1969.
[21] G. Grimmett and C.J.H. McDiarmid, «On Colouring Random Graphs», *Mathematical Proceedins of the Cambridge Philosophical Society* 77, 313 - 324, 1975.
[22] M. Haimovich, «The Simplex Method is Very Good! - On the Expected Number of Pivot Steps and Related Properties of Random Linear Programs», *Columbia University, New York*, 1983.
[23] M. Haimovich, *Ph.D. Thesis, M.I.T.*, 1984.

[24] M. Haimovich and A.H.G. Rinnooy Kan, «Bounds and Heuristics for Capacitated Routing Problems», *Mathematics of Operations Research* (to appear).
[25] D.S. Hochbaum and J.M. Steele, «Steinhaus' Geometric Location Problem for Random Samples in the Plane», *Advances in Applied Probability* 14, 56 - 67, 1981.
[26] O.H. Ibarra and S. Moran, «Deterministic and Probabilistic Algorithms for Maximum Bipartite Marching via Fast Matrix Multiplications», Information Processing Letters 13, 12 - 15, 1981.
[27] A. Itai and M. Rodeh, «Covering a Graph by Circuits», in G. Ausiello and C. Böhm, eds., *Automata, Languages and Programming*, Lecture Notes in Computer Science 62, Springer, Berlin 289 - 299, 1978.
[28] N. Karmarkar and R.M. Karp, «The Differencing Method of Set Partitioning», *Mathematics of Operations Research* (to appear).
[29] N. Karmarkar, R.M. Karp, G.S. Lueker and A. Odlyzko, «The Probability Distribution of the Optimal Value of a Partitioning Problem», *Bell Laboratories, Murray Hil, NJ*, 1984.
[30] R.M. Karp, «Probabilistic Analysis of Partitioning Algorithms for the Traveling Salesman Problem in the Plane», *Mathematics of Operation Research* 2, 209 - 224, 1977.
[31] R.M. Karp, «A Patching Algorithm for the Nonsymmetric Traveling Salesman Problem», *SIAM Journal on Computing* 8, 561 - 573, 1979.
[32] R.M. Karp, Private communication, 1984.
[33] R.M. Karp, J.K. Lenstra, C.J.H. McDiarmid and A.H.G. Rinnooy Kan, Chapter 6 in M. OhEigeartaigh, J.K. Lenstra and A.H.G. Rinnooy Kan, eds., *Combinatorial Optimization: Annotated Bibliographies*, Wiley, Chichester, 1985.
[34] R.M. Karp and J.M. Steele, Chapter 6 in E.L. Lawler, J.K. Lenstra, A.H.G. Rinnooy Kan and D.B. Shmoys, eds., *The Traveling Salesman Problem*, Wiley, Chichester, 1985.
[35] S. Kirkpatrick, C.D. Gelat Jr. and M.P. Vecchi, «Optimization by Simulated Annealing», *Science* 220, 671 - 680, 1983.
[36] W. Knödel, «A Bin Packing Algorithm with Complexity $O(n \log n)$ and Performance 1 in the Stochastic Limit», in J. Gruska and M. Chytil, eds., *Mathematical Foundations of Computer Science 1981*, Lecture Notes in Computer Science 118, Springer, Berlin, 369 - 378, 1981.
[37] J. Komlós and E. Szemerédi, «Limit Distribution for the Existence of Hamiltonian Cycles in Random Graphs», Discrete Mathematics 43, 55 - 63, 1983.
[38] A.D. Korsunov, «The Chromatic Number of n-vertex Graphs», *Metody Diskret. Analiz.* 35, 14 - 44, 104 (in Russian), 1980.
[39] J.C. Lagarias and A.M. Odiyzko, «Solving Low-Density Subset Sum Problems», *Proceedings of the 24th Annual IEEE Symposium on the Foundations of Computer Science*, 1 - 10, 1983.
[40] E.L. Lawler, J.K. Lenstra, A.H.G. Rinnooy Kan and D.B. Shmoys, eds., *The Traveling Salesman Problem*, Wiley, Chichester, 1985.
[41] L.A. Levin, «Problems, Complete in 'Average' Instance», *Proceedings of the 16th Annual ACM Symposium on the Theory of Computing*, 465, 1984.
[42] F. Maffioli, M.G. Speranza and C. Vercellis, Chapter 7 in M. OhEigeartaigh, J.K. Lenstra and A.H.G. Rinnooy Kan, eds., *Combinatorial Optimization: Annotated Bibliographies*, Wiley, Chichester, 1985.
[43] D.W. Matula, «The Largest Clique Size in a Random Graph», *Technical report CS7608, Department of Computer Science, Southern Methodist University, Dalles TX*, 1976.
[44] C.J.H. McDiarmid, «Colouring Random Graph Badly», in R.J. Wilson, ed., *Graph Theory and Combiatorics*, Pitman Research Notes in Mathematics 34, Pitman, London, 76 - 86, 1979.
[45] M. Meanti, A.H.G. Rinnooy Kan, L. Stougle and C. Vercellis, «A Probabilistic Analysis of the Multiknapsack Value Function», *Econometric Institute, Erasmus University, Rotterdam*, 1984.
[46] C.H. Papadimitriou, «The Complexity of the Capacitated Tree Problem», Networks 8, 217 - 230, 1978.
[47] Y. Perl, «Average Analysis of Simple Parth Algorithms», *Technical report UIUCDCS - R - 77 - 905, Department of Computer Science, University of Illinois at Urbana-Champaign*, 1977.
[48] L. Pósa, «Hamiltonian Circuits in Random Graphs», *Discrete Mathematics* 14, 359 - 364, 1976.
[49] M.O. Rabin, «Probabilistic Algorithms for Testing Primality», *Journal of Number Theory* 12, 128 - 138, 1980.
[50] P.W. Shor, «The Average Case Analysis of Some On-line Algorithms for Bin-Packing», *Proceedings of the 25th IEEE Symposium of the Foundations of Computer Science*, 1980.
[51] J.M. Steele, «Subadditive Euclidean Functionals and Nonlinear Growth in Geometric Probability», *Annals of Probability* 9, 365 - 376, 1981.
[52] J.M. Steele, «A Probabilistic Algorithm for the Directed Traveling Salesman Problem», *Mathematics of Operations Research* (to appear).
[53] L. Stougie, *Ph.D. Thesis, Econometric Institute, Erasmus University, Rotterdam*, 1985.
[54] M.J. Todd, «Polynomial Expected Behavior of a Pivoting Algorithm for Linear Complementarity and Linear Programming Problems», *Technical report 595, School of Operations Research and Industrial Engineering, Cornell University, Ithaca, NY*, 1983.

[55] D.W. Walkup, «On the Expected Value of a Random Assignment Problem», *SIAM Journal on Computing* 8, 440 - 442, 1979.
[56] E. Zemel, «Probabilistic Analysis of Geometric Location Problems», *Annals of Operations Reseach* I, 1984.

Alexander H.G. Rinnooy Kan
Econometric Institute
Erasmus University Rotterdam
P.O. Box 1738
3000 DR Rotterdam
The Netherlands

JUN 2 1 1989